Beyond Our Genes

Raffaele Teperino
Editor

Beyond Our Genes

Pathophysiology of Gene and Environment Interaction and Epigenetic Inheritance

Springer

Editor
Raffaele Teperino
Head of the Environmental Epigenetics Group
Institute of Experimental Genetics & German Center
for Diabetes Research (DZD), Helmholtz Zentrum
München GmbH – German Research Center
for Environmental Health
Neuherberg, Bayern, Germany

ISBN 978-3-030-35212-7 ISBN 978-3-030-35213-4 (eBook)
https://doi.org/10.1007/978-3-030-35213-4

This Springer imprint is published by the registered company Springer Nature Switzerland AG
The registered company address is: Gewerbestrasse 11, 6330 Cham, Switzerland

Preface

The incidence of chronic – multifactorial – diseases including diabetes, obesity, and cancer is exponentially increasing worldwide, and the most recent forecasts predict its rise will continue. The scientific community has been trying since more than a decade to understand the genetic underpinnings of complex diseases pathogenesis, heritability, and susceptibility through genome-wide association studies (GWAS). While providing an unprecedented genetic framework for our understanding of complex diseases, GWAS studies have failed in explaining heritability and individual susceptibility.

These results, though, have shed light on an unexpected hope. The relative failure of GWAS, indeed, underlines the existence of additional pathophysiological mechanisms and the need for the scientific community to start looking at the organism not as a genetic *monad* rather as the result of a complex interaction between his genetic blueprint and the environment he lives in. Epidemiological studies, indeed, suggest that our environment – including nutrition, lifestyle, circadian rhythm, and exposure to toxicants and pollutants among others – heavily impacts our health and, though still debated, that of our descendants. This new form of inheritance, called epigenetic inheritance as it relies on the existence of heritable epigenetic mechanisms, has been shown in several organisms and suggested by epidemiological studies in humans. We are just about to understand the mechanisms of this parent-to-offspring transmission, and yet we clearly see its relevance for human health.

The overarching goal of this volume is to introduce, with scientific accuracy, yet understandable terminology, the general reader to the importance of the environment for our health and the health of our offspring and, at the same time, to shed light on the current knowledge on epigenetic inheritance and open a window on future developments in the field. My hope is that the volume will inspire every category of readers: the nonspecialists to improve their health by improving their lifestyle and the specialists to improve our health by deepening their knowledge on the fascinating and promising field of epigenetic inheritance.

Enjoy!!

Neuherberg, Germany — Raffaele Teperino

Contents

Part I
The Physiology of Gene/Environment Interaction

Chapter 1
Food and Nutrition as Prime Environmental Factors

Immacolata Cristina Nettore, Paola Ungaro, and Paolo Emidio Macchia

Keywords Diet · Life style · Malnutrition · Nutrient deficiencies · Obesity · Hidden Hinger · Physical activity · Childhood · Adulthood · Elderly · Pregnancy

"Let food be thy medicine and medicine be thy food." This is one of the aphorisms that has been attributed to the Greek physician Hippocrates of Kos (460-between 375 and 351 BCE [1]), who is considered the father of Western medicine.

Although the attribution of this phrase to Hippocrates is almost certainly a historical misquotation, since the "Aphorism" does not appear in any of the volumes included in the Hippocratic Collection (Corpus Hippocraticum) [2], it clearly indicates the importance of food and nutrition in classical and many traditional forms of medicine [3].

Nutrition is the study of how food affects the body. The term "nutrition" defines the integrated processes by which cells, tissues, organs and the whole body acquire the energy and nutrients for their normal structure and function. This is achieved at body level through dietary supply, and thanks to the capacity of the body to transform the substrates and cofactors necessary for metabolism. Diet, metabolic capacity, body composition, level of demand for energy and nutrients are all influenced by physiological and pathological states and can vary according to different levels of physical activity [4].

Environmental factors, accordingly to the Alan Giplin's Dictionary of Environment and Sustainable Development [5], are factors abiotic or biotic that may influences living organisms, and therefore, food and life styles can be consid-

I. C. Nettore
Dipartimento di Medicina Clinica e Chirurgia, Università degli Studi di Napoli Federico II, Naples, Italy

P. Ungaro
Istituto di Endocrinologia ed Oncologia, Consiglio Nazionale delle Ricerche, Naples, Italy
e-mail: pungaro@ieos.cnr.it

P. E. Macchia (✉)
Dipartimento di Medicina Clinica e Chirurgia, Università degli Studi di Napoli Federico II, Naples, Italy
e-mail: pmacchia@unina.it

R. Teperino (ed.), *Beyond Our Genes*,
https://doi.org/10.1007/978-3-030-35213-4_1

ered among the most relevant and modifiable environmental factors that may influence the human health.

1.1 Diet Quality

In most developed countries over the past several decades, major shifts in body composition have been produced by the dramatic changes in consumption (both food and drink) and physical activity pattern. One of the consequences is the progressive increase in the prevalence of obesity [6]. In the USA, the annual medical cost of obesity was calculated in $147 billion in 2008 [7], with an estimated cost for obese in individual medical care of $3429/year higher than that reported in normal weight individuals [8].

If adequate nutrition will determine a global trend towards an increase in both height and weight, an increase in weight achieved before the height can lead to an increase in childhood overweight and adiposity as well as an increased risk of shortness/stunting and obesity.

Malnutrition affects both developing or transitioning countries and high income and industrialized countries. Indeed, despite well-publicized dietary guidance, data obtained in several European countries indicated that diet in the examined populations is not in line with the current advices [9–11], and, in addition to the weight increase of the population, several nutritional deficiencies can be observed in the population, across the different ages [12].

1.2 Obesity and Overweight

As already mentioned, the proportion of obese or overweight people has significantly risen in recent years [6]. Obesity has been linked to several health conditions widely recognized for many years, including cardiovascular disease (CVD), hypertension (HTN), type 2 diabetes mellitus (T2DM), hyperlipidemia, stroke, certain cancers, sleep apnea, liver and gall bladder disease, osteoarthritis, and gynecological problems [13, 14]. More recently, obesity and overweight have been linked to additional health problems, including periodontal disease, poor school performance, altered pre-pubertal hormones, and attention-deficit hyperactivity disorder in children [15], or chronic kidney disease which may result in early-onset disability [16–18].

The long-term physiological effects of obesity are numerous and potentiate each other. In addition, recent literature has also focused on the psychosocial consequences of obesity. Indeed, it has been reported that the experience of weight stigma or perceived weight discrimination is associated with depression, anxiety, bulimia, body dissatisfaction, and low body- and self-esteem [19–21]. This determined an increased risk for depression and other psychological disorders, which are less frequent among obese black girls and women, rather than white women [22]. These aspects are also worsened by negative attitudes that healthcare providers have more often towards obese patients, leading to perceptions of prejudice, ambivalence, and unsatisfactory healthcare treatment.

Clearly, psychosocial functioning may be affected on many levels among those with an increased BMI [23, 24].

1.3 Deficiency and Malnutrition

Malnutrition is a condition resulting from the unbalance between the nutritional requirements (calories and protein) and nutritional deficiencies or overconsumptions. This condition is a critical public health concern and it represents the largest single contributor to disease development in both developing and developed countries. Worldwide, ~805 million people are chronically undernourished [25].

Although deficiencies in vitamins and minerals are very common in the developing economies [26], micronutrient malnutrition is not a phenomenon limited to mothers, infants, and populations in developing economies. Inadequate micronutrient provision is a problem that pervades even into the most advanced economic areas, and it has been calculated that globally micronutrient malnutrition affects ~2 billion people [27, 28].

There are 50 known dietary nutrients essential for sustaining human life. These comprise water, carbohydrates, nine amino acids (histidine, isoleucine, leucine, lysine, methionine, phenylalanine, threonine, tryptophan, and valine), two fatty acids (linoleic acid and linolenic acid), eight macroelements (Na, N, K, Ca, Mg, S, P, and Cl), 16 microelements (Fe, Zn, Cu, Mn, I, F, B, Se, Mo, Ni, Cr, V, Si, As, Sn, and Co), and 13 vitamins (A, D, E, K, C, B1, B2, B3, niacin, B6, folate, biotin, and B12) [29].

Hidden hunger is defined as a chronic lack in any of the dietary micronutrients and vitamins listed above [30]. The term "hidden hunger" is commonly used to describe individuals who may have adequate energy consumption, but suboptimal micronutrient intakes, placing them at risk for nutrition-related diseases [30]. The health impact of such micronutrient malnutrition is very wild, and it ranges from hindered growth and development, through to persistent health issues and reduced lifespan [31].

Three interlinked factors supporting the hidden hunger: education, economics, and geography [26]. Education plays its role in terms of definition of the balanced dietary intake [32]; however, financial issues may be important preventing many people from accessing micronutrient-rich foods, which generally have higher in cost. Geography may contribute to the hidden hunger by determining the agro-environmental conditions of growing areas, in terms of soil element composition. In addition, geographical conditions influence the food distribution networks, with many people in rural and remote areas which may not benefit from an industrialized supply chain and are obligated to rely on local subsistence agriculture [33–35].

1.4 Interplay With Physical Activity

A critical aspect indirectly related to the topic of this chapter is the link between nutrition, physical activity levels and health.

Higher levels of physical activity (PA) were associated with better health-related quality of life while increased time of sedentary behaviour (SB) was linked to lower health-related quality of life among children and adolescents [36]. Children and adolescents who with increased levels of physical activities demonstrated better physical and mental health and psychosocial wellness than those with more inactive lifestyle [37, 38].

SBs is often characterized by the usage of screen-based media, including watching television (TV), computers/smartphones and playing video games. They are associated with various negative health consequences, such as obesity, coronary health disease and other health problems including a range of impaired psychological health. SB also contributes to a delay of cognitive development and a decrease in academic achievement of children and youth [39].

Intervening approach focus on reducing SB among school-age youth was conducted in the community, school, home, or clinic setting of several countries [37, 40]. The strategies include educational, health promotion, behavioral therapy, counseling, or management strategies at the individual and family levels. It should be noted that attending classes or playing a musical instrument, also represent behaviors that do not determine a significant energy expenditure. Overall, interventions that focused on decreasing SB were associated with reduction in time spent on SB and/or improvements in anthropometric measurements related to childhood obesity.

In the past, the beneficial effects of exercise in obese and type-2 diabetes patients on inflammation, lipid profile and glycemic factors have been largely studied. Recently, several studies demonstrate the benefit of High Intensity Interval Training (HIIT) for several groups of patients. The HIIT is a time-efficient methodology that allows improvements in cardiorespiratory fitness and work capacity, mitochondrial muscle biogenesis and GLUT-4 levels, insulin sensitivity and fasting glucose, reducing several cardiometabolic risk factors in overweight/obese populations [41, 42]. The association of HIIT along with low carbohydrate regimes in type 2 diabetic patients improves overall cardiovascular parameters, reduces pro-inflammatory markers and increases anti-inflammatory markers [43].

The original meta-analysis comparing the effects of HIIT vs moderate-intensity continuos training (MICT) on body composition in overweight and obese adults, suggested that both training cause a significant reduction in body fat mass and waist circumference, even if the HIIT required −40% less training time commitment each week. These results allow to HIIT the better time-efficient alternative for managing of patients [44].

The influence of physical activity in maintaining long-term weight loss has been also evaluated in the preparation of patients awaiting bariatric surgery (BS). A good physical activity program after BS provides benefits for health and fitness level in individuals with morbid obesity [45]. Similarly, a physical activity program prior to

BS (aerobic training with resistance training or only applying aerobic or resistance training) improves body weight, physical condition and quality of life related to health [46, 47]. More recently, a group of researchers demonstrated the beneficial effects of HIIT and resistance training in a small group of patients awaiting BS [48]. The most relevant effects were demonstrated on BMI and body composition, while small to moderate effects were determined on blood pressure and anthropometric measurements.

Work may contribute significantly to daily PA and SB. Physical inactivity is the fourth leading cause of global mortality [49, 50], being linked to major non-communicable diseases such as coronary heart disease, type 2 diabetes, and breast and colon cancers [49]. The World Health Organization estimates that around 3.2 million people die each year because of physical inactivity, and the work by Lee and coworkers estimated that the elimination of physical inactivity can determine a median potential gain of 0.68 years of life worldwide [49]. Physical inactivity and SB at work might be two major risk factors for premature morbidity. There is strong causal evidence linking PA and SB at work with late cardiovascular and cerebrovascular disease. In a cohort formed by retired adults aged >65 years, the risk for onset of myocardial infarction and stroke was lower among those who had a previous active work compared to those with previous sedentary work [51]. An increase in moderate-to-vigorous PA replacing SB and light PA (LPA) was also associated with a reduction in sarcopenia prevalence and better performance across muscle mass, handgrip strenght and gait speed, while LPA alone did not determined any significant effect [52].

1.5 Nutrition Across the Life Course

Nutritional requirements change accordingly to several factors that include age, sex, body weight, genotype, physical activity, physiological status (growth, pregnancy, lactation, …) and the presence or absence of disease. During the early years of life, the nutritional needs changes constantly, and an optimal nutrition plays a key role in lifelong health including in healthy ageing.

As suggested by Widdowson [53], both the quality (carbohydrate, lipid, amino acids, minerals, vitamins, trace elements, water, oxygen) and quantity (energy from macronutrients, carbohydrate, lipid [fat], protein) of nutrients are important for ensuring a healthy growth and the optimal functional capacity. In addition, in this equation should be also considered the physical activity, stresses as well as underlying pathological situations which modulate nutrient availability. Imbalances in nutrient will lead to abnormalities in body composition and determine either underweight or overweight and obesity.

On the basis of such criteria, a good nutrition is not simply the absence of nutrient deficiencies, but the intake of nutrients necessary to the appropriate growth and development across the entire life course. A good nutrition will properly modulate physiological processes, cognitive function and immune response, influencing, late in the life, aging and frailty. Moreover, adequate nutritional status can also affect resilience, disease susceptibility and response to therapy. It is not surprising, therefore, that a poor nutritional status is one of the major risk factors for several chronic diseases.

1.6 Nutrition During Pregnancy

The quantity and quality of nutrition during prenatal life plays a critical role in the long-term programming of health and disease.

The macronutrient dense but micronutrient poor diets could alter cellular metabolism. Nutrient interactions affect all stages of foetal development, influencing endocrine programming, organ development and the epigenetic programming of gene expression [54].

In 1931, it has been described how yeast extract could be effective in preventing tropical macrocytic anaemia of late pregnancy thanks to their ability to provide folate. Folate is an essential cofactor in metabolic pathways (methyl donor) that influence DNA methylation patterns, DNA synthesis and cell proliferation. Its low levels may lead to breaks in DNA, also predisposing to cancer. Supplementation of the diet with folic acid has been adopted by many countries particularly during pregnancy, when folate requirements increase because of increased rate of cell division and growth during embryo development [55]. Foetal cell proliferation requires the production of large quantities of DNA and membrane lipids as well as the synthesis of new protein. The derivatives of folic acid are important intermediates in the methionine cycle, which is in turn linked to lipid and nucleotide metabolism.

In the 1980s a series of studies showed how periconceptual folate supplementation could prevent Neural Tube Defects (NTD) such as spina bifida. A two-fold increase in NTD rate has been associated with maternal obesity. Pregnant women with high body mass index (BMI) and gestational diabetes (GD) can have inadequate dietary intakes of folate and lower folate concentrations, which predispose to greater risks of preterm deliveries, neural tube defects and low birth weight [56]. In order to prevent NTD, national guidelines have recommended to prescribe high dose folate and B12 supplementation for women with a Body Mass Index (BMI) >29.9 kg/m^2 [57].

Recently, it was demonstrated that high BMI and gestational diabetes are differentially associated with changes in maternal serum folate in late gestation [58]. The maternal folate concentrations were reduced with raised maternal pre-pregnancy BMI at 34 weeks of pregnancy, while the GD women present the highest folate concentrations. In addition, in those women with raised pre-pregnancy BMI, there was no detectable reduction in cord blood folate and all infants were healthy at term. These results suggest a protective role by the placenta of the foetus, further supporting the need to ensure optimal dietary folate intake.

It is well known that in obese subjects other several deficiencies are frequent including shortage in vitamin B1, vitamin B12, vitamin A, vitamin D and minerals such as iron and zinc [59, 60]. It has been estimated that, during pregnancy additional 740 mg of iron are necessary. Therefore, since an ideal diet provides only approximately 5 mg iron daily, a woman must enter pregnancy with an iron store of (at least) 300 mg to avoid deficiencies by the time of delivery. Many women, and particularly those in developing countries, have insufficient reserves and become iron deficient by the time the baby is born. Foetal and postnatal iron deficiency results in a range of adverse consequences for mother and infant including low birth weight, impaired cognitive development and poor immune function [54, 61].

Maternal nutrition during pregnancy and lactation, among other environmental factors, is also suggested to have an impact on the development of allergic diseases. Among these, the Cows' milk allergy (CMA) has been analysed in Finnish study that evaluated the maternal intake and/or the supplementation of folate, folic acid and vitamin D during pregnancy and lactation and demonstrated the association between food intake and CMA [62]. Thus, starting from pregnancy, the food represents one of the main environmental factors which can play a critical role in disease development or prevention.

1.7 Nutrition From Childhood to Early Adulthood

The quality of children's diet is more important before age 2 than at any other time in life. Changes, both in quality and quantity of nutrients during this period may permanently influence the development and the function of organs and systems [63]. These effects are termed 'programming' and represent an important risk factor for non-communicable diseases during adulthood, including metabolic syndrome and coronary heart disease [64].

Experimental models confirmed that nutrition in early life has the capacity to permanently establish physiological and metabolic states determining the risk of diseases occurring with ageing [65]. Diseases have been demonstrated to be influenced by several anthropometric measurements at birth including birthweight [66], the thinness at birth (measured as ponderal index; weight/length^2)[67], or head circumference [68].

The recommended practices in this critical developmental period (the so-called 1000 days) include: initiation of breastfeeding within the first hours of life, exclusive breastfeeding for the first 6 months and continued breastfeeding at least until age 2. At 6 months of age, children should be introduced to their first foods, starting with soft foods, followed by semi-solid and then solid foods. These complementary foods should be safe, nutritionally adequate and provided in response to a child's needs and hunger signals. The timing and type of foods used in weaning are also associated to various diseases in later life.

It has been also demonstrated that these recommendations have beneficial effects in terms of allergic, immunologic, and cardiovascular diseases prevention [69].

In the last decade food allergy has continued to increase in frequency, however studies are still discordant regarding the association between exposure to solid foods in first months of life. Some studies have shown that early exposure to solid foods (under 4 months) increases the incidence of several allergies [70, 71], while the European Society of Pediatric Gastroenterology, Hepatology and Nutrition (ESPGHAN) Committee of Nutrition in 2008 recommended to start exposing infants to solids, including known allergens, between the ages of 17–26 weeks. Further studies demonstrated that delayed exposure to certain foods such as wheat, oats, cows' milk, fish, and eggs contributes to increased incidence of atopic dermatitis, asthma, allergic rhinitis, and allergic tendency [70, 72].

Exclusive breastfeeding, probably until the age of 4 months [73] is also associated with decreased prevalence of obesity in children. The poor nutrition in infants and children plays an important role, as well as epigenetic mechanisms in the context of maternal nutrition during pregnancy. The breastfeeding remains the preferred way to feed infants and to avoid rapid weight gain, but it also prevents cardiovascular disease later in life [74]. By contrast, the administration of protein-rich formulas (2.6–3 g of protein per kilogram per day) is associated with increased body fat by 30% at age 8–10 years.

Obesity in childhood is also a risk factor for the later development of metabolic syndrome and type 2 diabetes. Furthermore, early exposure to gluten (before the age of 3 months) in infants of parents affected by type 1 diabetes has been associated with higher risk of type 1 diabetes (five times more compared to those exposed after the age of 3 months) [75]. The duration of breastfeeding seems to not influence the development of diabetes, in contrast to what previously demonstrated [76].

The adolescence (age between 10 and 19 years by World Health Organization) is a period of rapid growth and behavioural changes. An adequate nutrition is crucial for achieving full growth potential, and failure to achieve optimal nutrition may lead to delayed and stunted linear growth and impaired organ remodelling [77]. Overweight and obesity affect one in every three adolescents worldwide [78]. It has been estimated that in 2011 7% of children younger than 5 years were overweight, with a 54% increase in prevalence of obesity in comparison to 1990 [79]. The appetite in adolescence increases, and sedentary individuals are more likely to accumulate fat if they consume too high-energy food. Thus, the physical activity among adolescent is an important factor to prevent several diseases later.

The prevalence of underweight among adolescent females between 13 and 17 years is generally less than 5%, even if it has been estimated that in some countries of Africa and Asia 10% or more of young adolescent girls are malnutrited. Deficiencies in multiple micronutrients are of particular importance to adolescent health because of their direct effects, such as iron-deficiency anaemia and iodine-deficiency disorders. Recently, it has been suggested that a micronutrient supplementation in adolescent can significantly decrease a prevalence of anaemia [80]. Other nutrients necessary in adolescent period include vitamin D and calcium useful to favour bone growth, and amino acids for growth of striated muscle.

Nowadays it is beneficial follow simple recommendations regarding the daily nutritional approach.

In children, it is recommended reduce to minimum the supply of trans-fatty acids and cholesterol, normally present in packed snacks and in foods with a long shelf life since they contribute to raise the levels of LDL-cholesterol and lower HDL-cholesterol. Polyunsaturated fat (PUFA) and Monounsaturated fat consumption should account for 10% of total daily calories. Found in vegetable oils and in olive oil respectively, the PUFA reduces the levels of LDL, while the monounsaturated fat raised the HDL levels.

Fruits and vegetables have low calorie content and contribute to satiety, thus decreasing total caloric intake. Moreover, a diet rich in whole grains and fibbers reduces risk for coronary heart disease and cardiovascular mortality, hypertension, diabetes, and obesity. It is recommended to eat the fruit and vegetable as native foods (rather than juice). The adolescent should consume between 5 and 8 servings of fruits or vegetables per day [81].

1.8 Nutrition in Elderly

The elderly population is defined as people aged 65 and over. Thanks to the improved general health conditions, the growth of the elderly segment in the general population increases significantly all over the world.

The nutritional intake in elderly represents a key factor of health problems, especially in hospital structure. To assess the nutritional status in elderly population, rather than calculate Body Mass Index (BMI) is more important to evaluate the functional status. The measurement of fluid intake is also critical because elderly is prone to dehydration.

The prevalence of malnutrition in elderly across the world shows an increasing trend in the last years, even in the well-developed countries [82]. The mainly cause of malnutrition is the aging process itself [83], characterized by loss of bone density and sarcopenia, changes in digestive systems and sensory changes. These last, can lead to dysregulation of appetite and thirst, and make the food less appealing. The digestive system can be modified with a reduction of the secretion of acid and a consequent limited absorption of iron and vitamin B-12. The sarcopenia can cause loss of strength, functional decline, and poor endurance, while the loss of bone density can increase the risk of osteoporosis.

The consumption of higher levels of essential amino acids has been suggested to have beneficial effects on muscle health in older population [84]. Moreover, several evidences demonstrated the specific role of dietary approaches for osteoporosis' prevention [85], and the contribute by vitamins and minerals to skeletal health [86].

There are several age-related health complications and, among these, the age-related macular degeneration (AMD) is one of the major causes of irreversible blindness in older adults [87]. The dietary intake represents an additionally risk factor for AMD [88], in particular, the consumption of fruits and vegetables containing antioxidant components such as vitamin C, α-carotene, and β-carotene seems to have a protective effect on AMD [89]. All these factors influence and can be influenced by nutritional status of elderly, inducing an increased risk of falling ill and malnourishment. Malnutrition itself causes many other health problems, such as poor immune functions, poor wound healing, and muscle weakness [82].

Unfortunately, the malnutrition among hospitalized elderly has always been underestimated, underdiagnosed, and undertreated. To date, several nutritional assessment tools can be used to estimate the degree of malnutrition in elderly. Some of them consist in record of clinical history and physical examinations; others are characterized by a list of questions associated with anthropometric measurements. The elderly who has been screened and has been identified to be at risk of malnutrition will then be directed to a hospital nutrition professional for further evaluation [90], since the treatment of malnutrition will allow a significant reduction in complication and a faster recovery of the patient.

1.9 Nutrition and Diet as Primary Environmental Factor for Disease Development

In conclusion, this chapter demonstrated how unhealthy diets and physical inactivity should be considered among the most important risk factors for major chronic, non-communicable diseases. They account for about 30% of the global disease burden, indeed it has been recently calculated that unhealthy diets (low in fruit and vegetables and/or high in sugar, processed foods or sodium) directly account for 37% of all deaths [91]. Low levels of physical activity accounted for another 5% of all deaths and 3.4% of disability-adjusted life years (DALYs).

To date, several evidences already demonstrated the positive impacts of a healthy diet on outcomes such as major cardiovascular events [92–95] as well as on the association between physical activity and mortality [96, 97].

Therefore, since both diet and life style are modifiable environmental factors, a great effort should be place in programs promoting healthy "habits", that will help to reduce not only the prevalence of several diseases, but also the economic impacts on health care.

References

1. Potter, P. and B. Gundert, Hippocrates [6] of Cos. Brill's New Pauly: Encyclopedia of the Ancient World, 2005. 6: p. 354–363.
2. Cardenas D. *Let not thy food be confused with thy medicine: The Hippocratic misquotation.* e-SPEN Journal. 2013;8(6):e260–2.
3. Georgiou NA, Garssen J, Witkamp RF. Pharma–nutrition interface: the gap is narrowing. Eur J Pharmacol. 2011;651(1):1–8.
4. Mahan LK, Escott-Stump S. *Krause's food, nutrition, & diet therapy*. 10th ed. Philadelphia: W.B. Saunders. xxxiv; 2000. p. 1194.
5. Gilpin A. *Dictionary of environment and sustainable development.* Chichester; New York, Wiley.. xiii; 1996. p. 247.

6. Bhupathiraju SN, Hu FB. Epidemiology of obesity and diabetes and their cardiovascular complications. Circ Res. 2016;118(11):1723–35.
7. Mozaffarian D, et al. Heart disease and stroke statistics--2015 update: a report from the American Heart Association. Circulation. 2015;131(4):e29–322.
8. Biener A, Cawley J, Meyerhoefer C. The impact of obesity on medical care costs and labor market outcomes in the US. Clin Chem. 2017;
9. Sacks G, Robinson E. Investing for health: potential mechanisms for the investment community to contribute to obesity prevention and improved nutrition. Curr Obes Rep. 2018;7(3):211–9.
10. McEvilly A. *Identifying and managing malnutrition in the community*. Br J Community Nurs. 2016. Suppl Nutrition:S14–21.
11. Nardone P, et al. Dietary behaviour of children attending primary school in Italy found by the surveillance system "OKkio alla salute". Epidemiol Prev. 2015;39(5–6):380–5.
12. Rippin HL, et al. *Child and adolescent nutrient intakes from current national dietary surveys of European populations*. Nutr Res Rev. 2018:1–32.
13. Guh DP, et al. The incidence of co-morbidities related to obesity and overweight: a systematic review and meta-analysis. BMC Public Health. 2009;9:88.
14. Centers for Disease Control and Prevention, C.D.P.a.H.P. *Obesity: Halting the epidemic by making health easier at a glance*. 2010 [cited 2018; Available from: https://obesity.procon.org/sourcefiles/CDCobesityatglance2009.pdf.
15. Pulgaron ER. Childhood obesity: a review of increased risk for physical and psychological comorbidities. Clin Ther. 2013;35(1):A18–32.
16. Hall ME, et al. Obesity, hypertension, and chronic kidney disease. Int J Nephrol Renovasc Dis. 2014;7:75–88.
17. Reilly JJ, Kelly J. Long-term impact of overweight and obesity in childhood and adolescence on morbidity and premature mortality in adulthood: systematic review. Int J Obes. 2011;35(7):891–8.
18. Wong E, et al. The role of obesity duration on the association between obesity and risk of physical disability. Obesity (Silver Spring). 2015;23(2):443–7.
19. Hatzenbuehler ML, Keyes KM, Hasin DS. Associations between perceived weight discrimination and the prevalence of psychiatric disorders in the general population. Obesity (Silver Spring). 2009;17(11):2033–9.
20. Puhl R, Suh Y. Stigma and eating and weight disorders. Curr Psychiatry Rep. 2015;17(3):552.
21. Vartanian LR, Novak SA. Internalized societal attitudes moderate the impact of weight stigma on avoidance of exercise. Obesity (Silver Spring). 2011;19(4):757–62.
22. Fowler-Brown AG, Ngo LH, Wee CC. The relationship between symptoms of depression and body weight in younger adults. Obesity (Silver Spring). 2012;20(9):1922–8.
23. Gouveia MJ, et al. Quality of life and psychological functioning in pediatric obesity: the role of body image dissatisfaction between girls and boys of different ages. Qual Life Res. 2014;23(9):2629–38.
24. Williams EP, et al. Overweight and obesity: prevalence, consequences, and causes of a growing public health problem. Curr Obes Rep. 2015;4(3):363–70.
25. Institute, I.F.P.R., *lobal nutrition report 2014: Actions and accountability to accelerate the world's progress on nutrition*. 2014, Washington DC (USA): International Food Policy Research Institute (IFPRI).
26. Rehman HM, et al. Legume biofortification is an underexploited strategy for combatting hidden hunger. Plant Cell Environ. 2018;
27. Burchi F, Fanzo J, Frison E. The role of food and nutrition system approaches in tackling hidden hunger. Int J Environ Res Public Health. 2011;8(2):358–73.
28. WHO. *International Food Policy Research Institute. 2015 global nutrition report: actions and accountability to advance nutrition and sustainable development*. 2015 [cited 2018; Available from: http://www.fao.org/fileadmin/user_upload/raf/uploads/files/129654.pdf.
29. Welch RM, Graham RD. Agriculture: the real nexus for enhancing bioavailable micronutrients in food crops. J Trace Elem Med Biol. 2005;18(4):299–307.

30. Eggersdorfer M, et al. *Hidden Hunger: Solutions for America's Aging Populations*. Nutrients. 2018;10(9)
31. Martins VJ, et al. Long-lasting effects of undernutrition. Int J Environ Res Public Health. 2011;8(6):1817–46.
32. Khandelwal S, Kurpad A. Nurturing public health nutrition education in India. Eur J Clin Nutr. 2014;68(5):539–40.
33. Debabrata D, Monika S, Bibekananda D. Food quality of rural poor in India: a case study of Orissa and Bihar. Int JInn Appl Studies. 2014;6:622–8.
34. Graham, R.D., et al., *Nutritious subsistence food systems*, in *Advances in agronomy, Vol 92*, D.L. Sparks, Editor. 2007, Elsevier Academic Press Inc: San Diego. p. 1–74.
35. Morton JF. The impact of climate change on smallholder and subsistence agriculture. Proc Natl Acad Sci U S A. 2007;104(50):19680–5.
36. Wu XY, et al. The influence of physical activity, sedentary behavior on health-related quality of life among the general population of children and adolescents: a systematic review. PLoS One. 2017;12(11):e0187668.
37. Leung MM, et al. Intervening to reduce sedentary behaviors and childhood obesity among school-age youth: a systematic review of randomized trials. J Obes. 2012;2012:685430.
38. Iannotti RJ, et al. Patterns of adolescent physical activity, screen-based media use, and positive and negative health indicators in the U.S. and Canada. J Adolesc Health. 2009;44(5):493–9.
39. Tremblay MS, et al. Systematic review of sedentary behaviour and health indicators in school-aged children and youth. Int J Behav Nutr Phys Act. 2011;8:98.
40. Wassenaar TM, et al. Effects of a programme of vigorous physical activity during secondary school physical education on academic performance, fitness, cognition, mental health and the brain of adolescents (Fit to Study): study protocol for a cluster-randomised trial. Trials. 2019;20(1):189.
41. Batacan RB Jr, et al. Effects of high-intensity interval training on cardiometabolic health: a systematic review and meta-analysis of intervention studies. Br J Sports Med. 2017;51(6):494–503.
42. Jelleyman C, et al. The effects of high-intensity interval training on glucose regulation and insulin resistance: a meta-analysis. Obes Rev. 2015;16(11):942–61.
43. Asle Mohammadi Zadeh M, et al. *Diets along with interval training regimes improves inflammatory & anti-inflammatory condition in obesity with type 2 diabetes subjects*. J Diabetes Metab Disord. 2018;17(2):253–67.
44. Wewege M, et al. The effects of high-intensity interval training vs. moderate-intensity continuous training on body composition in overweight and obese adults: a systematic review and meta-analysis. Obes Rev. 2017;18(6):635–46.
45. Shah M, et al. High-volume exercise program in obese bariatric surgery patients: a randomized, controlled trial. Obesity (Silver Spring). 2011;19(9):1826–34.
46. Kalarchian MA, et al. Preoperative lifestyle intervention in bariatric surgery: initial results from a randomized, controlled trial. Obesity (Silver Spring). 2013;21(2):254–60.
47. Bond DS, et al. Exercise improves quality of life in bariatric surgery candidates: results from the Bari-active trial. Obesity (Silver Spring). 2015;23(3):536–42.
48. Pico-Sirvent I, et al. *Effects of a Combined High-Intensity Interval Training and Resistance Training Program in Patients Awaiting Bariatric Surgery: A Pilot Study*. Sports (Basel). 2019;7(3)
49. Lee IM, et al. Effect of physical inactivity on major non-communicable diseases worldwide: an analysis of burden of disease and life expectancy. Lancet. 2012;380(9838):219–29.
50. Andersen LB, Mota J, Di Pietro L. Update on the global pandemic of physical inactivity. Lancet. 2016;388(10051):1255–6.
51. Hupin D, et al. Even a previous light-active physical activity at work still reduces late myocardial infarction and stroke in retired adults aged>65 years by 32%: the PROOF cohort study. Front Public Health. 2019;7:51.

52. Sanchez-Sanchez JL, et al. Sedentary behaviour, physical activity, and sarcopenia among older adults in the TSHA: isotemporal substitution model. J Cachexia Sarcopenia Muscle. 2019;10(1):188–98.
53. Widdowson EM. Harmony of growth. Lancet. 1970;1(7653):902–5.
54. Rees WD. *Interactions between nutrients in the maternal diet and the implications for the long-term health of the offspring*. Proc Nutr Soc. 2018:1–9.
55. Kim H, et al. Relationship between body-mass index and serum folate concentrations in pregnant women. Eur J Clin Nutr. 2012;66(1):136–8.
56. Oyama K, et al. Folic acid prevents congenital malformations in the offspring of diabetic mice. Endocr J. 2009;56(1):29–37.
57. O'Malley EG, et al. Folate and vitamin B12 levels in early pregnancy and maternal obesity. Eur J Obstet Gynecol Reprod Biol. 2018;231:80–4.
58. Martino J, et al. *The Impact of Maternal Pre-Pregnancy Body Weight and Gestational Diabetes on Markers of Folate Metabolism in the Placenta*. Nutrients. 2018;(11):10.
59. Kaidar-Person O, et al. Nutritional deficiencies in morbidly obese patients: a new form of malnutrition? Part A: vitamins Obes Surg. 2008;18(7):870–6.
60. Kaidar-Person O, et al. Nutritional deficiencies in morbidly obese patients: a new form of malnutrition? Part B: minerals Obes Surg. 2008;18(8):1028–34.
61. Lozoff B. Iron deficiency and child development. Food Nutr Bull. 2007;28(4 Suppl):S560–71.
62. Tuokkola J, et al. Maternal dietary folate, folic acid and vitamin D intakes during pregnancy and lactation and the risk of cows' milk allergy in the offspring. Br J Nutr. 2016;116(4):710–8.
63. Moreno Villares JM. Nutrition in early life and the programming of adult disease: the first 1000 days. Nutr Hosp. 2016;33(Suppl 4):337.
64. Langley-Evans SC. Nutrition in early life and the programming of adult disease: a review. J Hum Nutr Diet. 2015;28(Suppl 1):1–14.
65. Langley-Evans SC, McMullen S. Developmental origins of adult disease. Med Princ Pract. 2010;19(2):87–98.
66. Bavdekar A, et al. Insulin resistance syndrome in 8-year-old Indian children: small at birth, big at 8 years, or both? Diabetes. 1999;48(12):2422–9.
67. Eriksson J, et al. Size at birth, childhood growth and obesity in adult life. Int J Obes Relat Metab Disord. 2001;25(5):735–40.
68. Carrington LJ, Langley-Evans SC. Wheezing and eczema in relation to infant anthropometry: evidence of developmental programming of disease in childhood. Matern Child Nutr. 2006;2(1):51–61.
69. Elenberg Y, Shaoul R. The role of infant nutrition in the prevention of future disease. Front Pediatr. 2014;2:73.
70. Nakamura A, Osonoi T, Terauchi Y. Relationship between urinary sodium excretion and pioglitazone-induced edema. J Diabetes Investig. 2010;1(5):208–11.
71. Zutavern A, et al. Timing of solid food introduction in relation to atopic dermatitis and atopic sensitization: results from a prospective birth cohort study. Pediatrics. 2006;117(2):401–11.
72. Zutavern A, et al. Timing of solid food introduction in relation to eczema, asthma, allergic rhinitis, and food and inhalant sensitization at the age of 6 years: results from the prospective birth cohort study LISA. Pediatrics. 2008;121(1):e44–52.
73. Gillman MW, et al. Risk of overweight among adolescents who were breastfed as infants. JAMA. 2001;285(19):2461–7.
74. Singhal A. Early growth and later atherosclerosis. World Rev Nutr Diet. 2013;106:162–7.
75. Ziegler AG, et al. Early infant feeding and risk of developing type 1 diabetes-associated autoantibodies. JAMA. 2003;290(13):1721–8.
76. Gerstein HC, VanderMeulen J. The relationship between cow's milk exposure and type 1 diabetes. Diabet Med. 1996;13(1):23–9.
77. Story, M., *Nutritional requirements during adoles-cence*, in *Textbook of Adolescent Medicine*, E.R. McAnarney, et al., Editors. 1992, Saunders: London, UK p 75-84.
78. WHO, *World Health Statistics 2014*, WHO, Editor. 2014, World Health Organization: Geneva.

79. Black RE, et al. Maternal and child undernutrition and overweight in low-income and middle-income countries. Lancet. 2013;382(9890):427–51.
80. Salam RA, et al. Interventions to improve adolescent nutrition: a systematic review and meta-analysis. J Adolesc Health. 2016;59(4S):S29–39.
81. Gidding SS, et al. Dietary recommendations for children and adolescents: a guide for practitioners. Pediatrics. 2006;117(2):544–59.
82. Abd Aziz NAS, et al. Assessing the nutritional status of hospitalized elderly. Clin Interv Aging. 2017;12:1615–25.
83. Amarya S, Singh K, Sabharwal M. Changes during aging and their association with malnutrition. Journal of Clinical Gerontology and Geriatrics. 2015;6(3):78–84.
84. Baum JI, Kim IY, Wolfe RR. *Protein Consumption and the Elderly: What Is the Optimal Level of Intake*. Nutrients. 2016;8(6)
85. Savanelli MC, et al. Preliminary results demonstrating the impact of Mediterranean diet on bone health. J Transl Med. 2017;15(1):81.
86. Prentice A. Diet, nutrition and the prevention of osteoporosis. Public Health Nutr. 2004;7(1A):227–43.
87. Foran S, Wang JJ, Mitchell P. Causes of visual impairment in two older population cross-sections: the Blue Mountains eye study. Ophthalmic Epidemiol. 2003;10(4):215–25.
88. Koushan K, et al. The role of lutein in eye-related disease. Nutrients. 2013;5(5):1823–39.
89. Kim EK, et al. Associations between fruit and vegetable, and antioxidant nutrient intake and age-related macular degeneration by smoking status in elderly Korean men. Nutr J. 2017;16(1):77.
90. Chern CJH, Lee S-D. Malnutrition in hospitalized Asian seniors: an issue that calls for action. J Clin Gerontol Geriatr. 2015;6(3):73–7.
91. Collaborators GBDRF. Global, regional, and national comparative risk assessment of 79 behavioural, environmental and occupational, and metabolic risks or clusters of risks, 1990-2015: a systematic analysis for the global burden of disease study 2015. Lancet. 2016;388(10053):1659–724.
92. Tong TY, et al. Prospective association of the Mediterranean diet with cardiovascular disease incidence and mortality and its population impact in a non-Mediterranean population: the EPIC-Norfolk study. BMC Med. 2016;14(1):135.
93. Liyanage T, et al. Effects of the Mediterranean diet on cardiovascular outcomes-a systematic review and meta-analysis. PLoS One. 2016;11(8):e0159252.
94. Stefler D, et al. Mediterranean diet score and total and cardiovascular mortality in Eastern Europe: the HAPIEE study. Eur J Nutr. 2017;56(1):421–9.
95. Estruch R, et al. Primary prevention of cardiovascular disease with a Mediterranean diet supplemented with extra-virgin olive oil or nuts. N Engl J Med. 2018;378(25):e34.
96. Samitz G, Egger M, Zwahlen M. Domains of physical activity and all-cause mortality: systematic review and dose-response meta-analysis of cohort studies. Int J Epidemiol. 2011;40(5):1382–400.
97. Woodcock J, et al. Non-vigorous physical activity and all-cause mortality: systematic review and meta-analysis of cohort studies. Int J Epidemiol. 2011;40(1):121–38.

Chapter 2
Circadian Rhythms in Health and Disease

Silke Kiessling

Keywords Circadian clock · Health · Circadian disruption · Metabolism · Immune system · Cancer · Jetlag · Shift work · Inflammation · Metabolic syndrome · Microbiota

2.1 The Mammalian Circadian System

2.1.1 The Circadian Clock

The regular changes between day and night caused by the Earth's rotation around its axis influences all life on this planet, from single cells to higher life forms, including humans [1–3]. To anticipate these recurring environmental changes organisms have developed the so called circadian (lat. *Circa* and *dies*, approximately one day) timing system. Results obtained from cyanobacteria with genetically altered circadian clocks have demonstrated the evolutionary advantage of having a circadian clock [4]. Lesion experiments in rodents located the mammalian circadian clock to the ventrolateral hypothalamus, right above the optic chiasma, the suprachiasmatic nuclei (SCN) [5, 6]. The SCN drives the animal's rhythmic behavior and physiology including the sleep-wake cycle, food intake behavior, body temperature, hormonal secretion, cardiovascular activity, acuity of the sensory system, renal plasma flow, intestinal peristaltic and detoxification [7–12] (Fig. 2.1A). In addition to the circadian clock in the central nervous system, so called "peripheral circadian clocks" have been discovered in almost all tissues, organs and even in single cells throughout the body [13, 14]. Importantly, circadian rhythms in tissue culture from peripheral clocks and the SCN persist for days, weeks and even years, demonstrating that non-SCN cells, such as the liver, the adrenal gland and even single cells contain their own endogenous circadian oscillators [15–18].

S. Kiessling (✉)
Chair of Nutrition and Immunology, ZIEL – Institute of Food & Health, Technical University Munich, Freising, Germany
e-mail: silke.kiessling@tum.de

R. Teperino (ed.), *Beyond Our Genes*,
https://doi.org/10.1007/978-3-030-35213-4_2

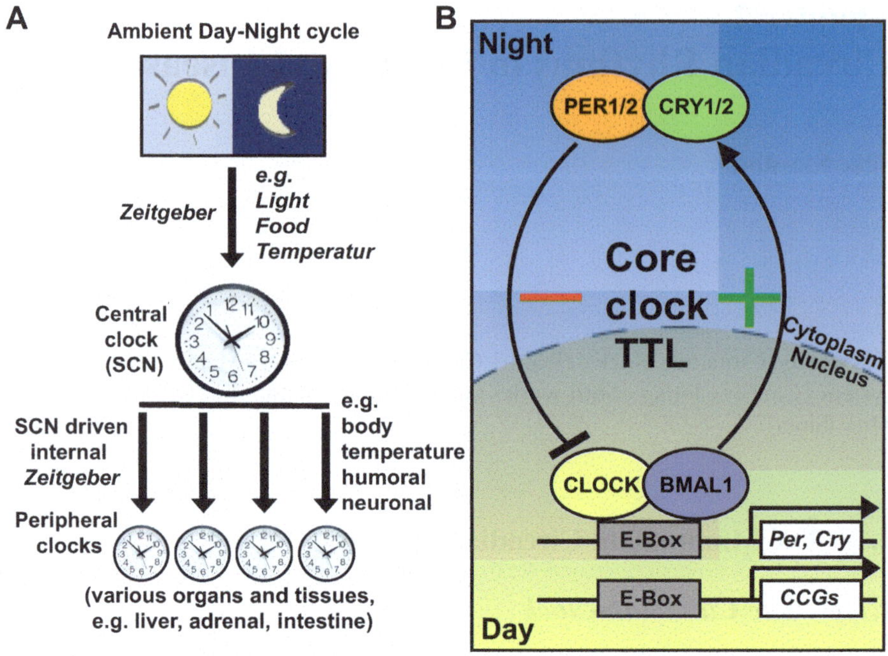

Fig. 2.1 The mammalian circadian clock. (**A**) Hierarchy of the circadian system: the central circadian clock located in the suprachiasmatic nucleus (SCN) is entrained to the environmental day by external *Zeitgeber*. The SCN orchestrates subordinated peripheral circadian clocks in various organs and tissues by internal signals. (**B**) Simplified molecular mechanism of the mammalian circadian clock: the core circadian clock is composed of a positive - negative feedback loop involving the genes Clock, Bmal1, Period and Cryptochrome. During the day BMAL1 and CLOCK proteins are expressed at high levels. The CLOCK::BMAL1 heterodimer is part of the positive arm of the transcription-translational feedback loop (TTL). The heterodimer binds to the E-Box in the promotor of clock genes, such as *Per1*, *Per2*, *Cry1*, *Cry2* and clock controlled genes (CCGs) and induces the transcription of these genes. The negative arm of the TTL is represented by the proteins from PER and CRY. During the night PER and CRY form dimers in the cytoplasm and translocate back to the nucleus where they block their own transcription

2.1.2 The Mammalian Molecular Clock Machinery

On the molecular level the mammalian circadian clock consists of a subset of clock genes, which form a cell-autonomous transcription-translational autoregulatory feedback loop [19]. Components of the positive limb are the two transcription factors: Circadian locomotor output cycles kaput (Clock) and Brain and muscle ARNT-like factor 1 (Bmal1). BMAL1 heterodimerizes with CLOCK by interaction of their basic helix-loop-helix (bHLH) domain [20] and binds to an E-Box enhancer element in the promotor of the Period (Per1/Per2/Per3) and Cryptochrome (Cry1/Cry2) genes. As part of the negative limb PER and CRY form a complex and after translocation to the nucleus the PER:CRY heterodimer represses their own transcription by inhibiting the enhancer activity of CLOCK:BMAL1 [14] (Fig. 2.1B).

In addition to the core loop, multiple accessory feedback loops have been discovered. The nuclear receptors (NRs) Rev-erb (Rev-erbα, Rev-erbβ) and retinoic acid receptor-like orphan receptor (RORα, RORβ, and RORγ), both activated through E-Box elements, rhythmically repress or activate the transcription of Bmal1 by binding to the RRE-element in the promotor of Bmal1 [21]. Another CLOCK:BMAL1-driven transcriptional loop involves the PAR-bZip (proline and acidic amino acid-rich basic leucine zipper) D-box-binding transcription factors (DBP), Thyrotroph embryonic factor (TEF), Hepatic leukemia factor (HLF) and E4 promoter–binding protein 4 (E4BP4), which feed back to the core loop e.g. by binding to a D-Box in the promotor of Cry1 [22]. Together, these accessory feedback loops are believed to stabilize the rhythm of the core loop [23]. In the past years a variety of posttranslational events acting on the (pre-) mRNA or proteins, including phosphorylation, sumoylation, ubiquitylation, intracellular transport, degradation or micro RNAs have been identified to be involved in the delay of several hours between mRNA and protein peaks and are thus critical for the fine-tuning of the 24-h oscillations (reviewed by [24]).

2.1.3 The Circadian Clock Regulates Overall Physiology

The basis for the circadian control of major physiological processes, including metabolism, immune functions and cell proliferation is that apart from clock genes, many non-clock genes contain E-boxes, D-boxes and RRE elements. Thereby, circadian clock components, such as CLOCK:BMAL1, DBP or REV-ERBα can rhythmically regulate the transcriptional activity of these so called clock-controlled genes (CCGs) [25] (Fig. 2.1B). Comparison of transcriptome profiling between various peripheral organs and the SCN has demonstrated that in any given tissue or organ thousands of genes are oscillating with a frequency of 24 hours and thus are controlled by circadian clocks [9, 26, 27]. Altogether, the results from these studies lead to the estimation that 50% of the whole transcriptome oscillates in at least one organ [26]. However, while the rhythmic core components of the circadian clock are conserved among tissues, the CCGs rarely overlap between tissues, and thus reflect the physiological function of each organ [28]. For example, in case of the liver, genes relevant for oxidative metabolism, mitochondrial functions, and amino acid turnover show circadian rhythms [29]. Experiments on mice with tissue-specific clock disruption in various peripheral clocks, such as the liver, the adrenal gland, the pancreas, the retina etc. further indicate that tissue-specific peripheral circadian clocks are the major circadian driver of their own specific functions [30–33]. For example, in rodents it has been demonstrated that the adrenal circadian clock regulates the rhythmic secretion of glucocorticoids [31, 34]. Similarly, mice with a liver-specific disruption of the circadian clock show abnormal glucose homeostasis, leading to defects in metabolic responses [30]. Even functions of single cells are controlled by their own clocks. For example the autonomous circadian clock in macrophages and T cells, two different immune cell subtypes, regulate the rhythmic expression of inflammatory markers, cytokine release or rhythmic response to antigen presentation [18, 35, 36].

2.2 The Circadian Clock in Relation to the Environment

2.2.1 The Circadian System Is in Synchrony with the Environment

Although circadian rhythms are generated intrinsically, organisms are influenced by environmental changes e.g. light, temperature, food availability, humidity, social cues or sound, which act on the circadian system as environmental timing cues, so called "*Zeitgeber*" (Fig. 2.1A). As a rule of thumb, in nocturnal animals (such as mice) endogenous clocks runs a little faster than 24 hours, while in diurnal animals (such as humans) clocks "tick" slightly slower than 24 hours [37]. Consequently, to maintain periodicity, the circadian clock needs to reach a stable phase relationship, not just between organ clocks, but also with the environmental time. The synchronization of the internal circadian clock time to these external *Zeitgebers* is referred to as entrainment and the process to adjust as resetting [38].

Light is the dominant *Zeitgeber* for all organisms, but how does the circadian system sense the day and night information? The location of the SCN above the optic nerve is ideal to directly receive the light information from the environment through the retino-hypothalamic tract [39]. Briefly, neurotransmitter release from retinal fibers induces a calcium dependent signaling cascade in SCN neurons (activation of calmodulin, MAP kinases and PKA) and the phosphorylation of the transcription factor CREB (cAMP response element binding protein) [40–42]. Activation of the cAMP response element in the promotor of clock genes, e.g. *Per* genes, induces their expression. Thereby, the SCN processes information about significant variations in the availability of light. To further adjust the body time with the environmental daytime, the circadian system is organized in a hierarchical manner. The light-entrainable central clock in the brain orchestrates subordinated peripheral clocks by neuronal, humoral and systemic cues (i.e. body temperature) [43] (Fig. 2.1A). For example, neural output from the SCN to other brain regions is responsible for autonomic and neuroendocrine regulation [44]. These pathways allow the SCN to coordinate daily variations in physiology and behavior in accordance with the environment.

Abolished rhythms in behavior and physiology have been demonstrated in rodents with SCN lesion and in transplantation studies [5, 6, 8]. Consequently the SCN was described as the "master clock" controlling circadian rhythms throughout the body. However, recent research on mice with a genetic disruption of the central clock indicated that peripheral clocks continue ticking, even in the absence of the SCN, although individual organ clocks gradually desynchronize from each other [45]. Accordingly, rather than a master clock, the SCN keeps peripheral circadian clocks in synchrony and thereby substantially enhances the physiology's efficiency. However, the molecular mechanisms underlying the SCN's controls over organ clocks are still not completely understood.

For a long time, it was believed that the SCN is the only clock capable of detecting environmental light information. However, these results were obtained from SCN lesion studies [46, 47]. Lesioning the SCN can damage the surrounding tissues including the retino-hypothalamic tract. Consequently the light input pathway to other peripheral tissues may have become defective. Later it was demonstrated in mice that the adrenal circadian clock can be activated by light via the autonomic nervous system [48] even at times when the SCN is irresponsive, indicating that a non-SCN pathway is involved [49]. Glucocorticoids, which are rhythmically produced in the adrenal gland, are potent synchronizing cues for other peripheral clocks [50, 51]. Therefore the adrenal gland may function as an internal synchronizer or 'body clock' acting to support the resetting role of the clock in the central nervous system [34].

2.2.2 Environmental Factors Can Disturb the Circadian Organization

Body clocks adjust the organism's behavior and physiology to recurring environmental changes related to day and night (such as food availability or the presence of predators). Consequently, virtually all aspects of behavior (e.g. sleep/wake cycle, fasting/feeding cycle), physiology (e.g. hormone secretion, immune response) and metabolism (e.g. glycolysis or fat-metabolism) show circadian rhythms [52]. However, in our modern society, frequent flyer, aircrew members and shift workers are exposed to artificial changes in environmental conditions.

What happens with the circadian system in people that no longer live in accordance with their inner clock time? In experiments on rodents exposed to an abrupt change in the light-dark cycle (so called Jetlag), it was found that (i) clock genes within the molecular clock as well as (ii) different tissue clocks adjust with different speed to the new light-dark conditions [34, 53]. For example, in mice the clock resides in the SCN and the adrenal reset within a few days to the new day-night conditions. This is not surprising, since both clocks are light responsive (see 3.2.1). In contrast, light-insensitive organs, e.g. the liver and the pancreas entrain weeks later [34]. As a consequence of the different resetting speeds of the circadian clocks throughout the body, even a single shift in the light-dark cycle globally disrupts the circadian system [34]. Although the disruption is temporary, it can affect the organism's behavior and physiology, e.g. hormone secretion, even for a few weeks.

In humans, shift work, in particular the night shift, is one of the most frequent reasons for the disruption of the circadian system. According to worldwide epidemiological data, up to 30% of the working population are employed in non-standard work hours e.g. evening or rotating shifts [54]. Similar to results obtained from rodents, modern life style disrupts circadian rhythms in humans [55]. Working in non-regular shifts causes significant alterations of sleep and biological functions, which, in turn can affect people's physical and psychological wellbeing. There is accumulating evidence that living in mismatch between your inner clock time and the external daytime time in the long-term provokes a wide range of pathologies including gastrointestinal diseases, metabolic syndrome, cardiovascular diseases, inflammation, mood disorders and even cancer [56–59]. The impact of the circadian clock on the development and progression of diseases, in particular during changing environmental conditions, will be introduced in the next section.

2.3 The Circadian Clock and Diseases

2.3.1 Circadian Regulation of Metabolic Functions

Emerging evidence closely links circadian clock function to metabolic homeostasis. Major regulators of energy homeostasis, such as glucose transporter Sglt1, Glut2 and Glut5, peptide transporter Pept1 or lipid regulator such as Pparγ, fluctuate between day and night in various organs e.g. liver, intestine, muscle and adipocytes [30]. Accordingly, disruption of circadian clock function may contribute to the development of metabolic diseases. Indeed, metabolic syndrome has frequently been found in people living in mismatch with their environment. A higher body mass index (BMI), increased blood pressure and enhanced triglycerides correlate with the time nurses spend working on rotating shifts and sets them at higher risk of developing Type-2-diabetes [60]. In accordance to symptoms found in shift workers, mice with tissue-wide genetic disruption of the circadian clock, e.g. by loss or mutation of the core clock genes Bmal1 and Clock, reveal disturbed energy homeostasis [61]. The diurnal variation in glucose and triglycerides as well as the rhythms in gluconeogenesis is blunted. In addition, overall glucose and triglycerides in the blood are enhanced. The mice become less tolerant for glucose and more sensitive for insulin and show increased fat mass [30, 61, 62]. Interestingly, similar results were obtained in mice with tissue-specific circadian clock dysfunction, e.g. in adipocytes or hepatocytes [30, 63], indicating that organ-specific peripheral circadian oscillators play a prominent role in energy regulation. Of note, not all metabolic alterations were found in all different genetic mouse models for clock disruption, in particular weight gain was not consistently observed. Nevertheless, wildtype mice with a generally functional circadian clock develop metabolic syndrome when kept under simulated shift work conditions [64, 65].

People who work unusual hours expose their circadian system to light during the night and in addition tend to consume fat-rich food at odd hours. In natural conditions, the daily fasting-feeding cycle is set by the central clock [66]. Although the dominant *Zeitgeber* for the central clock is light, peripheral circadian clocks, including the liver and the digestive tract, respond mostly to the timing of food intake [67]. However, when food is presented to rodents during rest hours, food-born signals can override the coordinated signals from the central clock and peripheral circadian clocks uncouple from the SCN [68]. Consequently, shift workers eating at unusual hours uncouple their peripheral circadian clocks, including those relevant for metabolism, from the SCN. In addition, artificial exposure to light at night in rodents can desynchronize all body clocks from each other [34] and has been shown to increase body mass by shifting the time of food intake [69]. Thus, shift work conditions and irregular feeding cycles cause circadian disturbance within and between the clocks in various organs, including the ones relevant for energy homeostasis, which is sufficient to induce metabolic malfunction. Moreover, shift workers eat at odd hours and at the same time, they tend to consume food with higher fat content. Metabolic processes can feed back to the circadian clock. A high fat diet (HFD) alters clock gene expression in the central clock, which in turn influences the response of the central clock to light [70] and disrupts behavioral rhythms in mice [71]. Mice under *ad libitum* HFD become more active during their usual resting time and show increased food uptake during the wrong time of the day [70]. Consequently, these mice exhibit circadian disruption similar to mice exposed to daytime-restricted food access (see section above). Taken together, light exposure at night, changes in food timing and diet may explain why metabolic disorders have frequently been linked to shift work.

2.3.2 Circadian Control of the Immune Response

A wide range of immunological functions, ranging from numbers of peripheral blood mononuclear cells as well as the level of cytokines, undergo daily fluctuations in humans and rodents [72, 73]. Interestingly, the highest number of immune cells (i.e. leukocytes, phagocytes) was detected in the circulation during the resting phase, namely during the night for humans and during the day in rodents [74–76]. Consequently, the susceptibility to infection likely underlies circadian variation. Indeed, experiments on mice demonstrated a higher inflammatory response to infection with *Salmonella typhimurium*, *Leishmania major* parasites and a higher pathogenicity of *Listeria monocytogenes* during their active phase [74, 76, 77]. A similar modulation of inflammatory response across the day has been observed in humans. For example, during the early morning the inflammatory response was strongest in people's reaction to allergic asthma [78] and in people suffering from sepsis an enhanced mortality was observed during the night [79].

Recent research indicated the existence of circadian clocks even in single cells of the immune system. For example, autonomous oscillations of clock gene expression has been found in T cells from mice and humans [36] and in macrophages [18]. Similar to the functionality of organs regulated by their intrinsic clocks, these single cell oscillators have been shown to mediate cell-type specific functions. For example the T cell clock was reported to control rhythmic cytokine release after stimulation of toll-like receptors (TLR) and gates a time of day-dependent immune response to immunization with antigen-loaded dendritic cells [35]. The macrophage clock continues ticking in culture and the clocks of macrophages obtained from mice sacrificed during the night induce an elevated cytokine release after LPS stimulation compared to macrophages harvested from animals euthanized during the day [18]. The physiological relevance of the circadian clock in immune cells, specifically in phagocytes, has been assessed in mice in the context of an infection with *Leishmania major* parasites [76]. The daytime-dependent differences in an inflammatory response to parasites were absent when experiments where performed in mice lacking the circadian clock in hematopoietic cells, indicating that the circadian clock in these immune cells is mediating the observed effect [76].

Other peripheral clocks may be involved in mediating an inflammatory response. For example lung epithelial cells can drive immune cell recruitment to an infection site by controlling the daytime-dependency of cytokine release after *Streptococcus pneumoniae* infection [80]. Consequently, besides circadian clocks residing in immune cells, an inflammatory response following pathogen stimulation seems to be driven by multiple body clocks. Accordingly, disruption of the circadian organization, as experienced during jetlag and shift work [34], was shown to relate to immune deficits. Severely reduced survival has been observed in mice undergoing simulated shift work conditions and subjected to endotoxic shock following LPS injection [81]. Similar to rodents, simulated night shift disrupts circadian rhythms of immune functions in humans [82] and thereby may enhance the susceptibility to infection and inflammation. Indeed, Boscolo and colleagues reviewed literature indicating an increased risk of night shift workers to develop autoimmune diseases [83]. Accordingly, results obtained by the research group of Kumar indicate that jetlag may enhance the susceptibility to infection with malaria due to disturbed circadian regulation of itching behavior [84].

In studies on shift workers, a higher prevalence to develop chronic gastrointestinal inflammation was observed, including inflammatory bowel diseases (IBD), a group of chronic inflammatory disorders of the gastrointestinal tract manifesting as Crohn's disease (CD) or ulcerative colitis [85] (reviewed by Swanson et al. [86]). For example, nurses on rotating shifts discover a higher prevalence of irritable bowel syndrome [87]. Since major players of the immune system are under circadian regulation and about 70% of the immune system is located in the gastrointestinal system [88], it is not surprising, that strong associations have been found between working on rotating shifts and gastrointestinal diseases. This association was further supported by experiments on rodents exposed to simulated shift work conditions. Enhanced development and progression of chronic inflammation within the gastrointestinal tract was observed [89, 90]. Altogether, these studies indicate the importance of the circadian system for a functional immune defense. However, future studies are required to further examine the molecular link between circadian disruption and inflammatory diseases following shift work. Nevertheless, experiments on rodents identified that a functional circadian clock maintains the intestinal barrier, a key function for gastrointestinal immune homeostasis [89], which protects against the invasion of foreign pathogens. Elevated translocation of pro-inflammatory bacterial products such as LPS from the intestinal lumen into systemic circulation elicits a strong pro-inflammatory response [89]. Since the barrier dysfunction is an underlying factor of IBD [88], circadian disruption may promote intestinal inflammation due to elevated intestinal permeability.

2.3.3 The Circadian Clock and Environmental Factors Control Microbiome Fluctuations

Amongst others, the microbiome forms an important aspect of nutrient supply and immune responses. Intestinal microbiota are strongly influenced by the host's immune system and in turn the immune system is constantly challenged to contain the microbes in the intestinal lumen. The first line of defense are intestinal epithelial cells (IECs), which form a biochemical and physical barrier that maintains segregation between luminal microbial communities and the mucosal immune system and thereby maintains immune homeostasis [88]. The second defense line is represented by the host's immune system. The presence of microbiota and pathogens in the gut is sensed in IECs e.g. by clock-controlled TLR [91]. Surface marker and metabolic end products from microbiota activate cytokine secretion to initiate an immune defense, which protects against potential pathogens [92]. Nevertheless, microbiota can become pathogenic and intensely attack the immune system, leading to inflammation and even carcinogenesis. Bacterial signaling in health and disease at the intestinal epithelial interface has been recently reviewed by Coleman & Haller [93].

Interestingly, the host's circadian clock controls the intestinal immune homeostasis by regulating the abundance of immune cells, such as lymphocytes and the barrier function of the intestine by controlling mucus and antimicrobial peptide secretion [91, 94–96]. In addition, major regulators of microbiota composition are under the control of the host's circadian clock (reviewed by [97]). Given the symbiotic relationship between humans and their resident gut bacteria, it is not surprising that daytime-dependent oscillations in microbiota composition and function have been reported in humans, and in mice [98]. Moreover, diurnal rhythms in microbiota have been demonstrated to rely on a functional circadian clock, since disruption of the circadian clock in mice either deficient for two major clock genes (Per1 and Per2), or kept under jetlag conditions, partly abolishes the oscillations in microbiota composition, e.g. in *Bacteroidales* and *Ruminococcaceae* respectively [98].

Environmental factors, such as changes in the light dark cycle, and irregular meal times, can strongly influence the diurnal fluctuations in the microbiome. Rotating light dark cycles in mice have been reported to cause a microbial imbalance, so called *dysbiosis* [98, 99]. Shift work conditions may likely cause dysbiosis through disturbance of the circadian system, since the circadian clock has been reported to – at least partly – control microbiota fluctuations. For example, the abundance of various operational taxonomic units (OTUs) becomes arrhythmic in mice kept in simulated jetlag conditions [98]. In addition, circadian disruption has been reported to cause an imbalance in the intestinal barrier [89, 99]. Since, the clock controls the expression of barrier markers, such as cytosolic occluding (OCLN), claudin-1 (CLDN1) and E-cadherin (CDH1) [89, 100, 101], this may constitute a mechanism how circadian disruption affects intestinal permeability. Although mainly harmless, microbiota can induce an immune response and lead to chronic inflammation, such as IBD, when the intestinal barrier function is disturbed. Indeed, mice exposed to rotating light-dark cycles, show increased sensitivity to LPS [81], and are more susceptible to IBD [90, 102, 103].

A side effect of rotating light schedules is an altered timing of food uptake, which may be an additional factor causing the observed dysbiosis in shift workers. The feeding rhythm and the content of the diet has been described to orchestrate diurnal microbiota composition in mice [104]. Accordingly, HFD, but not regular chow diet, promoted microbiota dysbiosis in mice during simulated shift work conditions [99]. On one hand, the timing of feeding behavior sets the phase of specific microbiota oscillations, i.e. the peak in the abundance of *Bacteroides* shifted by ~12 hours when mice were exposed to daytime restricted feeding [98]. On the other hand, the microbiota can manipulate the host's feeding behavior. Potential mechanism have been reviewed by Alcock et al. [105]. By influencing the circadian clock, i.e. as observed in mice fed a HFD, the microbiota may affect the host's metabolism [106]. Taken together, the balance between the circadian system and microbes presents another risk factor for metabolic and inflammatory disorders.

2.3.4 Environmental Changes Promote Cancer Through Clock Dysfunction

Cancer is one of the most common causes of death. Interestingly, developed countries exhibit a ten-fold higher tumor incidence compared to developing nations. This difference can be attributed to risk factors including lifestyle, diet, obesity, physical inactivity, alcohol consumption and smoking. In 2007 the World Health Organization has classified shift work as "probably carcinogenic" based on results from various experimental and epidemiological studies [107]. For example, night shift work resulted in a higher incidence of endometrial and colorectal cancer in nurses [108] and increased the risk to develop non-Hodgkin lymphoma [109]. Studies in humans were supported by experiments in rodents, e.g., chronic jetlag condition promoted the incidence of lung cancer in rats following injection of tumor cells [110] and enhanced the progression of Glasgow osteosarcoma in mice [111]. Jetlag and shift work disturb the circadian system. Consequently, another so far completely underestimated risk factor for carcinogenesis has been identified: circadian clock disruption. The molecular clock has been shown to rhythmically regulate cellular functions, including proliferation, DNA damage response, senescence, apoptosis, angiogenesis and metabolism. These functions can become hallmarks of cancer, when uncontrolled. Interestingly, disruption of the circadian clock is associated with a higher proliferation rate and enhanced tumor growth [112, 113]. This is not surprising, since several tumor-suppressor and key cell cycle genes, such as *Myc* and *Ccnd1* are directly controlled by the CLOCK-BMAL1 dimer [114].

Recently, alterations of circadian clock genes have been found in various cancer cell lines and tumor tissues from humans and mice (summarized by Kiessling et al. [115]). Suppression of the circadian clock in melanoma cells has been identified as causal for enhanced tumor growth in mice [116]. Thus, the circadian clock may become a target to develop novel strategies to treat cancer in people undergoing shift work or repeated jetlag. Indeed, resetting the clock in B16 melanoma cells inoculated in mice restored the rhythmic expression of clock and cell cycle genes, which in turn slowed down the speed of the cell cycle and dramatically reduced tumor growth [116].

Besides genetic alterations, the tumor microenvironment plays a determining role in tumorigenesis [117]. In this regard, inflammation and microbes were identified to be associated with tumorigenesis. For example, elevated intestinal permeability in a colorectal cancer mouse model resulted in increased infiltration of cytokines and chemokines, which induced an inflammatory response and thus may play a causative role in the development of several inflammatory and metabolic diseases, such as IBD and colitis-associated cancer [118]. In addition, increased intestinal permeability enhances the interaction between the intestinal microbiota and the host. It is already known that microorganism can trigger specific tumorigenic pathways to promote tumorigenesis by an increased frequency of gene mutations. For example, *Enterococcus faecalis* caused chromosomal instability in the host by extracellular superoxide production; *Escherichia coli* induced DNA double-strand breaks by their produced genotoxin or *Bacteroides fragilis* produces enterotoxin with an increased permeability, cellular proliferation and cytokine infiltration as consequence (reviewed by [119]). Frequent environmental changes, popular in people with a western lifestyle, likely promote cancer development or progression through disturbance of the circadian clock, induction of inflammation and microbiota dysbiosis. Taken together, environmental factors can severely influence the host's physiology, by acting on the circadian system, and thus promote the risk to develop various diseases, including chronic inflammation, metabolic disorders and even cancer (Fig. 2.2).

2.4 Chapter Conclusion

This chapter focuses on the basics of disturbances of the mammalian circadian system, such as during jetlag or shift work and related physiological changes affecting overall physiology and health. The described studies highlight the importance of

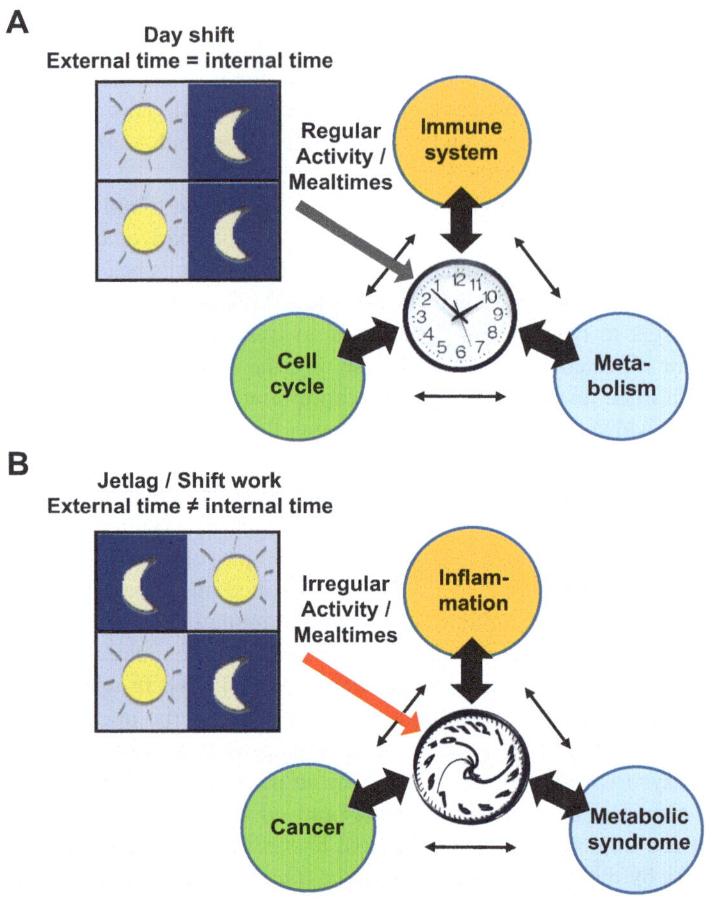

Fig. 2.2 Physiological consequences of circadian disruption. (**A**) Living in synchrony with the environmental day- night cycle results in regular light – dark exposure, mealtimes and rest-activity cycles and keeps the circadian system stably entrained with the environment. Several environmental factors can cause circadian disruption. (**B**) Changes in the environmental day-night cycle, such as occurs in shift workers or during jetlag, cause a disruption of the circadian system on multiple level. Dysfunction of circadian clocks can contribute to multifactorial diseases, such as inflammation, cancer and metabolic syndrome

the circadian system for a functional physiology and demonstrate its influence on metabolism, cell cycle and immune functions. Data obtained from mice exposed to circadian disruption illustrate the molecular and physiological consequences of environmental factors on the circadian system, which may lead to the development of physiological disorders. Consequently, the circadian clock may constitute a mechanisms by which the environment, diet, microbiota and the immune system affect multiple illnesses, and may represent a target for future therapies.

References

1. Edery I. Circadian rhythms in a nutshell. Physiol Genomics. 2000;3(2):59–74. https://doi.org/10.1152/physiolgenomics.2000.3.2.59.
2. Devlin PF, Kay SA. Circadian photoperception. Annu Rev Physiol. 2001;63:677–94. https://doi.org/10.1146/annurev.physiol.63.1.677.
3. Perez-Roger I, Solomon DL, Sewing A, Land H. Myc activation of cyclin E/Cdk2 kinase involves induction of cyclin E gene transcription and inhibition of p27(Kip1) binding to newly formed complexes. Oncogene. 1997;14(20):2373–81.
4. Ma P, Woelfle MA, Johnson CH. An evolutionary fitness enhancement conferred by the circadian system in cyanobacteria. Chaos, solitons, and fractals. 2013;50:65–74. https://doi.org/10.1016/j.chaos.2012.11.006.
5. Moore RY, Eichler VB. Loss of a circadian adrenal corticosterone rhythm following suprachiasmatic lesions in the rat. Brain Res. 1972;42(1):201–6.
6. Stephan FK, Zucker I. Circadian rhythms in drinking behavior and locomotor activity of rats are eliminated by hypothalamic lesions. Proc Natl Acad Sci U S A. 1972;69(6):1583–6.
7. Ralph MR, Foster RG, Davis FC, Menaker M. Transplanted suprachiasmatic nucleus determines circadian period. Science. 1990;247(4945):975–8.
8. Sollars PJ, Kimble DP, Pickard GE. Restoration of circadian behavior by anterior hypothalamic heterografts. J Neurosci. 1995;15(3 Pt 2):2109–22.
9. Panda S, Antoch MP, Miller BH, Su AI, Schook AB, Straume M, et al. Coordinated transcription of key pathways in the mouse by the circadian clock. Cell. 2002;109(3):307–20.
10. Schibler U, Sassone-Corsi P. A web of circadian pacemakers. Cell. 2002;111(7):919–22.. S0092867402012254 [pii]
11. Storch KF, Lipan O, Leykin I, Viswanathan N, Davis FC, Wong WH, et al. Extensive and divergent circadian gene expression in liver and heart. Nature. 2002;417(6884):78–83. https://doi.org/10.1038/nature744.
12. Sujino M, Masumoto KH, Yamaguchi S, van der Horst GT, Okamura H, Inouye ST. Suprachiasmatic nucleus grafts restore circadian behavioral rhythms of genetically arrhythmic mice. Curr Biol 2003;13(8):664–668. .doi:S0960982203002227 [pii].
13. Mohawk JA, Green CB, Takahashi JS. Central and peripheral circadian clocks in mammals. Annu Rev Neurosci. 2012;35:445–62. https://doi.org/10.1146/annurev-neuro-060909-153128.
14. Dibner C, Schibler U, Albrecht U. The mammalian circadian timing system: organization and coordination of central and peripheral clocks. Annu Rev Physiol. 2010;72:517–49. https://doi.org/10.1146/annurev-physiol-021909-135821.
15. Balsalobre A, Damiola F, Schibler U. A serum shock induces circadian gene expression in mammalian tissue culture cells. Cell. 1998;93(6):929–37.. S0092-8674(00)81199-X [pii]
16. Yoo SH, Yamazaki S, Lowrey PL, Shimomura K, Ko CH, Buhr ED, et al. PERIOD2::LUCIFERASE real-time reporting of circadian dynamics reveals persistent circadian oscillations in mouse peripheral tissues. Proc Natl Acad Sci U S A. 2004;101(15):5339–46. https://doi.org/10.1073/pnas.03087091010308709101.. [pii]
17. Welsh DK, Logothetis DE, Meister M, Reppert SM. Individual neurons dissociated from rat suprachiasmatic nucleus express independently phased circadian firing rhythms. Neuron. 1995;14(4):697–706.
18. Keller M, Mazuch J, Abraham U, Eom GD, Herzog ED, Volk HD, et al. A circadian clock in macrophages controls inflammatory immune responses. Proc Natl Acad Sci U S A. 2009;106(50):21407–12. https://doi.org/10.1073/pnas.0906361106.
19. Takahashi JS. Transcriptional architecture of the mammalian circadian clock. Nat Rev Genet. 2017;18(3):164–79. https://doi.org/10.1038/nrg.2016.150.
20. Huang N, Chelliah Y, Shan Y, Taylor CA, Yoo SH, Partch C, et al. Crystal structure of the heterodimeric CLOCK:BMAL1 transcriptional activator complex. Science. 2012;337(6091):189–94. https://doi.org/10.1126/science.1222804.

21. Guillaumond F, Dardente H, Giguere V, Cermakian N. Differential control of Bmal1 circadian transcription by REV-ERB and ROR nuclear receptors. J Biol Rhythm. 2005;20(5):391–403. https://doi.org/10.1177/0748730405277232.
22. Ukai-Tadenuma M, Yamada RG, Xu H, Ripperger JA, Liu AC, Ueda HR. Delay in feedback repression by cryptochrome 1 is required for circadian clock function. Cell. 2011;144(2):268–81. https://doi.org/10.1016/j.cell.2010.12.019.
23. Hastings MH. Circadian clockwork: two loops are better than one. Nat Rev Neurosci. 2000;1(2):143–6. https://doi.org/10.1038/35039080.
24. Preussner M, Heyd F. Post-transcriptional control of the mammalian circadian clock: implications for health and disease. Pflugers Archiv: European journal of physiology. 2016;468(6):983–91. https://doi.org/10.1007/s00424-016-1820-y.
25. Albrecht U. Timing to perfection: the biology of central and peripheral circadian clocks. Neuron. 2012;74(2):246–60. https://doi.org/10.1016/j.neuron.2012.04.006.
26. Zhang R, Lahens NF, Ballance HI, Hughes ME, Hogenesch JB. A circadian gene expression atlas in mammals: implications for biology and medicine. Proc Natl Acad Sci U S A. 2014;111(45):16219–24. https://doi.org/10.1073/pnas.1408886111.
27. Hughes ME, DiTacchio L, Hayes KR, Vollmers C, Pulivarthy S, Baggs JE, et al. Harmonics of circadian gene transcription in mammals. PLoS Genet. 2009;5(4):e1000442. https://doi.org/10.1371/journal.pgen.1000442.
28. Panda S, Hogenesch JB. It's all in the timing: many clocks, many outputs. J Biol Rhythm. 2004;19(5):374–87. https://doi.org/10.1177/0748730404269008.
29. Bass J, Takahashi JS. Circadian integration of metabolism and energetics. Science. 2010;330(6009):1349–54. https://doi.org/10.1126/science.1195027.
30. Lamia KA, Storch KF, Weitz CJ. Physiological significance of a peripheral tissue circadian clock. Proc Natl Acad Sci U S A. 2008;105(39):15172–7. https://doi.org/10.1073/pnas.0806717105.
31. Oster H, Damerow S, Kiessling S, Jakubcakova V, Abraham D, Tian J, et al. The circadian rhythm of glucocorticoids is regulated by a gating mechanism residing in the adrenal cortical clock. Cell Metab. 2006;4(2):163–73. https://doi.org/10.1016/j.cmet.2006.07.002.
32. Storch KF, Paz C, Signorovitch J, Raviola E, Pawlyk B, Li T, et al. Intrinsic circadian clock of the mammalian retina: importance for retinal processing of visual information. Cell. 2007;130(4):730–41. https://doi.org/10.1016/j.cell.2007.06.045.
33. Sadacca LA, Lamia KA, deLemos AS, Blum B, Weitz CJ. An intrinsic circadian clock of the pancreas is required for normal insulin release and glucose homeostasis in mice. Diabetologia. 2011;54(1):120–4. https://doi.org/10.1007/s00125-010-1920-8.
34. Kiessling S, Eichele G, Oster H. Adrenal glucocorticoids have a key role in circadian resynchronization in a mouse model of jet lag. J Clin Invest. 2010;120(7):2600–9. https://doi.org/10.1172/JCI41192.
35. Fortier EE, Rooney J, Dardente H, Hardy MP, Labrecque N, Cermakian N. Circadian variation of the response of T cells to antigen. J Immunol. 2011;187(12):6291–300. https://doi.org/10.4049/jimmunol.1004030.
36. Bollinger T, Leutz A, Leliavski A, Skrum L, Kovac J, Bonacina L, et al. Circadian clocks in mouse and human CD4+ T cells. PLoS One. 2011;6(12):e29801. https://doi.org/10.1371/journal.pone.0029801.
37. Daan S, Pittendrigh C. A functional analysis of circadian pacemakers in nocturnal rodents. J Comp Physiol. 1976;106(3):253–66. https://doi.org/10.1007/bf01417857.
38. Aschoff J, Pohl H. Phase relations between a circadian rhythm and its zeitgeber within the range of entrainment. Naturwissenschaften. 1978;65(2):80–4.
39. Moore RY, Lenn NJ. A retinohypothalamic projection in the rat. J Comp Neurol. 1972;146(1):1–14. https://doi.org/10.1002/cne.901460102.
40. Hirota T, Fukada Y. Resetting mechanism of central and peripheral circadian clocks in mammals. Zool Sci. 2004;21(4):359–68.
41. Xia Z, Dudek H, Miranti CK, Greenberg ME. Calcium influx via the NMDA receptor induces immediate early gene transcription by a MAP kinase/ERK-dependent mechanism. J Neurosci. 1996;16(17):5425–36.

42. Antle MC, Silver R. Orchestrating time: arrangements of the brain circadian clock. Trends Neurosci. 2005;28(3):145–51. https://doi.org/10.1016/j.tins.2005.01.003.
43. Schibler U, Ripperger J, Brown SA. Peripheral circadian oscillators in mammals: time and food. J Biol Rhythm. 2003;18(3):250–60.
44. Kalsbeek A, Perreau-Lenz S, Buijs RM. A network of (autonomic) clock outputs. Chronobiol Int. 2006;23(3):521–35. https://doi.org/10.1080/07420520600651073.
45. Husse J, Zhou X, Shostak A, Oster H, Eichele G. Synaptotagmin10-Cre, a driver to disrupt clock genes in the SCN. J Biol Rhythm. 2011;26(5):379–89. https://doi.org/10.1177/0748730411415363.
46. Niijima A, Nagai K, Nagai N, Nakagawa H. Light enhances sympathetic and suppresses vagal outflows and lesions including the suprachiasmatic nucleus eliminate these changes in rats. J Auton Nerv Syst. 1992;40(2):155–60.
47. Mutoh T, Shibata S, Korf HW, Okamura H. Melatonin modulates the light-induced sympathoexcitation and vagal suppression with participation of the suprachiasmatic nucleus in mice. J Physiol 2003;547(Pt 1):317–332. doi:https://doi.org/10.1113/jphysiol.2002.028001.
48. Ishida A, Mutoh T, Ueyama T, Bando H, Masubuchi S, Nakahara D, et al. Light activates the adrenal gland: timing of gene expression and glucocorticoid release. Cell metabolism. 2005;2(5):297–307. doi:S1550–4131(05)00269-X [pii]. https://doi.org/10.1016/j.cmet.2005.09.009.
49. Kiessling S, Sollars PJ, Pickard GE. Light stimulates the mouse adrenal through a retinohypothalamic pathway independent of an effect on the clock in the suprachiasmatic nucleus. PLoS One. 2014;9(3):e92959. https://doi.org/10.1371/journal.pone.0092959.
50. Balsalobre A, Brown SA, Marcacci L, Tronche F, Kellendonk C, Reichardt HM, et al. Resetting of circadian time in peripheral tissues by glucocorticoid signaling. Science. 2000;289(5488):2344–7.. doi:8856[pii]
51. Pezuk P, Mohawk JA, Wang LA, Menaker M. Glucocorticoids as entraining signals for peripheral circadian oscillators. Endocrinology. 2012;153(10):4775–83. https://doi.org/10.1210/en.2012-1486.
52. Reppert SM, Weaver DR. Coordination of circadian timing in mammals. Nature. 2002;418(6901):935–41. https://doi.org/10.1038/nature00965.
53. Reddy AB, Field MD, Maywood ES, Hastings MH. Differential resynchronisation of circadian clock gene expression within the suprachiasmatic nuclei of mice subjected to experimental jet lag. J Neurosci. 2002;22(17):7326–30.
54. Parent-Thirion A, Fernández Macías E, Hurley J, Vermeylen G. Fourth European Working Conditions Survey. 2007.
55. Haus EL, Smolensky MH. Shift work and cancer risk: potential mechanistic roles of circadian disruption, light at night, and sleep deprivation. Sleep Med Rev. 2012;17(4):273–84. https://doi.org/10.1016/j.smrv.2012.08.003.
56. Bechtold DA, Gibbs JE, Loudon AS. Circadian dysfunction in disease. Trends Pharmacol Sci. 2010;31(5):191–8. https://doi.org/10.1016/j.tips.2010.01.002.
57. Thomas C, Power C. Shift work and risk factors for cardiovascular disease: a study at age 45 years in the 1958 British birth cohort. Eur J Epidemiol. 2010;25(5):305–14. https://doi.org/10.1007/s10654-010-9438-4.
58. Karlsson B, Knutsson A. Lindahl B. is there an association between shift work and having a metabolic syndrome? Results from a population based study of 27,485 people. Occup Environ Med. 2001;58(11):747–52.
59. Selvi Y, Gulec M, Agargun MY, Besiroglu L. Mood changes after sleep deprivation in morningness-eveningness chronotypes in healthy individuals. J Sleep Res. 2007;16(3):241–4. https://doi.org/10.1111/j.1365-2869.2007.00596.x.
60. Pan A, Schernhammer ES, Sun Q, Hu FB. Rotating night shift work and risk of type 2 diabetes: two prospective cohort studies in women. PLoS Med. 2011;8(12):e1001141. https://doi.org/10.1371/journal.pmed.1001141.
61. Rudic RD, McNamara P, Curtis AM, Boston RC, Panda S, Hogenesch JB, et al. BMAL1 and CLOCK, two essential components of the circadian clock, are involved in glucose homeostasis. PLoS Biol. 2004;2(11):e377. https://doi.org/10.1371/journal.pbio.0020377.

62. Turek FW, Joshu C, Kohsaka A, Lin E, Ivanova G, McDearmon E, et al. Obesity and metabolic syndrome in circadian clock mutant mice. Science. 2005;308(5724):1043–5. https://doi.org/10.1126/science.1108750.
63. Paschos GK, Ibrahim S, Song WL, Kunieda T, Grant G, Reyes TM, et al. Obesity in mice with adipocyte-specific deletion of clock component Arntl. Nat Med. 2012;18(12):1768–77. https://doi.org/10.1038/nm.2979.
64. Kettner NM, Voicu H, Finegold MJ, Coarfa C, Sreekumar A, Putluri N, et al. Circadian homeostasis of liver metabolism suppresses Hepatocarcinogenesis. Cancer Cell. 2016;30(6):909–24. https://doi.org/10.1016/j.ccell.2016.10.007.
65. Kettner NM, Mayo SA, Hua J, Lee C, Moore DD, Fu L. Circadian dysfunction induces Leptin resistance in mice. Cell Metab. 2015;22(3):448–59. https://doi.org/10.1016/j.cmet.2015.06.005.
66. Coomans CP, van den Berg SA, Lucassen EA, Houben T, Pronk AC, van der Spek RD, et al. The suprachiasmatic nucleus controls circadian energy metabolism and hepatic insulin sensitivity. Diabetes. 2013;62(4):1102–8. https://doi.org/10.2337/db12-0507.
67. Polidarova L, Sladek M, Sotak M, Pacha J, Sumova A. Hepatic, duodenal, and colonic circadian clocks differ in their persistence under conditions of constant light and in their entrainment by restricted feeding. Chronobiol Int. 2011;28(3):204–15. https://doi.org/10.3109/07420528.2010.548615.
68. Damiola F, Le Minh N, Preitner N, Kornmann B, Fleury-Olela F, Schibler U. Restricted feeding uncouples circadian oscillators in peripheral tissues from the central pacemaker in the suprachiasmatic nucleus. Genes Dev. 2000;14(23):2950–61.
69. Fonken LK, Workman JL, Walton JC, Weil ZM, Morris JS, Haim A, et al. Light at night increases body mass by shifting the time of food intake. Proc Natl Acad Sci U S A. 2010;107(43):18664–9. https://doi.org/10.1073/pnas.1008734107.
70. Mendoza J, Pevet P, Challet E. High-fat feeding alters the clock synchronization to light. J Physiol. 2008;586(Pt 24):5901–5910. doi:jphysiol.2008.159566 [pii]https://doi.org/10.1113/jphysiol.2008.159566.
71. Kohsaka A, Laposky AD, Ramsey KM, Estrada C, Joshu C, Kobayashi Y, et al. High-fat diet disrupts behavioral and molecular circadian rhythms in mice. Cell metabolism. 2007;6(5):414–21. doi:S1550–4131(07)00266–5 [pii]. https://doi.org/10.1016/j.cmet.2007.09.006.
72. Boivin DB, James FO, Wu A, Cho-Park PF, Xiong H, Sun ZS. Circadian clock genes oscillate in human peripheral blood mononuclear cells. Blood. 2003;102(12):4143–5. https://doi.org/10.1182/blood-2003-03-0779.
73. Cermakian N, Westfall S, Kiessling S. Circadian clocks and inflammation: reciprocal regulation and shared mediators. Arch Immunol Ther Exp. 2014;62:303–18. https://doi.org/10.1007/s00005-014-0286-x.
74. Nguyen KD, Fentress SJ, Qiu Y, Yun K, Cox JS, Chawla A. Circadian gene Bmal1 regulates diurnal oscillations of Ly6C(hi) inflammatory monocytes. Science. 2013;341(6153):1483–8. https://doi.org/10.1126/science.1240636.
75. Scheiermann C, Kunisaki Y, Frenette PS. Circadian control of the immune system. Nat Rev Immunol. 2013;13(3):190–8. https://doi.org/10.1038/nri3386.
76. Kiessling S, Dubeau-Laramee G, Ohm H, Labrecque N, Olivier M, Cermakian N. The circadian clock in immune cells controls the magnitude of Leishmania parasite infection. Sci Rep. 2017;7(1):10892. https://doi.org/10.1038/s41598-017-11297-8.
77. Bellet MM, Deriu E, Liu JZ, Grimaldi B, Blaschitz C, Zeller M, et al. Circadian clock regulates the host response to Salmonella. Proc Natl Acad Sci U S A. 2013;110(24):9897–902. https://doi.org/10.1073/pnas.1120636110.
78. Turner-Warwick M. Epidemiology of nocturnal asthma. Am J Med. 1988;85(1B):6–8.
79. Sam S, Corbridge TC, Mokhlesi B, Comellas AP, Molitch ME. Cortisol levels and mortality in severe sepsis. Clin Endocrinol. 2004;60(1):29–35.
80. Gibbs J, Ince L, Matthews L, Mei J, Bell T, Yang N, et al. An epithelial circadian clock controls pulmonary inflammation and glucocorticoid action. Nat Med. 2014;20(8):919–26. https://doi.org/10.1038/nm.3599.

81. Castanon-Cervantes O, Wu M, Ehlen JC, Paul K, Gamble KL, Johnson RL, et al. Dysregulation of inflammatory responses by chronic circadian disruption. J Immunol. 2010;185(10):5796–805. https://doi.org/10.4049/jimmunol.1001026.
82. Cuesta M, Cermakian N, Boivin DB. Glucocorticoids entrain molecular clock components in human peripheral cells. FASEB journal: official publication of the Federation of American Societies for Experimental Biology. 2015;29(4):1360–70. https://doi.org/10.1096/fj.14-265686.
83. Boscolo P, Youinou P, Theoharides TC, Cerulli G, Conti P. Environmental and occupational stress and autoimmunity. Autoimmun Rev. 2008;7(4):340–3. https://doi.org/10.1016/j.autrev.2007.12.003.
84. Kumar CJ, Sharma VK, Kumar A. Jet lag and enhanced susceptibility to malaria. Med Hypotheses. 2006;66(3):671. https://doi.org/10.1016/j.mehy.2005.10.002.
85. Sandborn WJ. Current directions in IBD therapy: what goals are feasible with biological modifiers? Gastroenterology. 2008;135(5):1442–7. https://doi.org/10.1053/j.gastro.2008.09.053.
86. Swanson GR, Burgess HJ, Keshavarzian A. Sleep disturbances and inflammatory bowel disease: a potential trigger for disease flare? Expert Rev Clin Immunol. 2011;7(1):29–36. https://doi.org/10.1586/eci.10.83.
87. Nojkov B, Rubenstein JH, Chey WD, Hoogerwerf WA. The impact of rotating shift work on the prevalence of irritable bowel syndrome in nurses. Am J Gastroenterol. 2010;105(4):842–7. https://doi.org/10.1038/ajg.2010.48.
88. Peterson LW, Artis D. Intestinal epithelial cells: regulators of barrier function and immune homeostasis. Nat Rev Immunol. 2014;14(3):141–53. https://doi.org/10.1038/nri3608.
89. Summa KC, Voigt RM, Forsyth CB, Shaikh M, Cavanaugh K, Tang Y, et al. Disruption of the circadian clock in mice increases intestinal permeability and promotes alcohol-induced hepatic pathology and inflammation. PLoS One. 2013;8(6):e67102. https://doi.org/10.1371/journal.pone.0067102.
90. Pagel R, Bar F, Schroder T, Sunderhauf A, Kunstner A, Ibrahim SM, et al. Circadian rhythm disruption impairs tissue homeostasis and exacerbates chronic inflammation in the intestine. FASEB journal: official publication of the Federation of American Societies for Experimental Biology. 2017;31(11):4707–19. https://doi.org/10.1096/fj.201700141RR.
91. Mukherji A, Kobiita A, Ye T, Chambon P. Homeostasis in intestinal epithelium is orchestrated by the circadian clock and microbiota cues transduced by TLRs. Cell. 2013;153(4):812–27. https://doi.org/10.1016/j.cell.2013.04.020.
92. Cario E. Microbiota and innate immunity in intestinal inflammation and neoplasia. Curr Opin Gastroenterol. 2013;29(1):85–91. https://doi.org/10.1097/MOG.0b013e32835a670e.
93. Coleman OI, Haller D. Bacterial signaling at the intestinal epithelial interface in inflammation and cancer. Front Immunol. 2017;8:1927. https://doi.org/10.3389/fimmu.2017.01927.
94. Froy O, Chapnik N, Miskin R. Mouse intestinal cryptdins exhibit circadian oscillation. FASEB journal: official publication of the Federation of American Societies for Experimental Biology. 2005;19(13):1920–2. https://doi.org/10.1096/fj.05-4216fje.
95. Larsen KR, Moore JG, Dayton MT. Circadian rhythms of gastric mucus efflux and residual mucus gel in the fasting rat stomach. Dig Dis Sci. 1991;36(11):1550–5.
96. Born J, Lange T, Hansen K, Molle M, Fehm HL. Effects of sleep and circadian rhythm on human circulating immune cells. J Immunol. 1997;158(9):4454–64.
97. Rosselot AE, Hong CI, Moore SR. Rhythm and bugs: circadian clocks, gut microbiota, and enteric infections. Curr Opin Gastroenterol. 2016;32(1):7–11. https://doi.org/10.1097/MOG.0000000000000227.
98. Thaiss CA, Zeevi D, Levy M, Zilberman-Schapira G, Suez J, Tengeler AC, et al. Transkingdom control of microbiota diurnal oscillations promotes metabolic homeostasis. Cell. 2014;159(3):514–29. https://doi.org/10.1016/j.cell.2014.09.048.
99. Voigt RM, Summa KC, Forsyth CB, Green SJ, Engen P, Naqib A, et al. The circadian clock mutation promotes intestinal Dysbiosis. Alcohol Clin Exp Res. 2016;40(2):335–47. https://doi.org/10.1111/acer.12943.

100. Oh-Oka K, Kono H, Ishimaru K, Miyake K, Kubota T, Ogawa H, et al. Expressions of tight junction proteins Occludin and Claudin-1 are under the circadian control in the mouse large intestine: implications in intestinal permeability and susceptibility to colitis. PLoS One. 2014;9(5):e98016. https://doi.org/10.1371/journal.pone.0098016.
101. Hoogerwerf WA, Hellmich HL, Cornelissen G, Halberg F, Shahinian VB, Bostwick J, et al. Clock gene expression in the murine gastrointestinal tract: endogenous rhythmicity and effects of a feeding regimen. Gastroenterology. 2007;133(4):1250–60. https://doi.org/10.1053/j.gastro.2007.07.009.
102. Preuss F, Tang Y, Laposky AD, Arble D, Keshavarzian A, Turek FW. Adverse effects of chronic circadian desynchronization in animals in a "challenging" environment. Am J Physiol Regul Integr Comp Physiol. 2008;295(6):R2034–40. https://doi.org/10.1152/ajpregu.00118.2008.
103. Yu X, Rollins D, Ruhn KA, Stubblefield JJ, Green CB, Kashiwada M, et al. TH17 cell differentiation is regulated by the circadian clock. Science. 2013;342(6159):727–30. https://doi.org/10.1126/science.1243884.
104. Zarrinpar A, Chaix A, Yooseph S, Panda S. Diet and feeding pattern affect the diurnal dynamics of the gut microbiome. Cell Metab. 2014;20(6):1006–17. https://doi.org/10.1016/j.cmet.2014.11.008.
105. Alcock J, Maley CC. Aktipis CA. Is eating behavior manipulated by the gastrointestinal microbiota? Evolutionary pressures and potential mechanisms. BioEssays: news and reviews in molecular, cellular and Dev Biol. 2014;36(10):940–9. https://doi.org/10.1002/bies.201400071.
106. Leone V, Gibbons SM, Martinez K, Hutchison AL, Huang EY, Cham CM, et al. Effects of diurnal variation of gut microbes and high-fat feeding on host circadian clock function and metabolism. Cell Host Microbe. 2015;17(5):681–9. https://doi.org/10.1016/j.chom.2015.03.006.
107. Straif K, Baan R, Grosse Y, Secretan B, El Ghissassi F, Bouvard V, et al. Carcinogenicity of shift-work, painting, and fire-fighting. Lancet Oncol. 2007;8(12):1065–6.
108. Lie JA, Kjuus H, Zienolddiny S, Haugen A, Stevens RG, Kjaerheim K. Night work and breast cancer risk among Norwegian nurses: assessment by different exposure metrics. Am J Epidemiol. 2011;173(11):1272–9. https://doi.org/10.1093/aje/kwr014.
109. Lahti TA, Partonen T, Kyyronen P, Kauppinen T, Pukkala E. Night-time work predisposes to non-Hodgkin lymphoma. Int j cancer J int cancer. 2008;123(9):2148–51. https://doi.org/10.1002/ijc.23566.
110. Logan RW, Zhang C, Murugan S, O'Connell S, Levitt D, Rosenwasser AM, et al. Chronic shift-lag alters the circadian clock of NK cells and promotes lung cancer growth in rats. J Immunol. 2012;188(6):2583–91. https://doi.org/10.4049/jimmunol.1102715.
111. Filipski E, Delaunay F, King VM, Wu MW, Claustrat B, Grechez-Cassiau A, et al. Effects of chronic jet lag on tumor progression in mice. Cancer Res. 2004;64(21):7879–85. https://doi.org/10.1158/0008-5472.CAN-04-0674.
112. Masri S, Cervantes M, Sassone-Corsi P. The circadian clock and cell cycle: interconnected biological circuits. Curr Opin Cell Biol. 2013;25(6):730–4. https://doi.org/10.1016/j.ceb.2013.07.013.
113. Papagiannakopoulos T, Bauer MR, Davidson SM, Heimann M, Subbaraj L, Bhutkar A, et al. Circadian rhythm disruption promotes lung tumorigenesis. Cell Metab. 2016;24(2):324–31. https://doi.org/10.1016/j.cmet.2016.07.001.
114. Levi F, Okyar A, Dulong S, Innominato PF, Clairambault J. Circadian timing in cancer treatments. Annu Rev Pharmacol Toxicol. 2010;50:377–421. https://doi.org/10.1146/annurev.pharmtox.48.113006.094626.
115. Kiessling S, Cermakian N. Clock genes and Cancer. In: Kumar V, editor. Biological timekeeping: clocks, rhythms and behaviour. New Delhi: Springer India; 2017. p. 481–500.
116. Kiessling S, Beaulieu-Laroche L, Blum ID, Landgraf D, Welsh DK, Storch KF, et al. Enhancing circadian clock function in cancer cells inhibits tumor growth. BMC Biol. 2017;15(1):13. https://doi.org/10.1186/s12915-017-0349-7.

117. Hanahan D, Weinberg RA. Hallmarks of cancer: the next generation. Cell. 2011;144(5):646–74. https://doi.org/10.1016/j.cell.2011.02.013.
118. Puppa MJ, White JP, Sato S, Cairns M, Baynes JW, Carson JA. Gut barrier dysfunction in the Apc(min/+) mouse model of colon cancer cachexia. Biochim Biophys Acta. 2011;1812(12):1601–6. https://doi.org/10.1016/j.bbadis.2011.08.010.
119. Coleman OI, Nunes T. Role of the microbiota in colorectal Cancer: updates on microbial associations and therapeutic implications. BioResearch open access. 2016;5(1):279–88. https://doi.org/10.1089/biores.2016.0028.

Chapter 3
Environmental Factors' Interference in Endocrine Aspects of Male Reproduction

Claudia Pivonello, Cristina de Angelis, Francesco Garifalos, Rosario Pivonello, and Annamaria Colao

Keywords Environmental factors · Endocrine disruptors · Obesogens · Tributyltin · Male reproduction

3.1 Environmental Factors as Endocrine Disruptors: Sources and Milestones of Endocrine Disruptors' History

Starting from eighteenth century the anthropogenic impact on earth's ecosystem gradually resulted in a global climate and biodiversity change [1]. Indeed, since the industrial revolution, people were concentrated in cities and the human activities

C. Pivonello
Dipartimento di Medicina Clinica e Chirurgia, Sezione di Endocrinologia, Università "Federico II" di Napoli, Naples, Italy

C. de Angelis
FERTISEXCARES Centro di Andrologia, Medicina della Riproduzione e della Sessualità Maschile e Femminile, Università "Federico II" di Napoli, Naples, Italy

Dipartimento di Sanità Pubblica, Università "Federico II" di Napoli, Naples, Italy

F. Garifalos
FERTISEXCARES Centro di Andrologia, Medicina della Riproduzione e della Sessualità Maschile e Femminile, Università "Federico II" di Napoli, Naples, Italy

R. Pivonello · A. Colao (✉)
Dipartimento di Medicina Clinica e Chirurgia, Sezione di Endocrinologia, Università "Federico II" di Napoli, Naples, Italy

FERTISEXCARES Centro di Andrologia, Medicina della Riproduzione e della Sessualità Maschile e Femminile, Università "Federico II" di Napoli, Naples, Italy

Cattedra Unesco "Educazione alla salute e allo sviluppo sostenibile", Università "Federico II" di Napoli, Naples, Italy
e-mail: colao@unina.it

R. Teperino (ed.), *Beyond Our Genes*,
https://doi.org/10.1007/978-3-030-35213-4_3

substantially represented the main cause of the environment burden. The emission of anthropogenic atmospheric pollution, contributing to the effluence generated by the natural sources through the emission of several gases with a long atmospheric lifetime, well known as greenhouse gases, including carbon dioxide, methane, nitrous oxide and fluorinated gases, particles and aerosols has contributed to the global climate warming [1]. Moreover, a result of the intensive human activities is also represented by the production of many synthetic chemicals, which can perturb the endocrine system by disrupting the endocrine function; therefore, these substances are well known as Endocrine Disrupting Chemicals (EDCs) [2]. Over the past years, the expert committees of different international organizations, including the World Health Organization (WHO), the Agency for the Environmental Protection (EPA) and the Food and Drug Administration (FDA), proposed several definitions of EDCs, and, according to the definition of the Endocrine Society's second Scientific Statement, an EDC is: "an exogenous chemical, or mixture of chemicals, that interferes with any aspect of hormone action" [3].

The history of EDCs started in 1938 with the production of the first synthetic non-steroidal estrogen: diethylstilbestrol (DES). During 1950s–1980s this compound was prescribed in women who experienced previous miscarriage or threatened miscarriage, as treatment to prevent miscarriage or premature delivery. Only in the early 1970s, cases of young women, progeny of DES-exposed women, with vaginal and cervix clear cell adenocarcinoma started to be diagnosed [4–7].

Historically, an early expression of what the scientific community now consider as endocrine disruption was anticipated in 1958 by the endocrinologist Roy Hertz. Hertz was the first dissecting the potential deleterious role of some hormones used in the cattle feed lots, which, carried by faecal excretion after leaching in environmental matrices (soil and water), can reach the human body through the food chain, exerting severe consequences on the development, growth and reproductive function, introducing the concept of "steroid cycle" and anticipating the hypothesis of "bio-accumulation" [8, 9]. In 1962, the current concepts of bio-accumulation (the concentration of an EDC increases from one trophic level to the next within the food chain) and of bio-magnification (animals and humans that are on the highest trophic levels have the highest concentration of EDCs), were pointed out by the biologist Rachel Carson in one of the most important environmental book entitled "Silent spring" in which the author raised important questions about the anthropogenic impact on nature; in particular, Rachel Carson claimed that the chemical compound p,p'-Dichlorodiphenyltrichloroethane (DDT) and other pesticides entered the food chain and accumulated in fatty tissues of animals and humans, by causing genetic damage and cancer [10]. "Silent spring" message had important implications, inspiring several researchers in deepening the knowledge on EDCs. In 1966, the researchers of the National Institute of Environmental Health Sciences (NIEHS) started investigations with the mission to "reduce the burden of human illness by understanding how the environment influences the development and the progression of human disease", and, in 1971, they established that DES is a transplacental carcinogen whose toxicity involves the activation of estrogen receptors, by causing adverse effects in the offspring, up to the third generation, without necessarily affecting the mother [11]. In the same year, the FDA issued a drug bulletin urging medical physicians to stop prescribing DES, although DES was definitely banned by FDA only in 2000s.

Only in 1991, Theodora Colborn and other scientists coined the term "endocrine disruptor" during the meeting "Wingspread Congress". In the Wingspread Consensus Statement Colborn and colleagues described many observations acquired by studies on EDCs, and defined concepts such as "the critical window of exposure of susceptibility", the potential of bio-accumulation and the long latency between exposure and the appearance of the effects [12]. During the following years, many studies on humans and animal models were conducted on several EDCs, and, in 1998, during the Rotterdam Convention on the Prior Informed Consent Procedures for Certain Hazardous Chemicals and Pesticides, over 150 countries ratified a list of several different chemicals that included known EDCs [13]. The list was updated in 2001, during the Stockholm Convention, with the introduction of several persistent organic pollutants (POPs) [14].

EDCs are either man-made or naturally occurring chemicals, present in the environment (air, water and soil), food sources, personal care products, and manufactured products. EDCs include components of plastics, such as bisphenol A (BPA) and phthalates, and other compounds such as dioxins, polychlorinated biphenyls (PCBs), polybrominated biphenyls (PBBs), pesticides including p,p'-Dichlorodiphenyltrichloroethane (DDT) and its metabolites, heavy metals, industrial chemicals and fuels and many others [2].

3.2 EDCs as Obesogens

In 2002, Paula Baillie-Hamilton, in her article published in the Journal of Alternative and Complementary Medicine, reported the idea that some EDCs could be contributing to set obesity as an epidemic condition [15]; indeed, the author presented the hypothesis that the current obesity epidemic could be a consequence of human exposure to particular chemicals that are able to damage the mechanisms of human weight-control. In the early 2000s, the University of California biology professor Bruce Blumberg, during a meeting in Japan, heard for the first time that tin-based compounds like tributyltin (TBT), at that time used as wood preservative, caused sex reversal in several fish species, and hypothesized that this effect was a consequence of sex steroid receptor activation; nevertheless, Blumberg surprisingly discovered that TBT activate peroxisome proliferator-activated-receptor gamma (PPARγ), a master regulator of adipogenesis [16]. Moreover, Blumberg discovered that the offspring of pregnant mice exposed to TBT were predisposed to gain weight compared to offspring born by unexposed pregnant mice [17], showing that obesity can be linked to exposure to some EDCs, which act as risk factors. In 2006, the term "*obesogens*" was coined by Blumberg, referring to endocrine disruptors (*obesogens*) able to alter lipid homeostasis, and to promote adipogenesis and lipid accumulation, consequently resulting to be involved in weight gain and contributing to obesity epidemic. Among EDCs, TBT (a biocide in anti-fouling paint used as ship preservative), DES (a synthetic non-steroidal estrogen used to prevent threatened miscarriage), POPs (most of them used as insecticides), BPA and phthalates (used in the manufacture of polycarbonate plastics and epoxy resins), parabens (used as antimicrobial agents), 4-Nonyiphenol (used as surfactant in industrial and domestic chemicals), phytoestrogens (naturally produced by several plants), and polybrominated diphenylethers (PBDEs) (used as flame retardants), have been shown to possess obesogenic properties.

3.3 Endocrine System as Physiological Interface With the Environment: Obesogens Perturbations of Hypothalamic-Pituitary-Gonadal (HPG) Axis

The endocrine system consists of ductless endocrine glands, which maintain homeostasis and long-term control of the human body, by using chemical messengers, involving hormones released into the blood stream. Homeostasis is the ability to keep a constant internal environment; therefore, it refers to stability, balance, or equilibrium within a cell or the entire organism. The release of hormones into the blood is controlled by a stimulus, and the response to a stimulus, in turn, changes the internal conditions, and may itself become a new stimulus. This self-adjusting mechanism is called feedback regulation; a negative feedback occurs when the response to a stimulus reduces the original stimulus, whereas a positive feedback occurs when the response to a stimulus increases the original stimulus.

In general homeostasis, the HPG axis is regulated by the hypothalamus, comprising neuroendocrine cells, which synthesize and secrete gonadotropin-releasing hormone (GnRH) in the hypophysial portal circulation, in a pulsatile fashion, dependent on calcium influx. In response to GnRH, the gonadotropic cells of the anterior pituitary release gonadotropins: follicle-stimulating hormone (FSH) and luteinizing hormone (LH). Both gonadotropins regulate gonadal function, in both sexes, by controlling the synthesis and release of androgens and spermatogenesis, in males, and oestrogens, follicuologenesis and ovulation, in females. Hypothalamic and pituitary functions are strictly controlled by negative feedback loops, sustained by gonadal hormones [18, 19]. Moreover, the secretion of gonadotropins is also influenced by steroid hormones, as well as by glycoprotein hormones such as inhibins and activins, produced by the gonads [19].

Male gonads, testes, hold in the scrotum, consist in tiny U-shaped tubules, called seminiferous tubules, organized and packed in several lobules [19]. The seminiferous tubules converge into a series of uncoiled, interconnected, channels that form the rete testis. The efferent ducts connect the rete testis to a tightly coiled duct, the epididymis, which, in turn, terminates into a larger duct: the vas deferens. Two fibrous layers envelop the entire dense network of ducts: the outer layer is known as tunica vaginalis and the inner layer is known as tunica albuginea. Testes fulfil two pivotal reproductive functions: the production of male gametes, or spermatogenesis, and the secretion of male reproductive hormones, in particular, testosterone [19]. Spermatogenesis includes mitosis and meiosis to generate haploid spermatids from undifferentiated diploid precursors, the spermatogonia, followed by differentiation in mature spermatozoa, through the process of spermiogenesis [19]. Sertoli cells are somatic cells lining the seminiferous tubules, which surround the proliferating and differentiating germ cells, and provide nutrients and mechanistic support to spermatogenesis., Moreover, the tight junctions of Sertoli cells jointly constitute the blood-testis barrier, a tightly sealed unique structure segregating meiotic and post-meiotic germ cells within the apical compartment of the seminiferous tubule, by preventing the passage of antigenic products of germ cell maturation into blood circulation, and, therefore, autoimmunity against germ cells [20]. Blood-testis barrier is not a static ultrastructure, but undergoes massive remodelling during spermatogenesis in order to permit the transit of spermatocytes, meanwhile maintaining the barrier protecting from toxic and immunological factors [20]. Lastly, Sertoli cells, stimulated by FSH, produce testicular fluid that includes the androgen-binding protein. Androgen-binding protein is secreted into the lumen of seminiferous tubules where binds and concentrates testosterone, essential for the production of spermatozoa. Sertoli cells also produce inhibin, a dimeric glycoprotein existing in two bioactive forms, inhibin A and B; inhibin B, in particular, suppresses FSH secretion from the pituitary, and positively correlates with Sertoli cells function and semen quality in adult male [19, 21]. Leydig cells are stromal cells also defined as "interstitial cells" since they are interspersed among the seminiferous tubules. Leydig cells are mainly involved in the production and secretion of testosterone, in response to LH from the pituitary; testosterone promotes spermatogenesis and acquisition and maintenance of male secondary sex characteristics [19, 22].

Across lifespan, ECDs, including obesogens, can dramatically affect homeostasis, having the ability to mimic or block the effects of endogenous hormones, by directly binding to steroid hormone receptors, therefore interfering with the normal function of HPG axis. In the last decades, it has been well established that nutrition and reproduction are tightly coordinated: both reduced and excessive energy intake, consequence of an incorrect nutritional state, may have a negative impact on the human reproductive physiology. Indeed, as mounting evidences have demonstrated, obesogens can also indirectly perturb the homeostasis of HPG through a prolonged disturbance of adipose tissue functions, by binding to the nuclear receptor, PPARγ. Definitely, obesogens, being lipophilic toxic compounds, may accumulate in fatty tissues, and may be mostly bioaccumulated in obese subjects, rather than in lean individuals [23, 24], therefore exacerbating their deleterious effects.

The mammalian foetus is able to adapt its growth and development to the environmental stimuli provided by the mother. This phenomenon is defined as "developmental plasticity" [25]. Developmental plasticity is high during early foetal life, and the mammalian foetus is very sensitive to exposure to environmental factors within this timeframe since, during this period, pluripotent stem cells can differentiate into specific cell types to form different tissues. Hormones and other signalling molecules control these developmental events and their timing; therefore, the exposure to obesogens during the "in utero" window can alter functioning of the endocrine system, by interfering with the developmental processes. Consequently, the exposure to obesogens, in particular during "in utero" time window, but also during perinatal and postnatal periods, can dramatically influence the reproductive function in the offspring, at adulthood, and the effects of exposure may persist up to the F3 generation (transgenerational effect) [11, 17].

Considering that obesogens can participate to HPG axis function regulation at different levels, within different critical windows of exposure, and with different concentration-dependent intensity, and multiple mechanisms of action, it is difficult to define how exactly obesogens may hamper reproductive physiology [26].

3.4 Effects of Obesogens: Mechanisms of Regulation of Male Reproductive System Physiology

3.4.1 Obesogens and Hormonal Disorders

Obesity relies on the cooperation of a large number of hormones, which centrally control food intake behaviour (among them leptin produced by adipose tissue) and peripherally maintain the homeostasis of glucose blood levels (insulin, produced by pancreatic ß-cells). Indeed, one of the first steps in the aetiology of obesity is insulin resistance, a condition in which insulin no longer controls glucose production, by resulting in higher glucose blood levels. Consequently, glucose uptake in peripheral tissues is reduced, and lipolysis is enhanced, leading to increased free fatty acids blood levels. Gradually, a vicious circle is activated, with hyperglycaemia aggravating hyperinsulinemia, and, in turn, hyperinsulinemia aggravating hyperglycaemia and hypertriglyceridemia [27]. Epidemiological evidences in humans linked exposure to obesogens, assessed by the analysis of urinary levels of obesogens metabolites, with obesity and metabolic syndrome-associated disorders [28–32]. These observations were supported by studies performed in humans and animal models, demonstrating that obesogens, such as TBT, triphenyltin, phthalates, parabens and BPA, promote adipogenesis, lipid accumulation [16, 33–38], and differentiation of adipose stromal mesenchymal stem cells or fibroblast in adipocytes [33, 35, 36, 38, 39]. Obesogens are potent agonists of nuclear PPARs (PPARα, −δ, and -γ) and retinoid X receptor (RXR) at environmentally relevant levels (ppb – parts per billion). The binding of obesogens to PPARs in adipose tissue induces the formation of RXR-PPARs heterodimers, by triggering molecular pathways involved in increasing adipocytes volume, regulating lipid biosynthesis, proliferation and differentiation [40]. Indirect effects of obesogens on male reproductive system are mediated by the interference with the nuclear PPARs on adipose tissue, which results in impaired reproductive function (Fig. 3.1).

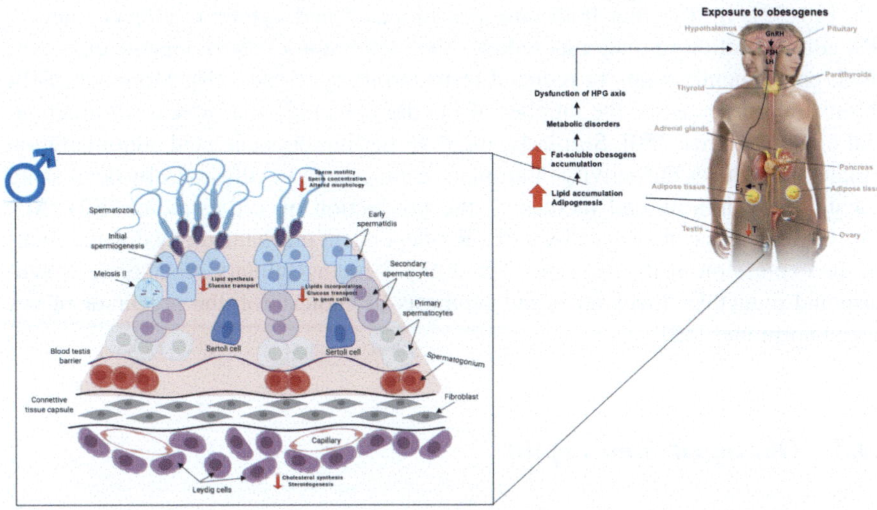

Fig. 3.1 Effects of the exposure to obesogens on male reproductive system. The exposure to obesogens, mainly occurring through the diet, predisposes to weight gain increasing the mass of adipose tissue. Being lipophilic, the obesogens accumulate in adipose tissue and activate adipogenic pathways. Adipocytes express high levels of aromatase enzyme that convert testosterone (T) in 17β-estradiol (E2). Peripheral E2 levels rise in response to an increased adipose tissue mass, by inducing a negative feedback on HPG axis, with a blockade of LH and FSH release from the pituitary, which, in turn, affect spermatogenesis

3.4.2 *Obesogens and Testosterone*

One of the most well characterized comorbidities of obesity is hypogonadism, which in turn leads to typical clinical signs and symptoms consistent with androgen deficiency [41]. Low circulating total testosterone levels in obese men have been previously linked to reduced SHBG levels [42]; nevertheless, subsequent studies reported low free testosterone levels in obese men [43], particularly in those with BMI > 40 kg/m^2 [44, 45]. In obese men, a condition defined as male obesity-related hypogonadotropic hypogonadism might be diagnosed, with low total and/or free testosterone levels associated to low/normal gonadotropins levels, along with signs and symptoms of hypogonadism, in absence of an organic impairment of the hypothalamus–pituitary axis, and of other known causes of hypogonadism [46]. Male obesity-related hypogonadotropic hypogonadism itself might worsen obesity, by enhancing fat mass increase, which may in turn contribute to hypogonadal profile [47]. Testis and epididymis exhibit high lipid content; therefore, these tissues may accumulate lipophilic toxic compounds, being exposed to the effects of obesogens across lifespan. Epidemiological studies focusing on the effects of obesogens on testicular steroidogenesis are lacking, and the main evidences on such an effect derive from several studies performed on animal models; the results of these studies proposed that obesogens might directly act on the testis, particularly, on Leydig cells [48–52], by exerting both anti-steroidogenic and cytotoxic effects. Indeed, BPA administration to male rats reduces the expression of steroidogenic enzymes, which consequently leads to reduced testosterone synthesis [48]. Moreover, BPA administration decreases the number of Leydig cells [49] and increases mitochondrial oxidative stress [50]. Similarly, the POP perfluorooctanic acid administration to male rats induces the activity of aromatase enzyme in Leydig cells, by decreasing testosterone synthesis and increasing the production of 17β-estradiol (E2) [51]. Several pesticides can also act by a PPARα-dependent mechanism, by downregulating the expression of the enzymes 3-hydroxy-3-methyl-glutaryl-coenzyme A synthase and reductase, involved in the synthesis of cholesterol, the precursor of sex steroid hormones [52].

3.4.3 Obesogens and Leptin

Besides storing and mobilizing energy, the adipose tissue, as endocrine organ, produces and secretes several factors known as adipokines, which regulate the functions of other endocrine tissues. The first adipokine described in literature was leptin, a 16-kDa anorexigenic protein produced mainly by adipose tissue, but also by placenta, stomach and skeletal muscle [53], exerting its effects by means of six isoforms of leptin receptors (ObR) [54].

Leptin is an adipokine messenger regulating feeding behaviour and energy expenditure, and a crucial cytokine regulating different physiological processes including reproduction [55]. Indeed, in recent years, the role of leptin in the regulation of reproductive neuroendocrine axis has been pointed out, although the underlying mechanisms have not been fully elucidated. It is clear that an adequate concentration of leptin is required for normal reproductive function, since leptin has been shown to act at different levels within the HPG axis, in both males and females [56, 57].

Circulating levels of leptin, positively correlated with adiposity, are gender-dependent with higher leptin serum concentrations observed in women than in men [58]. The gender-difference in leptin serum levels persists even after correction for variables including subcutaneous fat and visceral fat, which may influence circulating leptin levels. In normal homeostasis, gonadal and adipose tissue functions are regulated by a complex and reciprocal interplay. Physiological concentrations of leptin, produced by adipose tissue, regulates male and female reproductive function through the binding to ObR: indirectly stimulating the production of GnRH by hypothalamus and of gonadotropins by pituitary [59] after crossing the blood-brain barrier, and directly acting on testis [56]. In males, circulating androgens control adipocyte size and adipose mass, whereas plasmatic leptin indirectly increases steroidogenesis in Leydig cells, by stimulating LH, and promotes spermatogenesis by the FSH-dependent stimulation of Sertoli cells [60]. Moreover, leptin has been also detected in human seminiferous tubules, which express ObR, as well as in seminal plasma and spermatozoa [61, 62]; nevertheless, the direct positive effects of leptin on sperm motility, capacitation, and acrosome reaction are still a matter of debate [63–65].

In pathological conditions, such as obesity, the increased adipose mass induces the release of supraphysiological leptin levels. High serum leptin levels in obesity have been proposed to be secondary to "leptin resistance", whose mechanisms still remain unclear but potentially led back to the inability of circulating leptin to reach its target in the brain, reduction of ObR expression, and/or inhibition of the signalling events in specific brain regions [66]. Supraphysiological leptin levels act directly on the gonads, and, indirectly, on the hypothalamus and pituitary, in both males and females. In obese men, high leptin levels impact negatively on the reproductive potential by disrupting the HPG axis and reducing semen quality (sperm concentration, sperm motility and sperm morphology), as demonstrated by several observational studies [67–70]. In addition, the androgen response to hCG stimulation is impaired in obese men, and multivariate analysis demonstrated that leptin was the best hormonal predictor of the obesity-related reduction in androgen response. Taken together, these observations suggest that leptin excess might play an important role in the development of reduced androgen output in male obesity [53]. Indeed, the increase of adipose tissue mass, documented by the rise of body mass index (BMI), induces a reduction of sex hormone binding globulin (SHBG) and testosterone plasma concentrations, and a concomitant increase of estrogens plasma concentration [71]. Elevated estrogens in obese men may in part result from the increased mass of white adipose tissue in which the elevated activity of aromatase induces the conversion of testosterone in E2 [71]. The increased estrogens level triggers a negative feedback upon the HPG axis via kisspeptin neurons in the hypothalamus, resulting in a further downregulation of testosterone, and impaired spermatogenesis [71]. Moreover, in obese men, low levels of inhibin B have been also observed, suggesting Sertoli cells dysfunction, although a compensatory increase in FSH levels has not been demonstrated, therefore indicating a potential partial dysregulation of the HPG axis [72]. Lastly, leptin also exerts direct actions on the testis, by inhibiting steroidogenesis through interference with cAMP signalling in granulosa cells [63]. Obesogens, such as BPA and TBT, may induce downregulation of ObR, by attenuating the transport of leptin to the hypothalamus, via blood-brain barrier, therefore producing a condition of obesity [73], and may also contribute to induce leptin resistance, by leading to HPG axis dysfunction, and consequent reduction of testosterone synthesis [74]. The actions of leptin within the HPG axis are summarized in Fig. 3.2.

3.4.4 Obesogens and Insulin

One of the most relevant causative factors of obesity-induced hypogonadism is insulin resistance. Studies performed on animal models demonstrated that obesogens such as BPA and TBT might compromise glucose metabolism, by reducing the expression of glucose transporter 2 (GLUT2), insulin receptor substrate 1 (IRS1) and 2 (IRS2) and phosphatidylinositol 3 kinase (PI3K) within the testis [75, 76]. Considering the involvement of PI3K on glucose transporter 1 (GLUT1)

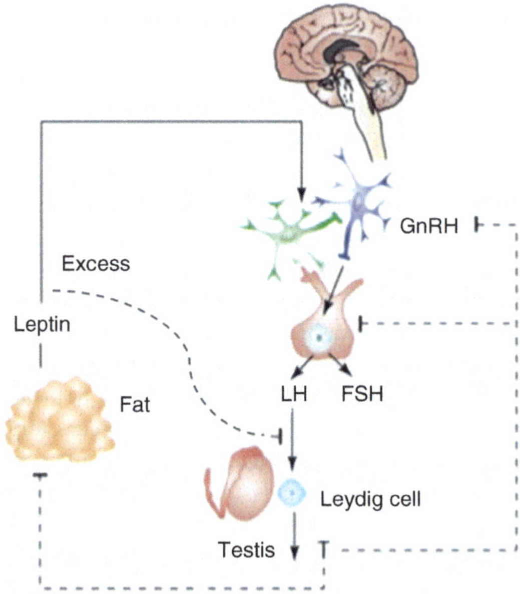

Fig. 3.2 Effects of leptin on the HPG axis. Leptin acts at various levels along the HPG axis, and its effects on the reproductive function seem to be elicited by different leptin blood thresholds, depending on the site of activity. High leptin concentrations stimulate hypothalamic GnRH, and, in turn, gonadotropin secretion; nevertheless, saturable leptin transport system modulates leptin transport across the blood–brain barrier, by preventing high concentrations of leptin from reaching hypothalamic leptin receptors. By contrast, excess of leptin may potentially act on peripheral leptin receptors and inhibit testicular steroidogenesis; testosterone, by inhibiting leptin synthesis and secretion by the adipose tissue closes the regulatory loop. Adapted from Caprio et al. [53]

translocation from cytoplasm to the membrane, the down-regulation of PI3K leads to a reduced glucose influx in Sertoli cells. The glucometabolic changes may trigger apoptosis of Sertoli cells, therefore affecting spermatogenesis [77]. In line with these observations, it has been demonstrated that TBT may affect GLUT1 translocation on the membrane, and decrease adenosine monophosphate-activated protein kinase (AMPK), by promoting insulin resistance and, consequently, impairing Sertoli and germ cells function [78].

3.4.5 Obesogens and Histopathology of the Testis and Sperm Parameters

Few in vivo studies have addressed the direct effects of obesogens on mammalian testis; nevertheless, such studies are unanimous in the idea that obesogens such as TBT and BPA may compromise testis histology and may damage spermatozoa membrane, by compromising sperm function [79–83]. Indeed, in animals treated with obesogens, severe degenerative alterations were observed within the testis: testis weight was reduced, seminiferous tubules showed degenerative changes in the germinal layer and markedly reduced number of spermatocytes and spermatids, with a consequently prejudiced spermatogenesis. Accordingly, sperm count and viability were decreased, and the proportion of spermatozoa with abnormal sperm morphology was increased.

Scarce epidemiological studies investigating the effects of obesogens on the human testis and semen quality were performed, by making it difficult to draw definitive conclusions on how obesogens may affect testis function [84, 85]. Interestingly, a recent epidemiological study pointed out that chronic exposure to an obesogens, such as BPA, in a pre-existing condition of obesity, might exacerbate the deleterious effect of BPA on semen quality [86]; the underlying molecular mechanisms reside in the condition of oxidative stress observed in obese patients, characterized by high levels of reactive oxygen species (ROS), which aggravate the toxic effect of BPA on semen quality, resulting in decreased sperm count [86]. Indeed, after treatment with BPA, histopathological examination of the testis showed a phenotype consistent with severe testicular trauma, and sperm count and concentration were significantly lower in obese mice, compared with lean mice [86]. Moreover, metabolomic analyses performed on the testis confirmed the hypothesis of an interaction between BPA and obesity, in affecting the male reproductive system; indeed, treatment with BPA in obese mice up-regulated oxidative stress metabolites associated with male reproductive dysfunction [86].

References

1. McMichael AJ, Woodruff RE, Hales S. Climate change and human health: present and future risks. Lancet. 2006;367(9513):859–69.
2. Gore AC, Chappell VA, Fenton SE, Flaws JA, Nadal A, Prins GS, et al. Executive summary to EDC-2: the Endocrine Society's second scientific statement on endocrine-disrupting chemicals. Endocr Rev. 2015;36(6):593–602.
3. Zoeller RT, Brown TR, Doan LL, Gore AC, Skakkebaek NE, Soto AM, et al. Endocrine-disrupting chemicals and public health protection: a statement of principles from the Endocrine Society. Endocrinology. 2012;153(9):4097–110.
4. Herbst AL, Ulfelder H, Poskanzer DC. Adenocarcinoma of the vagina. Association of maternal stilbestrol therapy with tumor appearance in young women. N Engl J Med. 1971;284(15):878–81.
5. Williams RR, Schweitzer RJ. Clear-cell adenocarcinoma of the vagina in a girl whose mother had taken diethylstilbestrol. Calif Med. 1973;118(6):53–5.
6. Puri S, Fenoglio CM, Richart RM, Townsend D. Clear cell carcinoma of cervix and vagina in progeny of women who received diethylstilbestrol: three cases with scanning and transmission electron microscopy. Am J Obstet Gynecol. 1977;128(5):550–5.
7. Robboy SJ, Scully RE, Welch WR, Herbst AL. Intrauterine diethylstilbestrol exposure and its consequences: pathologic characteristics of vaginal adenosis, clear cell adenocarcinoma, and related lesions. Arch Pathol Lab Med. 1977;101(1):1–5.
8. Gassner FX, Reifenstein Jr EC, Algeo JW, Mattox WE. Effects of hormones on growth, fattening, and meat production potential of livestock. Recent Prog Horm Res, 1958. 14: p. 183–210; discussion 210-7.

9. McLachlan JA. Environmental signaling: from environmental estrogens to endocrine-disrupting chemicals and beyond. Andrology. 2016;4(4):684–94.
10. Davis FR. Silent spring after 50 years. Endeavour. 2012;36(4):129–30.
11. Folkman J. Transplacental carcinogenesis by stilbestrol. N Engl J Med. 1971;285(7):404–5.
12. Colborn T, vom Saal FS, Soto AM. Developmental effects of endocrine-disrupting chemicals in wildlife and humans. Environ Health Perspect. 1993;101(5):378–84.
13. UNEP and FAO Rotterdam Convention – http://www.pic.int 1998.
14. UNEP and FAO Stockholm Convention – http://www.chm.pops.int. 2001.
15. Baillie-Hamilton PF. Chemical toxins: a hypothesis to explain the global obesity epidemic. J Altern Complement Med. 2002;8(2):185–92.
16. Grun F, Watanabe H, Zamanian Z, Maeda L, Arima K, Cubacha R, et al. Endocrine-disrupting organotin compounds are potent inducers of adipogenesis in vertebrates. Mol Endocrinol. 2006;20(9):2141–55.
17. Chamorro-Garcia R, Sahu M, Abbey RJ, Laude J, Pham N, Blumberg B. Transgenerational inheritance of increased fat depot size, stem cell reprogramming, and hepatic steatosis elicited by prenatal exposure to the obesogen tributyltin in mice. Environ Health Perspect. 2013;121(3):359–66.
18. Plant TM. 60 years of Neuroendocrinology: the hypothalamo-pituitary-gonadal axis. J Endocrinol. 2015;226(2):T41–54.
19. Weinbauer GF, LC, Simoni M, Nieschlag E, Physiology of testicular function, In: H.M. Behre Nieschlag E, Nieschlag S, editors. Andrology, 2010, Berlin: Springer.
20. Cheng CY, Mruk DD. The blood-testis barrier and its implications for male contraception. Pharmacol Rev. 2012;64(1):16–64.
21. O'Connor AE, De Kretser DM. Inhibins in normal male physiology. Semin Reprod Med. 2004;22(3):177–85.
22. Corradi PF, Corradi RB, Greene LW. Physiology of the hypothalamic pituitary gonadal Axis in the male. Urol Clin North Am. 2016;43(2):151–62.
23. Pelletier C, Despres JP, Tremblay A. Plasma organochlorine concentrations in endurance athletes and obese individuals. Med Sci Sports Exerc. 2002;34(12):1971–5.
24. Pereira-Fernandes A, Dirinck E, Dirtu AC, Malarvannan G, Covaci A, Van Gaal L, et al. Expression of obesity markers and persistent organic pollutants levels in adipose tissue of obese patients: reinforcing the obesogen hypothesis? PLoS One. 2014;9(1):e84816.
25. West-Eberhard MJ. Developmental plasticity and the origin of species differences. Proc Natl Acad Sci U S A. 2005;102(Suppl 1):6543–9.
26. Janesick AS, Blumberg B. Obesogens: an emerging threat to public health. Am J Obstet Gynecol. 2016;214(5):559–65.
27. Redinger RN. The pathophysiology of obesity and its clinical manifestations. Gastroenterol Hepatol. 2007;3(11):856–63.
28. Hatch EE, Nelson JW, Stahlhut RW, Webster TF. Association of endocrine disruptors and obesity: perspectives from epidemiological studies. Int J Androl. 2010;33(2):324–32.
29. La Merrill M, Birnbaum LS. Childhood obesity and environmental chemicals. Mt Sinai J Med. 2011;78(1):22–48.
30. Tang-Peronard JL, Andersen HR, Jensen TK, Heitmann BL. Endocrine-disrupting chemicals and obesity development in humans: a review. Obes Rev. 2011;12(8):622–36.
31. Lang IA, Galloway TS, Scarlett A, Henley WE, Depledge M, Wallace RB, et al. Association of urinary bisphenol a concentration with medical disorders and laboratory abnormalities in adults. JAMA. 2008;300(11):1303–10.
32. Trasande L, Attina TM, Blustein J. Association between urinary bisphenol a concentration and obesity prevalence in children and adolescents. JAMA. 2012;308(11):1113–21.
33. Masuno H, Kidani T, Sekiya K, Sakayama K, Shiosaka T, Yamamoto H, et al. Bisphenol a in combination with insulin can accelerate the conversion of 3T3-L1 fibroblasts to adipocytes. J Lipid Res. 2002;43(5):676–84.
34. Kanayama T, Kobayashi N, Mamiya S, Nakanishi T, Nishikawa J. Organotin compounds promote adipocyte differentiation as agonists of the peroxisome proliferator-activated receptor gamma/retinoid X receptor pathway. Mol Pharmacol. 2005;67(3):766–74.

35. Kirchner S, Kieu T, Chow C, Casey S, Blumberg B. Prenatal exposure to the environmental obesogen tributyltin predisposes multipotent stem cells to become adipocytes. Mol Endocrinol. 2010;24(3):526–39.
36. Yin L, Yu KS, Lu K, Yu X. Benzyl butyl phthalate promotes adipogenesis in 3T3-L1 preadipocytes: a high content Cellomics and metabolomic analysis. Toxicol In Vitro. 2016;32:297–309.
37. Feige JN, Gelman L, Rossi D, Zoete V, Metivier R, Tudor C, et al. The endocrine disruptor monoethyl-hexyl-phthalate is a selective peroxisome proliferator-activated receptor gamma modulator that promotes adipogenesis. J Biol Chem. 2007;282(26):19152–66.
38. Li X, Ycaza J, Blumberg B. The environmental obesogen tributyltin chloride acts via peroxisome proliferator activated receptor gamma to induce adipogenesis in murine 3T3-L1 preadipocytes. J Steroid Biochem Mol Biol. 2011;127(1–2):9–15.
39. Hu P, Chen X, Whitener RJ, Boder ET, Jones JO, Porollo A, et al. Effects of parabens on adipocyte differentiation. Toxicol Sci. 2013;131(1):56–70.
40. Grun F, Blumberg B. Environmental obesogens: organotins and endocrine disruption via nuclear receptor signaling. Endocrinology. 2006;147(6 Suppl):S50–5.
41. Basaria S. Male hypogonadism. Lancet. 2014;383(9924):1250–63.
42. Laaksonen DE, Niskanen L, Punnonen K, Nyyssonen K, Tuomainen TP, Salonen R, et al. Sex hormones, inflammation and the metabolic syndrome: a population-based study. Eur J Endocrinol. 2003;149(6):601–8.
43. Dhindsa S, Prabhakar S, Sethi M, Bandyopadhyay A, Chaudhuri A, Dandona P. Frequent occurrence of hypogonadotropic hypogonadism in type 2 diabetes. J Clin Endocrinol Metab. 2004;89(11):5462–8.
44. Giagulli VA, Kaufman JM, Vermeulen A. Pathogenesis of the decreased androgen levels in obese men. J Clin Endocrinol Metab. 1994;79(4):997–1000.
45. Vermeulen A, Kaufman JM, Deslypere JP, Thomas G. Attenuated luteinizing hormone (LH) pulse amplitude but normal LH pulse frequency, and its relation to plasma androgens in hypogonadism of obese men. J Clin Endocrinol Metab. 1993;76(5):1140–6.
46. Saboor Aftab SA, Kumar S, Barber TM. The role of obesity and type 2 diabetes mellitus in the development of male obesity-associated secondary hypogonadism. Clin Endocrinol. 2013;78(3):330–7.
47. Isidori AM, Giannetta E, Greco EA, Gianfrilli D, Bonifacio V, Isidori A, et al. Effects of testosterone on body composition, bone metabolism and serum lipid profile in middle-aged men: a meta-analysis. Clin Endocrinol. 2005;63(3):280–93.
48. Jin P, Wang X, Chang F, Bai Y, Li Y, Zhou R, et al. Low dose bisphenol a impairs spermatogenesis by suppressing reproductive hormone production and promoting germ cell apoptosis in adult rats. J Biomed Res. 2013;27(2):135–44.
49. Nakamura D, Yanagiba Y, Duan Z, Ito Y, Okamura A, Asaeda N, et al. Bisphenol a may cause testosterone reduction by adversely affecting both testis and pituitary systems similar to estradiol. Toxicol Lett. 2010;194(1–2):16–25.
50. El-Beshbishy HA, Aly HA, El-Shafey M. Lipoic acid mitigates bisphenol A-induced testicular mitochondrial toxicity in rats. Toxicol Ind Health. 2013;29(10):875–87.
51. Biegel LB, Liu RC, Hurtt ME, Cook JC. Effects of ammonium perfluorooctanoate on Leydig cell function: in vitro, in vivo, and ex vivo studies. Toxicol Appl Pharmacol. 1995;134(1):18–25.
52. Harada Y, Tanaka N, Ichikawa M, Kamijo Y, Sugiyama E, Gonzalez FJ, et al. PPARalpha-dependent cholesterol/testosterone disruption in Leydig cells mediates 2,4-dichlorophenoxyacetic acid-induced testicular toxicity in mice. Arch Toxicol. 2016;90(12):3061–71.
53. Caprio M, Fabbrini E, Isidori AM, Aversa A, Fabbri A. Leptin in reproduction. Trends Endocrinol Metab. 2001;12(2):65–72.
54. Friedman JM, Halaas JL. Leptin and the regulation of body weight in mammals. Nature. 1998;395(6704):763–70.
55. Tong Q, Xu Y. Central leptin regulation of obesity and fertility. Curr Obes Rep. 2012;1(4):236–44.
56. Mammi C, Calanchini M, Antelmi A, Cinti F, Rosano GM, Lenzi A, et al. Androgens and adipose tissue in males: a complex and reciprocal interplay. Int J Endocrinol. 2012;2012:789653.

57. Hausman GJ, Barb CR, Lents CA. Leptin and reproductive function. Biochimie. 2012;94(10):2075–81.
58. Hickey MS, Israel RG, Gardiner SN, Considine RV, McCammon MR, Tyndall GL, et al. Gender differences in serum leptin levels in humans. Biochem Mol Med. 1996;59(1):1–6.
59. Hill JW, Elmquist JK, Elias CF. Hypothalamic pathways linking energy balance and reproduction. Am J Physiol Endocrinol Metab. 2008;294(5):E827–32.
60. Gnessi L, Fabbri A, Spera G. Gonadal peptides as mediators of development and functional control of the testis: an integrated system with hormones and local environment. Endocr Rev. 1997;18(4):541–609.
61. Ishikawa T, Fujioka H, Ishimura T, Takenaka A, Fujisawa M. Expression of leptin and leptin receptor in the testis of fertile and infertile patients. Andrologia. 2007;39(1):22–7.
62. Glander HJ, Lammert A, Paasch U, Glasow A, Kratzsch J. Leptin exists in tubuli seminiferi and in seminal plasma. Andrologia. 2002;34(4):227–33.
63. Li HW, Chiu PC, Cheung MP, Yeung WS, W.S. O. Effect of leptin on motility, capacitation and acrosome reaction of human spermatozoa. Int J Androl. 2009;32(6):687–94.
64. Lampiao F, du Plessis SS. Insulin and leptin enhance human sperm motility, acrosome reaction and nitric oxide production. Asian J Androl. 2008;10(5):799–807.
65. Guo J, Zhao Y, Huang W, Hu W, Gu J, Chen C, et al. Sperm motility inversely correlates with seminal leptin levels in idiopathic asthenozoospermia. Int J Clin Exp Med. 2014;7(10):3550–5.
66. Munzberg H, Myers MG Jr. Molecular and anatomical determinants of central leptin resistance. Nat Neurosci. 2005;8(5):566–70.
67. Jensen TK, Andersson AM, Jorgensen N, Andersen AG, Carlsen E, Petersen JH, et al. Body mass index in relation to semen quality and reproductive hormones among 1,558 Danish men. Fertil Steril. 2004;82(4):863–70.
68. Hammoud AO, Wilde N, Gibson M, Parks A, Carrell DT, Meikle AW. Male obesity and alteration in sperm parameters. Fertil Steril. 2008;90(6):2222–5.
69. Hofny ER, Ali ME, Abdel-Hafez HZ, Kamal Eel D, Mohamed EE, Abd El-Azeem HG, et al. Semen parameters and hormonal profile in obese fertile and infertile males. Fertil Steril. 2010;94(2):581–4.
70. Paasch U, Grunewald S, Kratzsch J, Glander HJ. Obesity and age affect male fertility potential. Fertil Steril. 2010;94(7):2898–901.
71. Du Plessis SS, Cabler S, McAlister DA, Sabanegh E, Agarwal A. The effect of obesity on sperm disorders and male infertility. Nat Rev Urol. 2010;7(3):153–61.
72. Palmer NO, Bakos HW, Fullston T, Lane M. Impact of obesity on male fertility, sperm function and molecular composition. Spermatogenesis. 2012;2(4):253–63.
73. Ben-Jonathan N, Hugo ER, Brandebourg TD. Effects of bisphenol A on adipokine release from human adipose tissue: implications for the metabolic syndrome. Mol Cell Endocrinol. 2009;304(1–2):49–54.
74. Zuo Z, Chen S, Wu T, Zhang J, Su Y, Chen Y, et al. Tributyltin causes obesity and hepatic steatosis in male mice. Environ Toxicol. 2011;26(1):79–85.
75. D'Cruz SC, Jubendradass R, Jayakanthan M, Rani SJ, Mathur PP. Bisphenol A impairs insulin signaling and glucose homeostasis and decreases steroidogenesis in rat testis: an in vivo and in silico study. Food Chem Toxicol. 2012;50(3–4):1124–33.
76. D'Cruz SC, Jubendradass R, Mathur PP. Bisphenol A induces oxidative stress and decreases levels of insulin receptor substrate 2 and glucose transporter 8 in rat testis. Reprod Sci. 2012;19(2):163–72.
77. Dias TR, Rato L, Martins AD, Simoes VL, Jesus TT, Alves MG, et al. Insulin deprivation decreases caspase-dependent apoptotic signaling in cultured rat sertoli cells. ISRN Urol. 2013;2013:970370.
78. Galardo MN, Riera MF, Pellizzari EH, Cigorraga SB, Meroni SB. The AMP-activated protein kinase activator, 5-aminoimidazole-4-carboxamide-1-b-D-ribonucleoside, regulates lactate production in rat Sertoli cells. J Mol Endocrinol. 2007;39(4):279–88.
79. Chen Y, Zuo Z, Chen S, Yan F, Chen Y, Yang Z, et al. Reduction of spermatogenesis in mice after tributyltin administration. Toxicology. 2008;251(1–3):21–7.

80. Yu WJ, Lee BJ, Nam SY, Kim YC, Lee YS, Yun YW. Spermatogenetic disorders in adult rats exposed to tributyltin chloride during puberty. J Vet Med Sci. 2003;65(12):1331–5.
81. Yu WJ, Nam SY, Kim YC, Lee BJ, Yun YW. Effects of tributyltin chloride on the reproductive system in pubertal male rats. J Vet Sci. 2003;4(1):29–34.
82. Wisniewski P, Romano RM, Kizys MM, Oliveira KC, Kasamatsu T, Giannocco G, et al. Adult exposure to bisphenol A (BPA) in Wistar rats reduces sperm quality with disruption of the hypothalamic-pituitary-testicular axis. Toxicology. 2015;329:1–9.
83. Rahman MS, Kwon WS, Lee JS, Yoon SJ, Ryu BY, Pang MG. Bisphenol-A affects male fertility via fertility-related proteins in spermatozoa. Sci Rep. 2015;5:9169.
84. Goldstone AE, Chen Z, Perry MJ, Kannan K, Louis GM. Urinary bisphenol A and semen quality, the LIFE study. Reprod Toxicol. 2015;51:7–13.
85. Adoamnei E, Mendiola J, Vela-Soria F, Fernandez MF, Olea N, Jorgensen N, et al. Urinary bisphenol A concentrations are associated with reproductive parameters in young men. Environ Res. 2018;161:122–8.
86. Hu W, Dong T, Wang L, Guan Q, Song L, Chen D, et al. Obesity aggravates toxic effect of BPA on spermatogenesis. Environ Int. 2017;105:56–65.

Chapter 4
Prenatal Exposure to Endocrine Disrupting Chemicals and Their Effect on Health Later in Life

Elin Engdahl and Joëlle Rüegg

Keywords Endocrine disrupting chemicals, EDCs · Hormones · Nuclear receptors · Epigenetics · DNA methylation · Fertility problems · Neurodevelopmental disorders · Metabolic disorders

4.1 Endocrine Disrupting Chemicals and Their Effects on the Endocrine System

4.1.1 Rapid Increase of Chemical Exposure During the Last Decades

During the last decades the production of industrial chemicals has increased enormously, and is estimated to increase even more in the future [1, 2]. Global sales of chemicals increased from €1029 billion in 1996 to €3360 billion in 2016 [1], and > 80,000 chemicals are registered for commercial use in USA and Europe (reviewed in [3]). Chemicals are primarily used in industries producing rubber and plastics, pulp and paper, as well as in construction and in the automotive industry [1]. In the plastic industry, chemicals used as for example plasticizers and flame retardants, are added in order to enhance the desired properties of the material.

E. Engdahl (✉)
Unit of Integrative Toxicology, Institute of Environmental Medicine (IMM), Karolinska Institutet (KI), Stockholm, Sweden

Environmental Toxicology, Department of Organismal Biology, Uppsala University, Uppsala, Sweden
e-mail: Elin.Engdahl@ebc.uu.se

J. Rüegg
Environmental Toxicology, Department of Organismal Biology, Uppsala University, Uppsala, Sweden

R. Teperino (ed.), *Beyond Our Genes*,
https://doi.org/10.1007/978-3-030-35213-4_4

There has been a dramatic increase in plastic production from 1950 to 2015 and the sum of all nonfiber plastics (including additives) produced until 2015 has been estimated to 7300 million metric tons (reviewed in [4]).

One group of chemicals are the persistent organic pollutants (POPs). These are stable industrial chemicals that persist in the environment and accumulate in living organisms. Although some POPs (e.g. polychlorinated biphenols (PCBs) and dichlorodiphenyl-trichloroethane (DDT)) have been banned from use due to their toxicity and persistency, many of these substances can still be detected throughout the food chain and in human and wildlife tissue across the globe (reviewed in [5, 6]).

4.1.2 What is an EDC?

Some of these industrial chemicals are endocrine disrupting chemicals (EDCs), i.e. substances that interfere with any aspect of our natural hormone system [7, 8]. Mechanisms by which EDCs can interfere with hormonal action include binding to hormone receptors and thereby act similarly to the hormones themselves (agonistic) or block their function, i.e. act antagonistically. Furthermore, EDCs can interfere with the production, transport, release, metabolism or elimination of hormones. There are many different types of EDCs, in fact, close to 800 compounds in our environment are known or suspected to have endocrine disrupting properties (reviewed in [9]). In this chapter we have chosen to focus on a few well-studied ones in the context of human health (overview in Table 4.1) to exemplify the increasing evidence that EDC exposure early in life can affect human health throughout the lifespan.

EDCs are different from "classical" toxicants since their adverse effects may be subtle and not follow classical dose-responses curves. In fact, EDCs often have non-monotonic dose response curves [10] (Fig. 4.1) and can have different effects depending on concentration. Therefore, low doses can cause effects that are not predicted based on chemical testing using high doses [10]. In addition, the effects of EDCs may be much more severe during critical periods in life, for example during fetal development, when hormones have critical functions for development of the organism. Also, their effects are often different depending on the target tissue. It is thus concerning that tests on which chemical risk assessment is based today are most often not optimized to detect endocrine disrupting properties as (1) they are built on monotonic dose responses and hence the doses used may not be the doses giving adverse effects, (2) endpoint measurements are focused on more direct toxicity such as cell death, and (3) the time window of critical development of an organism is not taken into account. This lack of accurate risk assessment is one reason for the vast amount of chemicals around us that affect the endocrine system and thus human health.

Table 4.1 Overview of selected EDCs

Chemical	Primary usage	Predominant exposure route	% human exposure*
Bisphenols			
Bisphenol A	Hardener used in polycarbonate plastics (found in common consumer goods like food and beverage containers), epoxy resins (used to coat the inside of metal products like food cans, bottle tops and water supply lines).	Ingestion of packed food and drinks	96–97%
Phthalates			
High-molecular-weight phthalates[a]	Plasticizer in PVC plastics (found in e.g. toys, tubing, flooring and wall covering)	Ingestion of packed food and drinks, dust ingestion	Metabolite (chemical) MEHP (DEHP): 100%, MiNP (DiNP): 100%
Low-molecular-weight phthalates[a]	Added to personal care products (like nail polish, body lotions, cosmetics, shampoos, perfume etc.) to help lubricate other substances and to carry fragrances	Dermal absorption, dust ingestion, inhalation	Metabolite (chemical) MEP (DEP): 100%, MBP (DBP): 100%, MBzP (BBzP): 100%
Persistent organic compounds (POPs)[b]			
Polychlorinated biphenyls (PCBs)[a]	Previously widely used in industry in e.g. dielectric and coolant fluids in electrical apparatus, carbonless copy paper and in heat transfer fluids.	Food (especially fish consumption)	PCB-118: 77–100%, PCB-138: 95–100%, PCB-153: 98–100%, PCB-180: 95–96%
Perfluorinated alkylated substances (PFAS)[a]/[c]	Used as sufructant in fire-fighting foams, and for its ability to repel water and oil in e.g. stain repellents, impregnation agents, food packaging and non-stick pans	Drinking water, food, and use of consumer products	PFOS: 99–100%, PFOA: 99–100%
Polybrominated diphenyl ethers (PBDEs)[a]	Flame retardant in various industrial products such as electronics, textiles, furnishings, and building materials	Drinking water, food, and dust inhalation	PBDE-47: 99–100%, PBDE-99: 87–99%, PBDE-100: 98–99%, PBDE-153: 99–100%

[a]There are many different individual chemicals in these groups, but they all have similar properties

[b]Persistent organic pollutants (POPs) are organic compounds that are extremely resistant to environmental degradation and therefore persist in the environment for very long periods. They bioaccumulate in human and animal tissue and can harm human health and the environment. Sources of POPs are pesticides, industrial chemicals or unintentional production. The POPs listed in this table are all listed in the Stockholm Convention list where parties must take measures to eliminate or restrict the production and use of these chemicals

[c]PFOS is the only PFAS member on the Stockholm Convention list, but PFOA and PFHxS are under review for being added to the list

*In serum or urine of pregnant women, or in cord blood at birth, percent individuals with chemical level > Limit of detection (LOD). The most studied chemicals within a chemical group are presented. Min-max value from two studies are presented (BPA [17, 96]; phthalates [21, 118]; PCBs [17, 105]; PFAS [17, 195]; PBDEs [17, 23])

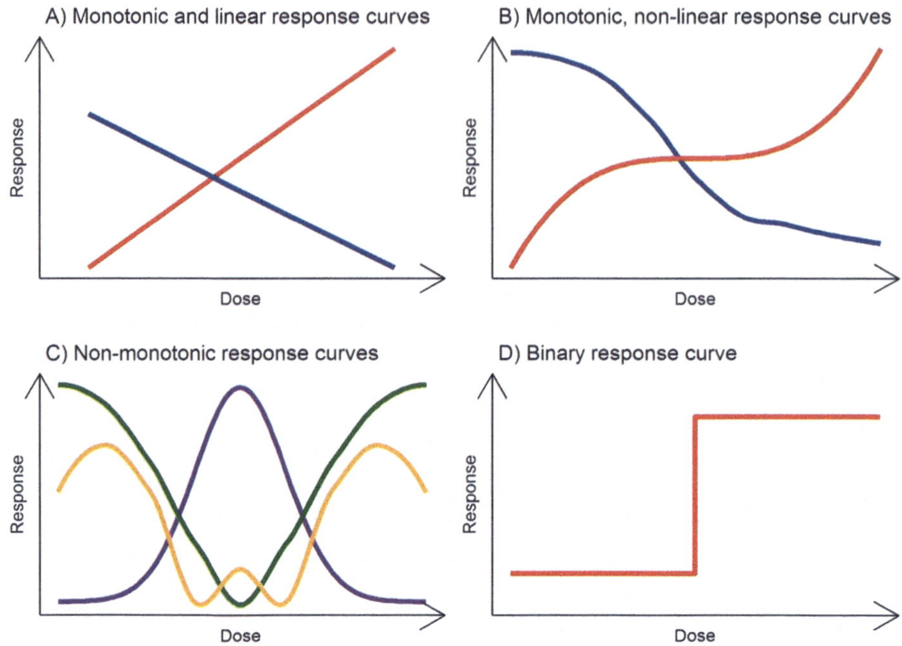

Fig. 4.1 Examples of dose response curves. (**A**) Linear dose response, where the response increases or decreases with the same rate as the dose, and the response to a certain chemical dose is therefore easy to predict; (**B**) monotonic but nonlinear dose response, i.e. the response increases or decreases in the same direction but the rate is not constant; (**C**) non-monotonic dose response where the direction of the slope changes at least once. With chemicals like these, it is impossible to predict the effect of a chemical dose based on knowledge regarding another dose; (**D**) Binary response where there is a threshold at which a dose elicits (or eliminates) a response

4.1.3 We Are Exposed to EDCs Throughout Our Life

We are constantly exposed to a cocktail of EDCs as numerous commonly used consumer products and articles contain and leak EDCs to the surrounding environment. A vast amount of EDCs are routinely found in food and water, as well as indoor air and dust, leading to constant human exposure (Table 4.1) [11–15], and EDCs or their metabolites are routinely detected in human urine, serum, breast milk and amniotic fluid (reviewed in [16]). Importantly, we are not exposed to one EDC at the time, but are constantly exposed to a mixture of EDCs [17], where the combined effects of the individual chemicals can be additive to each other [18, 19]. The large number of chemicals with potential or proven endocrine disrupting properties has thus raised global concern expressed in, for example, a recent WHO-report [9].

Although a united research field points at the risk of adverse health effects of EDC exposure during fetal development [20], many EDCs are found in >95% of pregnant women [17, 21–23] (Table 4.1). For example, Woodruff et al. [17] observed that when measuring 52 different environmental chemicals in pregnant women, the median number of chemicals detected in serum/urine was 37 and all women had chemicals from different chemical classes in their systems. The amount of chemicals present in pregnant women is of high concern as many of the environmental chemicals can cross the placenta [22, 24–29] and therefore reach the fetus. In an investigation by the Environmental Working Group (EWG) in 2005, 287 out of 413 measured environmental chemicals were detected in the cord blood of at least one of the 10 babies studied [30], indicating that the developing fetus is indeed exposed to hundreds of chemicals during sensitive developmental periods.

4.1.4 Fetuses are More Vulnerable to EDCs Than Adults

The process of fetal development, from one cell to a whole organism, is tightly regulated. Many developmental processes are controlled and fine-tuned by hormones, e.g. triggering cell proliferation and differentiation or regulating growth or retraction of morphological structures. Such hormonally regulated events are temporally restricted. Thus changes in hormonal balance during these developmental windows can result in permanent alterations with consequences for health later in life. Also, since hormone signaling differs between males and females, EDCs can cause different effects depending on sex. Examples of organs whose development is tightly regulated during sensitive time windows are the brain, reproductive tract and metabolic system. They are thus particularly receptive to environmental cues early in life (reviewed in [31–34]).

4.1.5 Endocrine Systems Affected by EDCs

Hormones are signaling molecules that enable communication between organs / tissues in order to regulate physiology and behavior. Hormones are most often secreted by endocrine glands and transported, via the blood stream, to a target organ where the hormone binds to its specific receptor and thereby affects important functions in the target cells. Hormones can have diverse chemical structures and regulate functions such as metabolism, sleep, mood, stress, reproduction, as well as growth and development. Since the endocrine system has many essential functions in the human body, changes in hormonal action (both increase and decrease) can have a variety of effects and these effects are different depending on target cell type and timing of the hormonal change. In adults, hormones have mostly temporary effects on their target cells regulating for example blood sugar levels or ovulation, whereas during fetal development, hormones have programming effects.

The most studied endocrine systems in the context of EDC effects are the thyroid hormone system and steroid hormones signaling pathways, in particular estrogen and androgen signaling. Steroid and thyroid hormones mediate their effects primarily through their nuclear receptors (NRs), i.e. ligand-induced transcription factors that, bound to the hormone, regulate expression of their target genes. In order to understand the diversity of EDC effects, it is important to emphasize that hormone signaling is complex. Depending on the cell type and tissue, NRs interact with different co-regulators that fine-tune their action and define the set of target genes specific for each cell type. Furthermore, many NRs come in different isoforms, like the thyroid hormone receptors (TRα and TRβ) or the estrogen receptors (e.g. ERα and ERβ). These receptor isoforms can have different temporal and spatial patterns of expression and differential effects on gene expression (reviewed in [9]).

4.2 The Effect of EDCs on Epigenetics as a Possible Explanation for Their Life Long Effects

There are many examples of that environmental factors that affect fetal development also affect health later in life (reviewed in [35–37]). Sensitivity to environmental information such as nutritional state or infections during development is thought to improve adaptation to the milieu after birth. This "developmental plasticity" in response to environmental cues is usually occurring during sensitive time windows (reviewed in [37]) and is thought to be, at least partly, due to epigenetic programming of the genetic material during early development.

Epigenetic information is defined as heritable information associated with the DNA but not encoded by the DNA sequence itself. Epigenetic mechanisms include DNA methylation, histone modifications and non-coding RNAs (including microRNAs; miRNA), which are all part of a complex regulatory network. Whereas DNA methylation and histone modifications determine the structure and accessibility of the DNA for transcription factors, and therefore regulate gene expression, miRNAs regulate gene expression at the posttranscriptional level (reviewed in [38]). Epigenetic processes are vital for the early stages of normal mammalian development where gene expression needs to be tightly regulated [39]. Changes in epigenetic information during development can permanently alter the functional genome, and subsequently cause adverse health outcomes that manifest in later life, or even in subsequent generations (reviewed in [40]).

4.2.1 EDC Exposure During Fetal Development Alters DNA Methylation

In the context of chemical exposure, the most studied epigenetic regulation is DNA methylation, where DNA methyltransferases (DNMT) add a methyl group to the cytosine base in the DNA strand. Generally, DNA methylation, particularly in a promoter region, decreases the accessibility of the DNA for transcription factors and therefore represses gene transcription. DNA methylation is mitotically heritable through the activity of DNMT1 that copies the methylation state of the mother strands to the daughter strands. Thus DNA methylation changes can be persistent over long periods in life.

Both experimental and epidemiological evidence is accumulating that exposure to EDCs, in particular during development, affects DNA methylation patterns (reviewed in [38, 41–43]). Mechanisms underlying such changes are not well studied but there is evidence that EDCs affect levels of the DNMTs or co-factors needed for the addition of the methyl group. Furthermore, hormone receptors have been shown to direct regulators of DNA methylation to specific genomic loci, hence interference with this function could also lead to EDC-induced DNA methylation (reviewed in [44]).

4.2.2 Transgenerational Effects

Chemical exposure in pregnancy results in exposure of three generations: the pregnant mother (F0), the fetus (F1) and the fetal germ cells that will give rise to another generation (F2). Environmental factors have been shown to affect disease susceptibility over generations, even reaching generations after F2. These so-called transgenerational effects are thought to be mediated through stable epigenetic marks inherited from one generation to the next.

Transgenerational effects of EDCs have not been assessed in humans yet due to the very long study design needed. Nevertheless, studies have been performed in rodents showing evidence of transgenerational effects of EDC exposure. For example, changes in behavior, reduced male fertility, as well as increased prevalence of pubertal abnormalities, testis and ovarian disease and obesity have been observed in F3 as an effect of exposure to EDCs such as bisphenol A (BPA), Bis(2-ethylhexyl) phthalate (DEHP), Dibutyl phthalate (DBP) and/or Vinclozolin exposure in F0 pregnant mice or rats [45–47]. Since there are many similarities between rodents and humans, and because EDC-induced effects have been observed in human F2 ("DES granddaughters" described in Sect. 3.1.2; [48]), it is possible that EDCs can affect disease susceptibility over generations in humans.

4.3 Rapid Increase of Chronic Common Diseases – A Possible Connection to Prenatal EDC Exposure?

During the last decades, there has been an increase in non-communicable diseases like infertility, neurodevelopmental disorders, obesity and diabetes (reviewed in [9]). During the same time period chemical production, and thus human exposure, has risen enormously (Fig. 4.2), which makes the connection between chemicals and disease prevalence interesting. However, these types of correlations need to be interpreted with care as also other life-style changes have occurred during the same time period. Nevertheless, epidemiological studies together with experimental

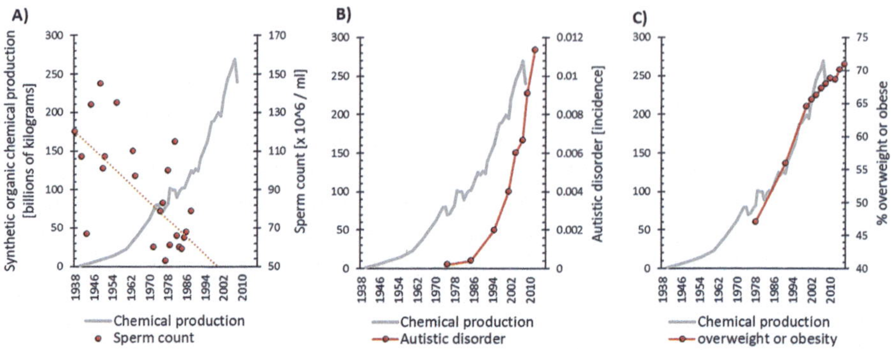

Fig. 4.2 Concurrent increase in chemical production and (**A**) reproductive, (**B**) neurodevelopmental and (**C**) metabolic diseases. One health outcome per disease category is visualized. Synthetic chemical production per year in USA (left y-axis) adapted from [150]. Sperm count determined in different cohorts from USA adapted from [196]. Autism disorder (AD) incidence in USA adapted from [197, 198], where the x axis represents study publication year. Obesity data was accessed from OECD [199], and presented as a percentage of the population aged ≥15 years with a BMI ≥ 25 (BMI values calculated from precise estimates of height and weight from health examinations) in USA

evidence and similar observations in wildlife species around the world (reviewed in [9]) clearly indicate a role for chemical exposure in the recent increase of non-communicable diseases. Indeed, when comparing different risk factors for developing complex diseases, environmental chemicals have been shown to be one major risk [49, 50]. In particular early life exposure to EDCs may be a major contributory factor in many later-life diseases such as fertility problems, cancer, obesity and neurodevelopmental disorders (reviewed in [6, 9]). It is likely that some chemically induced changes in the epigenome during fetal development can make the individual more susceptible to other triggers later in life, decreasing the threshold for disease onset.

To exemplify the role of EDCs in complex common diseases, we have chosen to focus on their role in infertility problems, neurodevelopmental disorders and metabolic disorders. Examples of EDCs associated with these diseases are discussed, as well as some of the known endocrine disrupting and, if studied, epigenetic mechanisms involved.

4.4 Fertility Problems

4.4.1 Introduction

Alarmingly, an over 50% decrease in sperm count between 1973 and 2011 has been reported in a meta-regression analysis of men from North America, Europe, Australia and New Zealand [51]. It is more difficult to study female fertility, but an increase in female reproductive organ diseases hand in hand with a decrease in female fertility have been suggested (reviewed in [32]). In addition, the overall decreased fertility can be reflected in the observed declining rates of natural conceptions [52, 53].

Infertility may be affected by, for example, diseases, infections, lifestyle related factors and female age [54, 55]. However, there are indications that the increase in synthetics chemicals, like EDCs, is linked to the decline in fertility around the world (Fig. 4.2) (reviewed in [32, 55–57]).

4.4.2 Associations Between Prenatal EDC Exposure and Fertility Problems in Adulthood

Due to the long study times, there are not many epidemiological studies investigating the effect of fetal EDC exposure on female fertility later in life. However, a lot has been learnt from exposure to the potent EDC diethylstilbestrol (DES) that was used between the 1940's and the early 1970's (reviewed in [58]). DES is a synthetic estrogen that was administered to pregnant women with the aim to reduce miscarriage and pregnancy complications. It is now clear that DES treatment during pregnancy increases infertility among the DES-exposed born girls ("DES daughters") in adulthood, where uterine and tubular problems were reported as a major cause of the female infertility [59]. Furthermore, pregnant DES daughters are less likely to have full-term live births compared to unexposed women [60]. Instead, the DES daughters had an increased risk of premature births, spontaneous pregnancy losses and ectopic pregnancies [60]. Moreover, daughters to DES daughters ("DES granddaughters") have been reported to have more irregular menstrual periods and fewer live births compared to unexposed women [48]. Although less potent than DES, BPA can also disrupt estrogen signaling, and *in utero* exposure to BPA has been suggested to cause similar effect as DES in female mice [61, 62], some of the effects being transgenerational [62]. If BPA affects sexual development and fertility in humans needs to be further studied.

Also DES sons have been studied, and although DES exposure does not seem to impair male fertility [63], congenital malformations of the genitalia, like cryptorchidism, has been reported to be increased in DES sons compared to sons of untreated women [63, 64]. Interestingly, the adverse effect of DES was more severe if DES exposure began before the 11th week of pregnancy [63, 64], supporting the notion that the timing of EDC exposure is important for its consequences on sexual development (reviewed in [57]).

Male fertility is epidemiologically easier to study compared to female reproduction since the proxy measurement anogenital distance (AGD; the distance between the anus and genitalia) is easily accessible already at birth. Shorter AGD is associated with male infertility and lower sperm count [65, 66]. In the context of male infertility, most evidence is available for some of the phthalate esters. For example,

DEHP and diisononyl phthalate (DiNP) metabolites in first trimester urine has been associated with shorter AGD in boys after birth [67, 68]. In addition, several phthalate ester metabolites measured during trimester two and three have been associated with shorter AGD [69–71], incomplete testicular descent [69] and smaller penile size [70]. Interestingly, the genetic background may play a role in susceptibility as white Americans and African Americans show differences regarding associations between phthalate ester exposure and anogenital measurements [71]. These epidemiological findings are corroborated by experimental data, in particular from rodent studies, where negative effects on male sexual development have been repeatedly reported after *in utero* phthalate ester exposure [72–74].

4.4.3 Endocrine Disrupting Properties & EDC-Induced DNA Methylation Associated With Reproductive Capacity

Reproductive development is highly dependent on steroid hormone signaling, in particular of the sexual hormones estrogens and androgens. Thus, prenatal exposure to chemicals that interfere with these hormones have the potential to affect sexual development in the exposed fetus, which may cause of infertility in reproductive age (reviewed in [32]).

A typical estrogenic compound is BPA whose estrogenic properties were discovered already in the 1930s. This was a desired property for pharmaceutical use, but when DES was synthesized and proven more potent, only DES was commercialized for this purpose. BPA has instead been extensively used during the past 60 years in plastics and epoxy resins (reviewed in [75]). It has been shown that the effects of DES on female reproductive abnormalities, like anovulation and cornification of the vaginal epithelium, are dependent on the presence of ERα in experimental studies [76, 77]. Interestingly, changes in the hormone balance *in utero* may also permanently alter the estrogen receptors balance for life, as observed in adult female sheep prenatally exposed to BPA, where ERα expression was increased and ERβ expression decreased in hypothalamus compared to controls [78]. This change in ER expression may be a consequence of BPA-induced non-monotonic sex-specific DNA methylation of ER genes, which has been observed in mouse brain [79]. Furthermore, prenatal DES exposure has been associated with altered concentrations of estradiol, follicle-stimulating hormone (FSH) and inhibin B in women aged 36–45 years [80], indicating that interfering with estrogen signaling *in utero* may have effects on hormone balance for life.

Phthalates, on the other hand, are well-known anti-androgens [81] and have been shown to reduce testosterone production in fetal rodent testis [82, 83]. There is a critical window for genital development (estimated to be 8–14 weeks of gestation in humans) where androgen action is crucial for masculinization of the organism [84]. Therefore, depending on the exposure timing of chemicals that interfere with androgen signaling, the effects on genital development will vary [85]. Apart from phthalates, a number of pesticides and fungicides are known to affect androgen signaling and, importantly, there is a dose-additive effect of different anti-androgenic chemicals on reproductive development (reviewed in [86]).

4.5 Neurodevelopmental Disorders

4.5.1 Introduction

Neurodevelopmental disorders are highly heterogeneous with etiologies involving impaired brain developmental processes. Examples for neurodevelopmental disorders are autism spectrum disorders (ASDs), attention-deficit/hyperactivity disorder (ADHD), intellectual disability, and schizophrenia. In children, impaired neurodevelopment can be expressed as impairments of cognitive functions (e.g. difficulties with language and speech, learning and memory), emotional behavior (e.g. increased anxiety and depression), and social behavior (e.g. conduct problems, aggressiveness).

Impaired neurodevelopment can be observed as a variety of symptoms, but not always diagnosed as a disease. Even so, neurodevelopmental diseases like ASD, ADHD, depression and schizophrenia have been reported to steeply increase during the last decades [87–92].

Although heredity is a well-documented risk factor, heritability is low for many neurodevelopmental disorders. For ASD, heritability is around 50% [93, 94], implicating that environmental factors also play a role in the disease etiology. In fact, in many cases, it is the interaction between genetic disposition and environmental factors that determine disease risk. An interesting example in this context is that genetic polymorphisms in genes involved in detoxification of environmental pollutants are more common in children diagnosed with ASD compared to in controls (reviewed in [95]).

4.5.2 Associations Between Prenatal EDC Exposure and Neurodevelopmental Health Outcomes

Epidemiological studies have established associations between prenatal exposure to EDCs and impaired neurodevelopment in children for, e.g., BPA [96–100], polybrominated diphenyl ethers (PBDEs) [100–104], PCBs [105–109] and phthalates [109–118]. These studies support that early life exposure to EDCs, measured during pregnancy, can affect neurodevelopment in domains primarily related to child behavior and cognition. This is corroborated by experimental studies in rodents, where neurodevelopmental impairments following prenatal EDC exposure is well documented (reviewed in [119, 120]).

It is important to note that different EDCs affect neurodevelopment differentially, depending on which hormonal system they affect, the timing of exposure, and not least the sex of the exposed individual. For example, prenatal PBDE (primarily PBDE-47) exposure mainly impairs mental development and lowers IQ in the offspring, whereas BPA exposure is associated with behavioral changes in the offspring, expressed as changes in both externalizing behavior (e.g. hyperactive and aggressive behavior) and internalizing behavior (e.g. anxiety and depression). In the case of BPA [96–100] as well as of exposure to phthalate esters [109–114], clear differences between males and females could be observed.

Considering that fetuses are exposed to a mixture of EDCs that affect development in different ways, it would be important to focus more on studying effects of mixtures. In a recent study, Birgerson et al. [121] have identified a mixture of EDCs in pregnant women associated with language delay at 30 months of age (an early marker for impaired neurodevelopment) in their children. Subsequently, this mixture was tested in experimental systems including concentrations found in the pregnant women. In human fetal primary neuronal stem cells, transcriptomic analyses revealed that this mixture affects expression of genes previously associated with neurodevelopmental disorders. In addition, the mixture altered behavior in tadpoles and zebra fish embryos, confirming the epidemiologically observed associations between EDC exposure and impaired neurodevelopment.

4.5.3 Endocrine Disrupting Properties & EDC-Induced DNA Methylation Associated With Neurodevelopment

Many hormonal systems are crucial for brain development (reviewed in [122]). In the context of EDCs, the best studied ones are thyroid hormone (TH) signaling, estrogen signaling and glucocorticoid signaling.

It is well known that TH signaling is essential for brain development in all vertebrates, including humans [123, 124]. Reduced levels of thyroid hormone levels during pregnancy can cause irreversible adverse effects on the child's neurodevelopment [124–127], observed as, e.g., IQ loss [125, 127]. The observed association between changes in maternal TH levels and impaired neurodevelopment may be, at least partly, due to that maternal TH levels affect brain morphology, more specifically the white matter/grey matter ratios, in the child [127]. Many EDCs can disrupt different parts of the thyroid signaling pathway (reviewed in [128]). For example, high maternal BPDE level is associated with changes in maternal thyroid hormone level [23], which may be a link between the reported association between prenatal BPDE exposure and impaired mental development and IQ loss [100–103].

Estrogen signaling, via ERα, ERβ, and membrane-bound ER, is involved in cell proliferation, differentiation and migration of cortical interneurons, and synaptogenesis in hippocampus and neocortex during development [129–131]. For example, estrogens can enhance N-methyl-D-aspartate receptor (NMDAR) mediated synaptic currents, possibly due to increased expression or recruitment of the NR2B subunit (reviewed in [132]). NMDARs are receptors for the excitatory neurotransmitter glutamate and are involved in synaptic plasticity, which is important for learning and memory (reviewed in [132, 133]). BPA exposure has been shown to alter NMDA receptor expression as well as synaptogenesis in mouse hippocampus, which leads to impaired learning behavior [134, 135]. Interestingly, in both human and rats, prenatal BPA exposure has been shown to alter methylation of *GRIN2B*, the gene encoding the NR2B subunit of the NMDA receptor, in females but not in males [136]. Genetic polymorphisms in *GRIN2B* have been associated with neurodevelopmental disorders such as ADHD and ASD [137, 138], thus the BPA-induced changes in regulation of this gene [136] may have a connection to the BPA-induced ADHD like symptoms observed in both mice and humans [139].

Furthermore, estrogen increases brain-derived neurotrophic factor (BDNF), a small secreted growth factor important in synaptic plasticity, in hippocampus and prefrontal cortex (reviewed in [130]). BDNF is essential for learning and memory, and its expression and methylation has been observed to be altered in neurodevelopmental diseases (reviewed in [140]). In both mice and human, prenatal BPA exposure is associated with sex-specific changes DNA methylation in the *BDNF* gene, measured in the cord blood of babies or in blood and hippocampus tissue of mice [141].

Glucocorticoid signaling is essential for the body's stress response, and an impaired regulation of stress response is involved in stress-related disorders such as depression and anxiety (reviewed in [142]). Glucocorticoids, like cortisol, are steroid hormones that bind glucocorticoid receptors (GR), where also co-regulatory proteins are needed. One such protein is FKBP51, a stress responsive protein encoded by the *FKBP5* gene. FKBP51 inhibits GR signaling by different mechanisms and increased *FKBP5* expression has been observed in, for example, major depressive disorder and schizophrenia patients (reviewed in [143]). Interestingly, prenatal BPA-exposure in rats increases methylation of *Fkbp5*, and lowers the FKBP51 protein levels, in the hippocampus of male rat offspring, mediated via an ER beta dependent pathway [144]. BPA induced *FKBP5* methylation may be one molecular link explaining the increased anxiety and depression observed in boys exposed to BPA *in utero* [97, 99].

Of note, these examples illustrate the sex-specific effects of BPA on *GRIN2B*, *BDNF* and *FKBP5* during fetal development, which are coherent with associations between BPA exposure and sex-specific neurodevelopmental alterations [96–100].

4.6 Metabolic Disorders

4.6.1 Introduction: Obesity and Diabetes

Both childhood and adult obesity has been reported to increase worldwide [89, 145, 146], even more pronounced in boys compared with girls [146]. Hand in hand, the related metabolic disease type 2 diabetes (T2D) is increasing [147, 148].

As for other complex diseases, there are many different factors involved in the etiology of obesity and diabetes, such as genetic predisposition and life style factors. Yet, the increase in human exposure to synthetic chemicals has been proposed to also play a role the etiology of obesity and diabetes, as many of these chemicals may interfere with the body's natural weight-control mechanisms [149, 150].

4.6.2 Associations Between Prenatal EDC Exposure and the Metabolic Disorders Obesity and Diabetes

Obese individuals have more fat cells than normal weight individuals. The fat cell number is programmed during fetal development and subsequently the number cannot be altered by diet or exercise [151], indicating the importance of not disturbing a metabolic set point during development. Some EDCs have been suggested to be "obesogens", i.e. having a causative role in the pathogenesis of obesity. EDCs induce susceptibility to obesity during development by, e.g., increasing the number of fat cells, shifting the energy balance to favor energy storage in fat tissue, altering basal metabolic rate, and/or by interfering with neuroendocrine control of appetite and satiety (reviewed in [31, 152]). Due to recent observations that EDCs also increase susceptibility to T2D, liver lipid abnormalities and other metabolic abnormatities, a new, broader name for this chemical group has been suggested, namely "metabolism disrupting chemicals" [152].

Of the EDCs mentioned in this chapter (Table 4.1), *in utero* exposure to PFASs [153–158], BPA [159–163], PCBs [164–168] and DEHP metabolites [158, 163] have been associated with metabolic outcomes like weight measurements, obesity and/or T2D in epidemiological studies. These studies indicate that early life exposure to EDCs can change the metabolic set-point and increase the susceptibility to obesity and diabetes later in life.

One chemical that is repeatedly associated with weight measurements is PFOA. In a meta-analysis of 9 epidemiological studies, high PFOA exposure *in utero* was shown to decrease birth weight [155]. Since low birth weight is associated with a fast weight gain after birth [169], it may not be surprising that prenatal serum PFOA levels were associated with a more rapid increase in BMI between 2–8 years, as well as greater adiposity at 8 years. These associations were most prominent when comparing the 2nd tertile with the 1st tertile, indicating a non-linear dose response for the PFOA effect on human adiposity [153]. Importantly, high PFOA exposure *in utero* has been associated with overweight or obesity in female, but not male, offspring at 20 years of age [154], supporting the notion that chemical exposure *in utero* can have lifelong effects.

Another chemical frequently associated with both obesity and glucose intolerance is BPA (reviewed in [170]). Although effects of BPA on metabolic outcomes have been repeatedly observed, the direction of the associations and exact outcomes do vary. For example, prenatal BPA exposure has been associated with both increased [160–163] and decreased [159, 160] body mass index, percent body fat, and waist circumference and/or skinfold thickness in childhood. In addition, these effects have been sex-specific and more pronounced in girls compared to boys [159, 162, 163].

4.6.3 Endocrine Disrupting Properties & EDC-Induced DNA Methylation Associated With Metabolic Disorders

PFOA can bind and activate PPARγ, a master regulator of fat cell differentiation, and promote adipocyte differentiation [171]. This ability of PFOA to increase the number of fat cells may be one link between prenatal PFOA exposure and obesity in adulthood [154]. On the molecular level, the metabolic effect of FFOA may also be due to methylation of Insulin-like growth factor 2 (*IGF2*). IGF2 is a protein hormone involved in fetal growth and development [172]. Maternal PFOA levels has been negatively correlated with cord blood *IGF2* methylation, where the reduced *IGF2* methylation explained around one-fifth of the observed association between PFOA exposure and reduced ponderal index at birth [173]. In cord blood, PFOA exposure has also been associated with expressional changes of metabolically relevant transcription factors [174]. These differences in methylation and expression at birth may be connected to differences in circulating metabolic hormone levels observed later in life. For example, PFOA exposure *in utero* has been positively associated with serum insulin and leptin levels and inversely associated with adiponectin in adult females [154], confirming that PFOA may be an obesogen.

Estrogen signaling is involved in the regulation of energy metabolism, for example through the modulation of feeding behavior, brown fat adipocyte function, glucose/insulin balance, body fat distribution, and energy expenditure (reviewed in [152, 175]). Since BPA can interfere with estrogen signaling, BPA has the potential to be an obesogen. Indeed, BPA has been shown to induce adipogenesis in human adipose and mesenchymal stromal/stem cells [176–178] through an ER-mediated pathway [177]. Interestingly, human prenatal BPA exposure has also been associated with decreased cord blood DNA methylation in the promoter of the obesity associated mesoderm specific transcript (*MEST*), and this methylation change was connected to MEST expression and BMI z scores of the children [176]. *MEST* expression correlates with adipocyte size and adipose tissue expansion [179]. The association between BPA and MEST was strengthen by demonstrating that also human mesenchymal stem cells exposed to BPA displayed decreased *MEST* methylation, increased MEST expression as well as increased lipid accumulation in the cells. In mice, BPA *in utero* has been shown to altered *Mest* and *Fggy* (an obesity-relevant enzyme that phosphorylates carbohydrates) promoter methylation and transcription of these genes, as well as increase in body weight, at >10 weeks of age

[176, 180]. These studies indicate that BPA induces methylation changes that are stable over time, which may play a role in regulation of metabolism long after the exposure was gone. In mice, there is also evidence of transgenerational effects on obesity since a low, but not high, dose of plastic-derived EDCs (BPA, DEHP and DBP) induced obesity in F3, but not in F1 or F2 [45].

BPA has also been suggested to play a role in glucose intolerance and diabetes. Rodent studies have shown that BPA during development induces sex-specific, non-monotonic dose-response relationships with metabolic outcomes like weight gain and impaired glucose tolerance in adult offspring [181–183]. Also, the BPA effects were accelerated and exacerbated when the offspring was fed a high fat diet [181, 182], indicating that BPA during development can shift the energy balance in favor for energy storage, which increases the susceptibility for obesity later in life.

4.7 Conclusion

In this chapter, we have illustrated the increasing evidence that EDC exposure, in particular during developmental phases, is contributing to the etiology of common non-communicable diseases. We have exemplified the role of EDC exposure for susceptibility to three disease groups but there are other common diseases that have been linked to prenatal EDC exposure, for example cancer [184, 185], cardiovascular disease [186] as well as inflammatory diseases like allergy [187] and asthma [188–191].

In conclusion, research over the last decades, both on the epidemiological and experimental level, has clearly demonstrated that EDCs are contributing to some of the most common diseases and thus to the related suffering and costs. Indeed, the annual costs for health problems related to EDC exposure are extensive [192–194], and have been estimated to €163 billion (1.28% of EU Gross Domestic Product) in Europe alone [194]. Yet, this might be an underestimation as most studies focus on one chemical at the time instead of investigating effects of real-life mixtures or combined effects of chemical exposure and other environmental factors (such as stress, nutrition, etc.) or genetic predisposition.

Also, we have only started to understand the mechanisms underlying the (life-) long lasting effects of EDCs. This will be needed, not only to causally link exposure to adverse health outcomes but also to identify early indicators for chemically induced disease risk that can be used for chemical testing. Because even though the use of some of the described chemicals (e.g. BPA and some phthalate esters) will be or has been restricted due to their demonstrated impact on human health, we are faced with thousands of newly produced chemicals that should be accurately assessed before they reach the market.

References

1. The European Chemical Industry Council (cefic), Facts and Figures 2017. 2017.
2. OECD. OECD environmental outlook to 2050: the consequences of inaction. Paris: OECD Publishing; 2012.
3. Grandjean P, Landrigan PJ. Developmental neurotoxicity of industrial chemicals. Lancet. 2006;368(9553):2167–78.
4. Geyer R, Jambeck JR, Law KL. Production, use, and fate of all plastics ever made. Sci Adv. 2017;3(7):e1700782.
5. Yang C, et al. Persistent organic pollutants as risk factors for obesity and diabetes. Curr Diab Rep. 2017;17(12):132.
6. Mallozzi M, et al. The effect of maternal exposure to endocrine disrupting chemicals on fetal and neonatal development: a review on the major concerns. Birth Defects Res C Embryo Today. 2016;108(3):224–42.
7. Zoeller RT, et al. A path forward in the debate over health impacts of endocrine disrupting chemicals. Environ Health. 2014;13:118.
8. Zoeller RT, et al. Endocrine-disrupting chemicals and public health protection: a statement of principles from the Endocrine Society. Endocrinology. 2012;153(9):4097–110.
9. WHO/UNEP, State of the science of endocrine disrupting chemicals – 2012. 2013.
10. Vandenberg LN, et al. Hormones and endocrine-disrupting chemicals: low-dose effects and nonmonotonic dose responses. Endocr Rev. 2012;33(3):378–455.
11. Rudel RA, Perovich LJ. Endocrine disrupting chemicals in indoor and outdoor air. Atmos Environ (1994), 2009. 43(1): p. 170–181.
12. Weiss JM, et al. Daily intake of phthalates, MEHP, and DINCH by ingestion and inhalation. Chemosphere. 2018;208:40–9.
13. Koch HM, et al. Identifying sources of phthalate exposure with human biomonitoring: results of a 48h fasting study with urine collection and personal activity patterns. Int J Hyg Environ Health. 2013;216(6):672–81.
14. Lopez-Cervantes J, Paseiro-Losada P. Determination of bisphenol A in, and its migration from, PVC stretch film used for food packaging. Food Addit Contam. 2003;20(6):596–606.
15. Hu XC, et al. Detection of poly- and Perfluoroalkyl Substances (PFASs) in U.S. drinking water linked to industrial sites, military fire training areas, and wastewater treatment plants. Environ Sci Technol Lett. 2016;3(10):344–50.
16. Faniband M, Lindh CH, Jonsson BA. Human biological monitoring of suspected endocrine-disrupting compounds. Asian J Androl. 2014;16(1):5–16.
17. Woodruff TJ, Zota AR, Schwartz JM. Environmental chemicals in pregnant women in the United States: NHANES 2003-2004. Environ Health Perspect. 2011;119(6):878–85.
18. Kortenkamp A. Ten years of mixing cocktails: a review of combination effects of endocrine-disrupting chemicals. Environ Health Perspect. 2007;115(Suppl 1):98–105.
19. Ramhoj L, et al. Perfluorohexane Sulfonate (PFHxS) and a mixture of endocrine disrupters reduce thyroxine levels and cause anti-androgenic effects in rats. Toxicol Sci. 2018;163:579–91.
20. Bergman A, et al. The impact of endocrine disruption: a consensus statement on the state of the science. Environ Health Perspect. 2013;121(4):A104–6.
21. Shu H, et al. Temporal trends of phthalate exposures during 2007-2010 in Swedish pregnant women. J Expo Sci Environ Epidemiol. 2018;28:437–47.
22. Sarcinelli PN, et al. Dietary and reproductive determinants of plasma organochlorine levels in pregnant women in Rio de Janeiro. Environ Res. 2003;91(3):143–50.
23. Vuong AM, et al. Maternal Polybrominated Diphenyl Ether (PBDE) exposure and thyroid hormones in maternal and cord sera: the HOME study, Cincinnati, USA. Environ Health Perspect. 2015;123(10):1079–85.
24. Needham LL, et al. Partition of environmental chemicals between maternal and fetal blood and tissues. Environ Sci Technol. 2011;45(3):1121–6.

25. Gutzkow KB, et al. Placental transfer of perfluorinated compounds is selective—a Norwegian mother and child sub-cohort study. Int J Hyg Environ Health. 2012;215(2):216–9.
26. Mazdai A, et al. Polybrominated diphenyl ethers in maternal and fetal blood samples. Environ Health Perspect. 2003;111(9):1249–52.
27. Guvenius DM, et al. Human prenatal and postnatal exposure to polybrominated diphenyl ethers, polychlorinated biphenyls, polychlorobiphenylols, and pentachlorophenol. Environ Health Perspect. 2003;111(9):1235–41.
28. Inoue K, et al. Perfluorooctane sulfonate (PFOS) and related perfluorinated compounds in human maternal and cord blood samples: assessment of PFOS exposure in a susceptible population during pregnancy. Environ Health Perspect. 2004;112(11):1204–7.
29. Schecter A, Kassis I, Papke O. Partitioning of dioxins, dibenzofurans, and coplanar PCBS in blood, milk, adipose tissue, placenta and cord blood from five American women. Chemosphere. 1998;37(9–12):1817–23.
30. Group EW. Body burden: the pollution in newborns. 2005.
31. Janesick AS, Blumberg B. Obesogens: an emerging threat to public health. Am J Obstet Gynecol. 2016;214(5):559–65.
32. Crain DA, et al. Female reproductive disorders: the roles of endocrine-disrupting compounds and developmental timing. Fertil Steril. 2008; 90(4):911–40.
33. Grandjean P, Landrigan PJ. Neurobehavioural effects of developmental toxicity. Lancet Neurol. 2014;13(3):330–8.
34. Rice D, Barone S Jr. Critical periods of vulnerability for the developing nervous system: evidence from humans and animal models. Environ Health Perspect. 2000;108(Suppl 3):511–33.
35. Bale TL, et al. Early life programming and neurodevelopmental disorders. Biol Psychiatry. 2010;68(4):314–9.
36. Barker DJ. The origins of the developmental origins theory. J Intern Med. 2007;261(5):412–7.
37. Gluckman PD, Hanson MA. Living with the past: evolution, development, and patterns of disease. Science. 2004;305(5691):1733–6.
38. Jacobs MN, et al. Marked for life: epigenetic effects of endocrine disrupting chemicals. Annu Rev Environ Resour. 2017;42:105–60.
39. Bird A. DNA methylation patterns and epigenetic memory. Genes Dev. 2002;16(1):6–21.
40. Skinner MK, Manikkam M, Guerrero-Bosagna C. Epigenetic transgenerational actions of environmental factors in disease etiology. Trends Endocrinol Metab. 2010;21(4):214–22.
41. Marczylo EL, Jacobs MN, Gant TW. Environmentally induced epigenetic toxicity: potential public health concerns. Crit Rev Toxicol. 2016;46(8):676–700.
42. Tran NQV, Miyake K. Neurodevelopmental disorders and environmental toxicants: epigenetics as an underlying mechanism. Int J Genomics. 2017;2017:7526592.
43. Walker DM, Gore AC. Epigenetic impacts of endocrine disruptors in the brain. Front Neuroendocrinol. 2017;44:1–26.
44. Alavian-Ghavanini A, Ruegg J. Understanding epigenetic effects of endocrine disrupting chemicals: from mechanisms to novel test methods. Basic Clin Pharmacol Toxicol. 2018;122(1):38–45.
45. Manikkam M, et al. Plastics derived endocrine disruptors (BPA, DEHP and DBP) induce epigenetic transgenerational inheritance of obesity, reproductive disease and sperm epimutations. PLoS One. 2013;8(1):e55387.
46. Wolstenholme JT, Goldsby JA, Rissman EF. Transgenerational effects of prenatal bisphenol A on social recognition. Horm Behav. 2013;64(5):833–9.
47. Anway MD, et al. Epigenetic transgenerational actions of endocrine disruptors and male fertility. Science. 2005;308(5727):1466–9.
48. Titus-Ernstoff L, et al. Menstrual and reproductive characteristics of women whose mothers were exposed in utero to diethylstilbestrol (DES). Int J Epidemiol. 2006;35(4):862–8.
49. Bellinger DC. A strategy for comparing the contributions of environmental chemicals and other risk factors to neurodevelopment of children. Environ Health Perspect. 2012;120(4):501–7.

50. Landrigan PJ, et al. The Lancet Commission on pollution and health. Lancet. 2018;391(10119):462–512.
51. Levine H, et al. Temporal trends in sperm count: a systematic review and meta-regression analysis. Hum Reprod Update. 2017;23(6):646–59.
52. Hamilton BE, Ventura SJ. Fertility and abortion rates in the United States, 1960-2002. Int J Androl. 2006;29(1):34–45.
53. Lassen TH, et al. Trends in rates of natural conceptions among Danish women born during 1960–1984. Hum Reprod. 2012;27(9):2815–22.
54. Vander Borght M, Wyns C. Fertility and infertility: definition and epidemiology. Clin Biochem. 2018;62:2–10.
55. Sharma R, et al. Lifestyle factors and reproductive health: taking control of your fertility. Reprod Biol Endocrinol. 2013;11:66.
56. Fowler PA, et al. Impact of endocrine-disrupting compounds (EDCs) on female reproductive health. Mol Cell Endocrinol. 2012;355(2):231–9.
57. Ho SM, et al. Environmental factors, epigenetics, and developmental origin of reproductive disorders. Reprod Toxicol. 2017;68:85–104.
58. Al Jishi T, Sergi C. Current perspective of diethylstilbestrol (DES) exposure in mothers and offspring. Reprod Toxicol. 2017;71:71–7.
59. Palmer JR, et al. Infertility among women exposed prenatally to diethylstilbestrol. Am J Epidemiol. 2001;154(4):316–21.
60. Kaufman RH, et al. Continued follow-up of pregnancy outcomes in diethylstilbestrol-exposed offspring. Obstet Gynecol. 2000;96(4):483–9.
61. Signorile PG, et al. Pre-natal exposure of mice to bisphenol A elicits an endometriosis-like phenotype in female offspring. Gen Comp Endocrinol. 2010;168(3):318–25.
62. Ziv-Gal A, et al. The effects of in utero bisphenol A exposure on reproductive capacity in several generations of mice. Toxicol Appl Pharmacol. 2015;284(3):354–62.
63. Wilcox AJ, et al. Fertility in men exposed prenatally to diethylstilbestrol. N Engl J Med. 1995;332(21):1411–6.
64. Palmer JR, et al. Urogenital abnormalities in men exposed to diethylstilbestrol in utero: a cohort study. Environ Health. 2009;8:37.
65. Eisenberg ML, et al. The relationship between anogenital distance, fatherhood, and fertility in adult men. PLoS One. 2011;6(5):e18973.
66. Mendiola J, et al. Shorter anogenital distance predicts poorer semen quality in young men in Rochester, New York. Environ Health Perspect. 2011;119(7):958–63.
67. Swan SH, et al. First trimester phthalate exposure and anogenital distance in newborns. Hum Reprod. 2015;30(4):963–72.
68. Bornehag CG, et al. Prenatal phthalate exposures and anogenital distance in Swedish boys. Environ Health Perspect. 2015;123(1):101–7.
69. Swan SH, et al. Decrease in anogenital distance among male infants with prenatal phthalate exposure. Environ Health Perspect. 2005;113(8):1056–61.
70. Bustamante-Montes LP, et al. Prenatal exposure to phthalates is associated with decreased anogenital distance and penile size in male newborns. J Dev Orig Health Dis. 2013;4(4):300–6.
71. Wenzel AG, et al. Influence of race on prenatal phthalate exposure and anogenital measurements among boys and girls. Environ Int. 2018;110:61–70.
72. Gray LE Jr, et al. Perinatal exposure to the phthalates DEHP, BBP, and DINP, but not DEP, DMP, or DOTP, alters sexual differentiation of the male rat. Toxicol Sci. 2000;58(2):350–65.
73. Macleod DJ, et al. Androgen action in the masculinization programming window and development of male reproductive organs. Int J Androl. 2010;33(2):279–87.
74. Foster PM. Disruption of reproductive development in male rat offspring following in utero exposure to phthalate esters. Int J Androl, 2006. 29(1): 140–147; discussion 181-5.
75. Vogel SA. The politics of plastics: the making and unmaking of bisphenol a "safety". Am J Public Health. 2009;99(Suppl 3):S559–66.

76. Nakamura T, et al. Estrogen receptor subtypes selectively mediate female mouse reproductive abnormalities induced by neonatal exposure to estrogenic chemicals. Toxicology. 2008;253(1–3):117–24.
77. Couse JF, Korach KS. Estrogen receptor-alpha mediates the detrimental effects of neonatal diethylstilbestrol (DES) exposure in the murine reproductive tract. Toxicology. 2004;205(1–2):55–63.
78. Mahoney MM, Padmanabhan V. Developmental programming: impact of fetal exposure to endocrine-disrupting chemicals on gonadotropin-releasing hormone and estrogen receptor mRNA in sheep hypothalamus. Toxicol Appl Pharmacol. 2010;247(2):98–104.
79. Kundakovic M, et al. Sex-specific epigenetic disruption and behavioral changes following low-dose in utero bisphenol a exposure. Proc Natl Acad Sci USA. 2013;110(24):9956–61.
80. Wise LA, et al. Prenatal diethylstilbestrol exposure and reproductive hormones in premenopausal women. J Dev Orig Health Dis. 2015;6(3):208–16.
81. Lee BM, Koo HJ. Hershberger assay for antiandrogenic effects of phthalates. J Toxicol Environ Health A. 2007;70(15–16):1365–70.
82. Wilson VS, et al. Phthalate ester-induced gubernacular lesions are associated with reduced insl3 gene expression in the fetal rat testis. Toxicol Lett. 2004;146(3):207–15.
83. Wu S, et al. Dynamic effect of di-2-(ethylhexyl) phthalate on testicular toxicity: epigenetic changes and their impact on gene expression. Int J Toxicol. 2010;29(2):193–200.
84. Welsh M, et al. Identification in rats of a programming window for reproductive tract masculinization, disruption of which leads to hypospadias and cryptorchidism. J Clin Invest. 2008;118(4):1479–90.
85. Martino-Andrade AJ, et al. Timing of prenatal phthalate exposure in relation to genital endpoints in male newborns. Andrology. 2016;4(4):585–93.
86. Gray LE Jr et al. Adverse effects of environmental antiandrogens and androgens on reproductive development in mammals. Int J Androl, 2006. 29(1): 96–104; discussion 105–8.
87. Newschaffer CJ, Falb MD, Gurney JG. National autism prevalence trends from United States special education data. Pediatrics. 2005;115(3):e277–82.
88. Hertz-Picciotto I, Delwiche L. The rise in autism and the role of age at diagnosis. Epidemiology. 2009;20(1):84–90.
89. (EPA), U.S.E.P.A., America's children and the environment. 2013.
90. McDonald ME, Paul JF. Timing of increased autistic disorder cumulative incidence. Environ Sci Technol. 2010;44(6):2112–8.
91. Mojtabai R, Olfson M, Han B. National Trends in the prevalence and treatment of depression in adolescents and young adults. Pediatrics. 2016;138(6):e20161878.
92. Chan KY, et al. Prevalence of schizophrenia in China between 1990 and 2010. J Glob Health. 2015;5(1):010410.
93. Hallmayer J, et al. Genetic heritability and shared environmental factors among twin pairs with autism. Arch Gen Psychiatry. 2011;68(11):1095–102.
94. Sandin S, et al. THe familial risk of autism. JAMA. 2014;311(17):1770–7.
95. Rossignol DA, Genuis SJ, Frye RE. Environmental toxicants and autism spectrum disorders: a systematic review. Transl Psychiatry. 2014;4:e360.
96. Braun JM, et al. Impact of early-life bisphenol A exposure on behavior and executive function in children. Pediatrics. 2011;128(5):873–82.
97. Harley KG, et al. Prenatal and early childhood bisphenol A concentrations and behavior in school-aged children. Environ Res. 2013;126:43–50.
98. Evans SF, et al. Prenatal bisphenol A exposure and maternally reported behavior in boys and girls. Neurotoxicology. 2014;45:91–9.
99. Perera F, et al. Bisphenol A exposure and symptoms of anxiety and depression among inner city children at 10–12 years of age. Environ Res. 2016;151:195–202.
100. Braun JM, et al. Prenatal environmental chemical exposures and longitudinal patterns of child neurobehavior. Neurotoxicology. 2017;62:192–9.
101. Chen A, et al. Prenatal polybrominated diphenyl ether exposures and neurodevelopment in U.S. children through 5 years of age: the HOME study. Environ Health Perspect. 2014;122(8):856–62.

102. Eskenazi B, et al. In utero and childhood polybrominated diphenyl ether (PBDE) exposures and neurodevelopment in the CHAMACOS study. Environ Health Perspect. 2013;121(2):257–62.
103. Herbstman JB, et al. Prenatal exposure to PBDEs and neurodevelopment. Environ Health Perspect. 2010;118(5):712–9.
104. Sagiv SK, et al. Prenatal and childhood polybrominated diphenyl ether (PBDE) exposure and attention and executive function at 9–12 years of age. Neurotoxicol Teratol. 2015;52(Pt B):151–61.
105. Forns J, et al. Prenatal exposure to polychlorinated biphenyls and child neuropsychological development in 4-year-olds: an analysis per congener and specific cognitive domain. Sci Total Environ. 2012;432:338–43.
106. Sagiv SK, et al. Prenatal organochlorine exposure and behaviors associated with attention deficit hyperactivity disorder in school-aged children. Am J Epidemiol. 2010;171(5):593–601.
107. Stewart PW, et al. The relationship between prenatal PCB exposure and intelligence (IQ) in 9-year-old children. Environ Health Perspect. 2008;116(10):1416–22.
108. Kyriklaki A, et al. Prenatal exposure to persistent organic pollutants in association with offspring neuropsychological development at 4years of age: the Rhea mother-child cohort, Crete, Greece. Environ Int. 2016;97:204–11.
109. Kim S, et al. Association between maternal exposure to major phthalates, heavy metals, and persistent organic pollutants, and the neurodevelopmental performances of their children at 1 to 2years of age- CHECK cohort study. Sci Total Environ. 2018;624:377–84.
110. Arbuckle TE, et al. Bisphenol A, phthalates and lead and learning and behavioral problems in Canadian children 6–11 years of age: CHMS 2007–2009. Neurotoxicology. 2016;54:89–98.
111. Whyatt RM, et al. Maternal prenatal urinary phthalate metabolite concentrations and child mental, psychomotor, and behavioral development at 3 years of age. Environ Health Perspect. 2012;120(2):290–5.
112. Kobrosly RW, et al. Prenatal phthalate exposures and neurobehavioral development scores in boys and girls at 6–10 years of age. Environ Health Perspect. 2014;122(5):521–8.
113. Olesen TS, et al. Prenatal phthalate exposure and language development in toddlers from the Odense Child Cohort. Neurotoxicol Teratol. 2018;65:34–41.
114. Kim Y, et al. Prenatal exposure to phthalates and infant development at 6 months: prospective Mothers and Children's Environmental Health (MOCEH) study. Environ Health Perspect. 2011;119(10):1495–500.
115. Lien YJ, et al. Prenatal exposure to phthalate esters and behavioral syndromes in children at 8 years of age: Taiwan Maternal and Infant Cohort Study. Environ Health Perspect. 2015;123(1):95–100.
116. Factor-Litvak P, et al. Persistent associations between maternal prenatal exposure to phthalates on child IQ at age 7 years. PLoS One. 2014;9(12):e114003.
117. Engel SM, et al. Prenatal phthalate exposure is associated with childhood behavior and executive functioning. Environ Health Perspect. 2010;118(4):565–71.
118. Engel SM, et al. Prenatal phthalates, maternal thyroid function, and risk of attention-deficit hyperactivity disorder in the Norwegian mother and child cohort. Environ Health Perspect. 2018;126(5):057004.
119. Wolstenholme JT, Rissman EF, Connelly JJ. The role of Bisphenol A in shaping the brain, epigenome and behavior. Horm Behav. 2011;59(3):296–305.
120. Rebuli ME, Patisaul HB. Assessment of sex specific endocrine disrupting effects in the prenatal and pre-pubertal rodent brain. J Steroid Biochem Mol Biol. 2016;160:148–59.
121. Birgersson L, et al. From cohorts to molecules: adverse impacts of endocrine disrupting mixtures. Deposited as preprint on BioRxiv 2017. https://doi.org/10.1101/206664
122. Miranda A, Sousa N. Maternal hormonal milieu influence on fetal brain development. Brain Behav. 2018;8(2):e00920.

123. Bernal J. Thyroid hormone receptors in brain development and function. Nat Clin Pract Endocrinol Metab. 2007;**3**(3):249–59.
124. Willoughby KA, McAndrews MP, Rovet JF. Effects of maternal hypothyroidism on offspring hippocampus and memory. Thyroid. 2014;24(3):576–84.
125. Haddow JE, et al. Maternal thyroid deficiency during pregnancy and subsequent neuropsychological development of the child. N Engl J Med. 1999;341(8):549–55.
126. Pop VJ, et al. Low maternal free thyroxine concentrations during early pregnancy are associated with impaired psychomotor development in infancy. Clin Endocrinol (Oxf). 1999;50(2):149–55.
127. Korevaar TI, et al. Association of maternal thyroid function during early pregnancy with offspring IQ and brain morphology in childhood: a population-based prospective cohort study. Lancet Diabetes Endocrinol. 2016;4(1):35–43.
128. Ghassabian A, Trasande L. Disruption in thyroid signaling pathway: a mechanism for the effect of endocrine-disrupting chemicals on child neurodevelopment. Front Endocrinol (Lausanne). 2018;9:204.
129. Bowers JM, Waddell J, McCarthy MM. A developmental sex difference in hippocampal neurogenesis is mediated by endogenous oestradiol. Biol Sex Differ. 2010;1(1):8.
130. Luine V, Frankfurt M. Interactions between estradiol, BDNF and dendritic spines in promoting memory. Neuroscience. 2013;239:34–45.
131. Wang L, et al. Estrogen receptor (ER)beta knockout mice reveal a role for ERbeta in migration of cortical neurons in the developing brain. Proc Natl Acad Sci USA. 2003;100(2):703–8.
132. Liu SB, Zhao MG. Neuroprotective effect of estrogen: role of nonsynaptic NR2B-containing NMDA receptors. Brain Res Bull. 2013;93:27–31.
133. Baez MV, Cercato MC, Jerusalinsky DA. NMDA receptor subunits change after synaptic plasticity induction and learning and memory acquisition. Neural Plast. 2018;2018:5093048.
134. Xu X, et al. Sex-specific effects of bisphenol-A on memory and synaptic structural modification in hippocampus of adult mice. Horm Behav. 2013;63(5):766–75.
135. Xu XH, et al. Perinatal exposure to bisphenol-A impairs learning-memory by concomitant down-regulation of N-methyl-D-aspartate receptors of hippocampus in male offspring mice. Horm Behav. 2010;58(2):326–33.
136. Alavian-Ghavanini A, et al. Prenatal bisphenol A exposure is linked to epigenetic changes in glutamate receptor subunit gene Grin2b in female rats and humans. Sci Rep. 2018;8(1):11315.
137. Dorval KM, et al. Association of the glutamate receptor subunit gene GRIN2B with attention-deficit/hyperactivity disorder. Genes Brain Behav. 2007;6(5):444–52.
138. Hu C, et al. Human GRIN2B variants in neurodevelopmental disorders. J Pharmacol Sci. 2016;132(2):115–21.
139. Rochester JR, Bolden AL, Kwiatkowski CF. Prenatal exposure to bisphenol A and hyperactivity in children: a systematic review and meta-analysis. Environ Int. 2018;114:343–56.
140. Boulle F, et al. Epigenetic regulation of the BDNF gene: implications for psychiatric disorders. Mol Psychiatry. 2012;17(6):584–96.
141. Kundakovic M, et al. DNA methylation of BDNF as a biomarker of early-life adversity. Proc Natl Acad Sci USA. 2015;112(22):6807–13.
142. Holsboer F, Ising M. Stress hormone regulation: biological role and translation into therapy. Annu Rev Psychol, 2010. 61: 81–109, C1–11.
143. Matosin N, Halldorsdottir T, Binder EB. Understanding the molecular mechanisms underpinning gene by environment interactions in psychiatric disorders: the FKBP5 model. Biol Psychiatry. 2018;83(10):821–30.
144. Kitraki E, et al. Developmental exposure to bisphenol A alters expression and DNA methylation of Fkbp5, an important regulator of the stress response. Mol Cell Endocrinol. 2015;417:191–9.
145. Flegal KM, et al. Overweight and obesity in the United States: prevalence and trends, 1960–1994. Int J Obes Relat Metab Disord. 1998;22(1):39–47.

146. Hamrik Z, et al. Trends in overweight and obesity in Czech Schoolchildren from 1998 to 2014. Cent Eur J Public Health. 2017;25(Suppl 1):S10–4.
147. Fang M. Trends in the prevalence of diabetes among U.S. adults: 1999–2016. Am J Prev Med. 2018;55(4)
148. Thomas MC, et al. Evidence of an accelerating increase in prevalence of diagnosed Type 2 diabetes in British men, 1978–2005. Diabet Med. 2009;26(8):766–72.
149. Baillie-Hamilton PF. Chemical toxins: a hypothesis to explain the global obesity epidemic. J Altern Complement Med. 2002;8(2):185–92.
150. Neel BA, Sargis RM. The paradox of progress: environmental disruption of metabolism and the diabetes epidemic. Diabetes. 2011;60(7):1838–48.
151. Spalding KL, et al. Dynamics of fat cell turnover in humans. Nature. 2008;453(7196):783–7.
152. Heindel JJ, et al. Metabolism disrupting chemicals and metabolic disorders. Reprod Toxicol. 2017;68:3–33.
153. Braun JM, et al. Prenatal perfluoroalkyl substance exposure and child adiposity at 8 years of age: the HOME study. Obesity (Silver Spring). 2016;24(1):231–7.
154. Halldorsson TI, et al. Prenatal exposure to perfluorooctanoate and risk of overweight at 20 years of age: a prospective cohort study. Environ Health Perspect. 2012;120(5):668–73.
155. Johnson PI, et al. The Navigation Guide – evidence-based medicine meets environmental health: systematic review of human evidence for PFOA effects on fetal growth. Environ Health Perspect. 2014;122(10):1028–39.
156. Maisonet M, et al. Maternal concentrations of polyfluoroalkyl compounds during pregnancy and fetal and postnatal growth in British girls. Environ Health Perspect. 2012;120(10):1432–7.
157. Mora AM, et al. Prenatal exposure to perfluoroalkyl substances and adiposity in early and mid-childhood. Environ Health Perspect. 2017;125(3):467–73.
158. Lenters V, et al. Prenatal phthalate, perfluoroalkyl acid, and organochlorine exposures and term birth weight in three birth cohorts: multi-pollutant models based on elastic net regression. Environ Health Perspect. 2016;124(3):365–72.
159. Harley KG, et al. Prenatal and postnatal bisphenol A exposure and body mass index in childhood in the CHAMACOS cohort. Environ Health Perspect. 2013;121(4):514–20.
160. Vafeiadi M, et al. Association of early life exposure to bisphenol A with obesity and cardiometabolic traits in childhood. Environ Res. 2016;146:379–87.
161. Valvi D, et al. Prenatal bisphenol a urine concentrations and early rapid growth and overweight risk in the offspring. Epidemiology. 2013;24(6):791–9.
162. Hoepner LA, et al. Bisphenol A and adiposity in an inner-city birth cohort. Environ Health Perspect. 2016;124(10):1644–50.
163. Yang TC, et al. Bisphenol A and phthalates in utero and in childhood: association with child BMI z-score and adiposity. Environ Res. 2017;156:326–33.
164. Agay-Shay K, et al. Exposure to endocrine-disrupting chemicals during pregnancy and weight at 7 years of age: a multi-pollutant approach. Environ Health Perspect. 2015;123(10):1030–7.
165. Govarts E, et al. Birth weight and prenatal exposure to polychlorinated biphenyls (PCBs) and dichlorodiphenyldichloroethylene (DDE): a meta-analysis within 12 European Birth Cohorts. Environ Health Perspect. 2012;120(2):162–70.
166. Maresca MM, et al. Prenatal exposure to phthalates and childhood body size in an urban cohort. Environ Health Perspect. 2016;124(4):514–20.
167. Valvi D, et al. Prenatal concentrations of polychlorinated biphenyls, DDE, and DDT and overweight in children: a prospective birth cohort study. Environ Health Perspect. 2012;120(3):451–7.
168. Casas M, et al. Prenatal exposure to PCB-153, p,p'-DDE and birth outcomes in 9000 mother-child pairs: exposure-response relationship and effect modifiers. Environ Int. 2015;74:23–31.
169. Kopec G, Shekhawat PS, Mhanna MJ. Prevalence of diabetes and obesity in association with prematurity and growth restriction. Diabetes Metab Syndr Obes. 2017;10:285–95.

170. Alonso-Magdalena P, Quesada I, Nadal A. Prenatal exposure to BPA and offspring outcomes: the diabesogenic behavior of BPA. Dose Response. 2015;13(2):1559325815590395.
171. Yamamoto J, et al. Perfluorooctanoic acid binds to peroxisome proliferator-activated receptor gamma and promotes adipocyte differentiation in 3T3-L1 adipocytes. Biosci Biotechnol Biochem. 2015;79(4):636–9.
172. Ghanipoor-Samami M, et al. Atlas of tissue- and developmental stage specific gene expression for the bovine insulin-like growth factor (IGF) system. PLoS One. 2018;13(7):e0200466.
173. Kobayashi S, et al. Effects of prenatal perfluoroalkyl acid exposure on cord blood IGF2/H19 methylation and ponderal index: the Hokkaido study. J Expo Sci Environ Epidemiol. 2017;27(3):251–9.
174. Remy S, et al. Metabolic targets of endocrine disrupting chemicals assessed by cord blood transcriptome profiling. Reprod Toxicol. 2016;65:307–20.
175. Xu Y, Lopez M. Central regulation of energy metabolism by estrogens. Mol Metab. 2018;15:104–15.
176. Junge KM, et al. MEST mediates the impact of prenatal bisphenol A exposure on long-term body weight development. Clin Epigenetics. 2018;10:58.
177. Ohlstein JF, et al. Bisphenol A enhances adipogenic differentiation of human adipose stromal/stem cells. J Mol Endocrinol. 2014;53(3):345–53.
178. Dong H, et al. Non-cytotoxic nanomolar concentrations of bisphenol A induce human mesenchymal stem cell adipogenesis and osteogenesis. Ecotoxicol Environ Saf. 2018;164:448–54.
179. Takahashi M, Kamei Y, Ezaki O. Mest/Peg1 imprinted gene enlarges adipocytes and is a marker of adipocyte size. Am J Physiol Endocrinol Metab. 2005;288(1):E117–24.
180. Taylor JA, et al. Prenatal exposure to bisphenol A disrupts naturally occurring bimodal DNA methylation at proximal promoter of fggy, an obesity-relevant gene encoding a carbohydrate kinase, in Gonadal White Adipose Tissues of CD-1 mice. Endocrinology. 2018;159(2):779–94.
181. Wei J, et al. Perinatal exposure to bisphenol A at reference dose predisposes offspring to metabolic syndrome in adult rats on a high-fat diet. Endocrinology. 2011;152(8):3049–61.
182. Mackay H, et al. Organizational effects of perinatal exposure to bisphenol-A and diethylstilbestrol on arcuate nucleus circuitry controlling food intake and energy expenditure in male and female CD-1 mice. Endocrinology. 2013;154(4):1465–75.
183. Angle BM, et al. Metabolic disruption in male mice due to fetal exposure to low but not high doses of bisphenol A (BPA): evidence for effects on body weight, food intake, adipocytes, leptin, adiponectin, insulin and glucose regulation. Reprod Toxicol. 2013;42:256–68.
184. Huo D, et al. Incidence rates and risks of diethylstilbestrol-related clear-cell adenocarcinoma of the vagina and cervix: update after 40-year follow-up. Gynecol Oncol. 2017;146(3):566–71.
185. Hilakivi-Clarke L. Maternal exposure to diethylstilbestrol during pregnancy and increased breast cancer risk in daughters. Breast Cancer Res. 2014;16(2):208.
186. Troisi R, et al. Medical conditions among adult offspring prenatally exposed to diethylstilbestrol. Epidemiology. 2013;24(3):430–8.
187. Stelmach I, et al. The effect of prenatal exposure to phthalates on food allergy and early eczema in inner-city children. Allergy Asthma Proc. 2015;36(4):72–8.
188. Buckley JP, et al. Associations of prenatal environmental phenol and phthalate biomarkers with respiratory and allergic diseases among children aged 6 and 7years. Environ Int. 2018;115:79–88.
189. Jahreis S, et al. Maternal phthalate exposure promotes allergic airway inflammation over 2 generations through epigenetic modifications. J Allergy Clin Immunol. 2018;141(2):741–53.
190. Gascon M, et al. Prenatal exposure to bisphenol A and phthalates and childhood respiratory tract infections and allergy. J Allergy Clin Immunol. 2015;135(2):370–8.
191. Shu H, et al. PVC flooring at home and development of asthma among young children in Sweden, a 10-year follow-up. Indoor Air. 2014;24(3):227–35.
192. Attina TM, et al. Exposure to endocrine-disrupting chemicals in the USA: a population-based disease burden and cost analysis. Lancet Diabetes Endocrinol. 2016;4(12):996–1003.

193. Legler J, et al. Obesity, diabetes, and associated costs of exposure to endocrine-disrupting chemicals in the European Union. J Clin Endocrinol Metab. 2015;100(4):1278–88.
194. Trasande L, et al. Burden of disease and costs of exposure to endocrine disrupting chemicals in the European Union: an updated analysis. Andrology. 2016;4(4):565–72.
195. Liew Z, et al. Prenatal exposure to perfluoroalkyl substances and IQ scores at age 5; a study in the Danish National Birth Cohort. Environ Health Perspect. 2018;126(6):067004.
196. Carlsen E, et al. Evidence for decreasing quality of semen during past 50 years. BMJ. 1992;305(6854):609–13.
197. Weintraub K. The prevalence puzzle: autism counts. Nature. 2011;479:22–4.
198. Demeneix B. Losing our minds: how environmental pollution impairs human intelligence and mental health (Oxford Series in Behavioral Neuroendocrinology). 1 ed. 2014: Oxford University Press.
199. OECD. Overweight or obese population (indicator). 05 September 2018.

Part II
Gene-Environment Interaction and Disease Susceptibility

Chapter 5
Gene-Environment Interaction and Individual Susceptibility to Metabolic Disorders

Ingrid Dahlman and Mikael Rydén

Keywords Obesity · Type 2 diabetes · Insulin resistance · Smoking · Diet · Physical activity

5.1 Introduction

Overweight and obesity (see Table 5.1 for definitions), particularly when fat is accumulated in the abdominal area, associate with a number of metabolic complications including hypertension, dyslipidemia, insulin resistance and type 2 diabetes (T2D). The close link to T2D constitutes a particular health threat given the increased risk for cardiovascular disease in diabetes. Worldwide, nearly 40% of the adult population is estimated to be overweight and 10–15% obese [1]. Furthermore, >400 million people worldwide are living with T2D [2], and an additional 10% of the global population are likely to develop the disease. The rapid increase in obesity/T2D is multifactorial but primarily due to life-style including caloric over-supply and sedentary habits. Ongoing urbanization is an important underlying factor [3]. Altogether, this makes understanding of the underlying mechanisms and development of preventive strategies prioritized research areas.

5.2 Genetic Susceptibility to Obesity and T2D

Genetic epidemiological studies have provided support for a strong hereditary impact on obesity and T2D. More recently, progress in genetic techniques has permitted mapping of hundreds of susceptibility (risk) gene loci for these diseases.

I. Dahlman (✉) · M. Rydén
Department of Medicine (H7), Karolinska Institutet, C2-94, Karolinska University Hospital, Stockholm, Sweden
e-mail: ingrid.dahlman@ki.se; mikael.ryden@ki.se

R. Teperino (ed.), *Beyond Our Genes*,
https://doi.org/10.1007/978-3-030-35213-4_5

Table 5.1 Classification of body weight. The degree of excess body weight is categorized using Body Mass Index (BMI) which is calculated by dividing the body weight in kilograms with the square of the height in meters. The corresponding value in kg/m^2 is used to categorize individuals as normal weight, over weight and obese

	Body Mass Index (kg/m^2)
Normal weight	18.5– < 25
Overweight	25.0– < 30
Obesity	≥30

Mendelian randomization, in which the risk factor obesity has been replaced by genetic risk loci for BMI, has established a causal link to T2D and coronary artery disease (CAD) [4]. Despite this, most susceptibility gene loci do not overlap between the different traits suggesting that obesity, T2D and CAD develop through distinct mechanisms [5]. This notion is supported by the clinical observation that a proportion of morbidly obese are relatively metabolically healthy [6] and conversely, some patients with T2D are lean. This is also the reason why obesity and T2D are discussed in separate paragraphs in this chapter. Importantly, despite progress in human genetics, for most genetic loci associated with BMI and/or T2D, the underlying causative genes and involved organs remain to be identified. At best, defining genes contributing to obesity and T2D can identify new therapeutic targets. However, there is also an expectation of "precision medicine", i.e. that genetic information can be used to highlight which primary prevention intervention strategies and/or therapies that are most effective in a specific individual.

5.3 The Relationship Between Genetic and Environmental Factors in Obesity and T2D

Life-style modifications are an integral part in obesity prevention/treatment and large randomized control trials (RCTs) have clearly shown that physical activity and dietary interventions can minimize the risk, or delay the onset, of T2D [7]. Furthermore, twin studies support the notion that genetic background influences change in body fat storage in response to dietary intervention [8], and that physical activity decreases the genetic impact on BMI and body fat distribution [9]. Thus, the response to life style factors is the result of an interaction with genetic background, and to design effective prevention and treatment for obesity and T2D it will be necessary to define the most critical causal environmental factors for each individual. For this to become possible researchers must define the interaction between genetic variants and environmental factors. Both obesity and T2D arise from the interactions between a genetic risk profile and obesogenic environmental factors which include not only physical inactivity and excessive caloric intake but may also involve factors such as medications, socioeconomic status, poor sleep quality, and the gastrointestinal microbiome [4]. Environmental factors may also share mechanisms with gene variants, and studies of gene-environment interactions can potentially highlight pathways underlying genetic susceptibility to disease.

5.4 Genetics of Obesity

Heredity has a substantial impact on obesity which has been confirmed in numerous genetic epidemiological studies. Early twin studies estimated the heritability of BMI, i.e. the proportion of variance in BMI controlled by additive genetic factors, to be between 40–70% [10]. The obesogenic environment underlying the obesity epidemic has not altered the overall heritability for BMI, which has remained unchanged in more recent studies [11].

The genetic causes behind obesity may in theory depend on variations/mutations in single (monogenic obesity) or a combination of genes (polygenic obesity). It is now known that monogenic obesity is a rare condition characterized by early onset severe obesity. The early successes in obesity genetics were obtained in this group of patients and identified causal mutations in several genes (e.g. *MC4R*, *BDNF*, *PCSK1*, *POMC*, *SH2B1*, *LEP*, *LEPR*, and *NTRK2)* implicated in hypothalamic pathways involved in central regulation of food intake and satiety. Studies of monogenic obesity highlighted the leptin–melanocortin pathway as a key regulator of energy intake [12].

Unlike monogenic obesity, the genetic risk of common obesity reflects the accumulation of multiple loci, each contributing a small portion of the total risk (polygenic obesity). Analyzes of multifactorial traits have been revolutionized by the genome-wide association study (GWAS) approach in which a large number of study subjects, cases and controls or a population based cohort, are genotyped for millions of single nucleotide polymorphisms (SNPs) covering the common variation in the genome. The SNPs are individually analyzed for association with the disease or trait of interest. Due to the large number of independent tests, a nominal $P < 5\text{x}10^{-8}$ is required for genome-wide significant association of <5%. In 2007, SNPs in the first intron of *FTO* were reported to be associated with BMI [13]. *FTO* remains to date the strongest susceptibility gene locus for common obesity and has been confirmed in multiple studies and ethnic groups. *FTO* encodes a 2-oxoglutarate-dependent nucleic acid demethylase that is ubiquitously expressed but most highly expressed in hypothalamic nuclei governing energy balance [14], and has been reported to affect Leptin signaling [15]. However, the exact mechanisms remain elusive and more recent studies have proposed that SNPs in the *FTO* gene could impact on BMI by controlling the expression of the nearby gene *IRX3* and by influencing the metabolic function of adipocytes [16].

Subsequent GWAS in larger cohorts have identified additional genetic risk loci for BMI. Interestingly, several of the genes in the vicinity of these loci have previously been implicated in monogenic obesity [4]. The most recent and largest meta-analysis encompassing ~500,000 study subjects of mainly European and

Japanese descent brought the total number of BMI-associated loci to >200 [5]. The effect sizes of individual BMI-loci are modest, ~0.06–0.4 kg/m^2 per BMI-increasing allele, with the *FTO* locus having the largest effect. In accordance with the modest effect of individual loci, the joint effect of known BMI loci explain in the order of 3% of the population variance in BMI [5]. As a consequence of the low explanatory power, genetic risk loci or a genetic risk score for BMI based on multiple unique loci are inefficient to predict obesity.

Genes encoded near BMI-associated genetic loci display enriched expression in the central nervous system, suggesting that genetic control of BMI primarily involves control of food intake. Nevertheless, while BMI-associated genetic variants are over-represented for tissue-specific enhancers active in the CNS, they are also present in immune cells and adipose tissue indicating the potential importance of additional cell types/organs for the development of obesity [5].

5.5 Genetics of Body Fat Distribution

An important aspect of body fat mass is the accumulation into different peripheral depots. Expansion of the abdominal subcutaneous and visceral depot is closely associated with metabolic complications. An estimate of body fat distribution is obtained from the waist-to-hip ratio (WHR). A large GWAS meta-analysis in >200,000 study subjects identified 49 loci linked to BMI-adjusted WHR [17]. Genes expressed near these loci are enriched in adipose tissue, suggesting that fat distribution is at least to some degree determined by regulation in the fat depots themselves. While the 49 WHR-loci together explain 1.4% of the population variance in WHR, there is a gender effect as the associations were stronger in women than in men. Importantly, most genetic loci associated with WHR do not overlap with those for BMI, suggesting that genetic variations in fat distribution and total adiposity are mediated via independent mechanisms [4]. Nonetheless, the central roles of the *FTO* and *MC4R* loci for adiposity are strengthened by the finding that these genetic loci are associated with multiple adiposity traits. Studies of gene-environment interaction have mainly focused on genetic loci associated with BMI, and not those exclusively associated with BMI-adjusted WHR. This makes sense since BMI, but not body fat distribution, is strongly influenced by behavioral factors. It is therefore mainly BMI that is discussed below in conjunction with potential factors interacting with genetic susceptibility.

5.6 Genetics of Type 2 Diabetes

The genetic architecture of T2D exhibits great similarities to that of obesity, although a distinct set of genetic risk loci are involved. The heritability of T2D has been calculated to be between 30–35% in genetic epidemiological studies [18]. A few risk loci for T2D were identified before the GWAS era. These include a locus on chromosome 10 encoding the transcription factor *TCF7L2*, as well as a common variant in the 5′ region of *PPARG* on chromosome 3. In GWAS, around 250 genetic risk loci for T2D have been mapped [19]. Together, these loci explain ~20% of the genetic risk of developing T2D [18]. Analyses of quantitative traits have identified three clusters of susceptibility loci for T2D; the first and largest cluster encodes genes primarily involved in insulin secretion (i.e. *GIPR, C2CDC4A, CDKAL1, GCK, TCF7L2, GLIS3, THADA,* and *IGF2BP2*). The second cluster is primarily associated with insulin sensitivity and encodes genes such as *PPARG*, *KLF14*, and *IRS1*, which encode proteins involved in the peripheral regulation of insulin signaling. The third cluster contains genes such as *NRXN3*, *CMIP, APOE*, and *MC4R* and is associated with BMI and lipid traits [20]. More recent exome sequencing in cases and controls has identified additional rare variants associated with T2D, but most often these are in loci already discovered by GWAS [21].

The chromosome 10 region encoding *TCF7L2* is the locus with strongest impact on the risk of developing T2D. Each risk-allele of *TCF7L2* increases the risk about ~1.3 fold [22]. *TCF7L2* is an effector in the Wnt signaling pathway, which has important functions in proliferation and differentiation processes. *TCF7L2* seems mainly to influence T2D risk by impact on beta cell insulin secretion [23]. However, *TCF7L2* has also been linked to adipogenesis [24] and liver insulin sensitivity [25].

5.7 Gene Environment Interaction

By *gene-environment interaction* we refer to situations with synergistic effects, that is where the joint effect of genotype and environment is less or greater than would be expected if the effect is additive. Numerous gene-environment interaction studies for obesity and T2D have been published. However, many findings have not been possible to reproduce, which has been attributed to small effects and sample sizes yielding low statistical power, as well as failure to account for multiple testing given the many genetic loci and potential environmental factors available for analysis. Herein, we have prioritized results from studies analyzing larger cohorts, even if they may not be the first to report the association with a specific environmental trigger. Furthermore, given that many findings in animal models cannot be translated into humans, we only report results from clinical studies. Most gene-environment studies for obesity and T2D are observational cross-sectional or cohort studies with information about disease incidence during a follow up period after exposure. A few RCTs are also mentioned. These have the advantage of testing the effects of specific environmental factors, while controlling for confounders; the drawback is low power due to limited sample size.

5.8 Gene Environment Interaction in Obesity

5.8.1 Obesogenic Environment

Overall, individuals with the greatest genetic predisposition to obesity seem to be more susceptible when exposed to today's obesogenic environment. This notion is based on several independent observations. Thus, a risk score based on 29 BMI-associated SNPs had stronger effects on BMI in those born more recently [26]. In concordance, studies of the *FTO* locus and high penetrant mutations in the *MC4R* gene have reported a stronger association between risk alleles and BMI in later birth cohorts [27, 28].

So what are the environmental factors that modify the impact of susceptibility gene loci? A comprehensive study based on the UK Biobank showed that a composite score of obesogenic environmental factors is of importance [29]. An index for social deprivation based on work and housing situation accentuates genetic susceptibility to high BMI. Here, the impact on BMI of a genetic risk score comprising 69 SNPs was larger in the group with the most relatively deprived situation. For the half living under the most deprived situations, carrying 10 additional BMI-raising alleles was associated with 3.8 kg extra weight, whereas for the half living under the least deprived situations, carrying 10 additional BMI-raising alleles was associated with 2.9 kg increase in weight. The same study reported interaction between genetic risk and physical activity, sedentary time, or TV watching in predicting BMI of similar effect sizes. Altogether, these analyses suggest that genetic predisposition to obesity is influenced, albeit to a minor degree, by an obesogenic environment.

5.8.2 Diet

A healthier diet is associated with lower BMI [30]. A number of studies have assessed if the diet also influences genetic predisposition to obesity. A genetic risk score comprising BMI-associated SNPs has been reported to be associated with a lower total energy intake as well as higher intake of fiber, but not with relative intake of other macronutrients [31]. By contrast, there seems to be no interaction between genetic risk score and dietary composition on BMI [29–31]. However, a genetic risk score based on WHR-associated SNPs showed nominally significant interaction with a favorable diet i.e. relative higher intake of whole grains, fish, fruits, vegetables, and nuts/seeds [30]. The healthier diet strengthened the association between the genetic risk score and WHR, thus not supporting the hypothesis that healthy diet offset genetic risk. Cooking method could, hypothetically, also influence genetic impact on BMI. Conflicting results have been reported as regards interaction between genetic risk score and intake of fried food in determining BMI [29, 32]. Thus, overall macronutrient intake and cooking method seem to have at most a modest impact on genetic susceptibility to obesity.

As *FTO* is the genetic locus with strongest effect on common obesity, numerous studies have investigated the association and interaction of the *FTO* locus with dietary factors [33]. The BMI increasing *FTO* allele has been reported to be associated with modestly lowered total energy intake and with relative higher dietary protein intake among adults, suggesting that the *FTO* gene could be involved in some aspects of food preference [33]. By contrast, in children and adolescents the BMI-increasing *FTO* allele is associated with increased total energy intake but not with macronutrient composition [34]. Nevertheless, the *FTO* gene variant interacts with protein intake i.e. the association between *FTO* genotype and BMI is stronger in individuals with high protein intake [34]. The *MCR4* gene, another important risk locus for obesity involved in central regulation of food intake, does not seem to associate with total energy or macronutrient composition [33].

Compelling evidence supports a link between the consumption of sugar-sweetened beverages [35] and an increased risk of obesity, and sugar-sweetened beverages may also adversely affect genetic susceptibility. In agreement with this, it has been reported that the effect of a genetic risk score of BMI-associated SNPs is twofold higher in those with the highest (≥1 serving/day) versus lowest (<1 serving/month) intake of sugar sweetened beverages, and the risk of developing obesity was fourfold higher [36]. On the other hand, in the UK Biobank no interaction was observed between consumption of fizzy drinks and a genetic risk score of BMI-associated SNPs on BMI [29]. However, the results are not directly comparable as drink intake was measured and grouped in different ways in the two studies. Intake of other beverages can also modify the association between genetic risk and BMI; increased alcohol intake has been reported to attenuate the association between a genetic risk score and BMI [37], and between *FTO* SNPs and BMI [38]. Thus, whereas intake of sugar-sweetened beverages seems to have an unfavorable impact on BMI in particular among those with a high genetic risk for obesity, this is not the case for alcohol intake.

Genetic susceptibility to obesity can be mediated via control of food intake but also the hedonic effects of food suggesting the potential importance of eating behavior. Eating behavior can be classified as emotional, uncontrolled, or cognitive constraint. A risk score for BMI has been reported to be positively associated with emotional eating behavior [39]. Furthermore, an interaction between cognitive constraint eating behavior and the genetic risk score on BMI was observed. The association was strongest in the lowest tertile of "cognitive constraint" supporting the notion that eating behavior could help to protect genetically susceptible individuals from weight gain.

5.8.3 Physical Activity

Physically active individuals have lower risk of obesity [40]. In genetic epidemiological studies physical activity reduces the genetic influence on BMI [9]. In agreement with this, in the European Prospective Investigation of Cancer (EPIC)-Norfolk cohort a genetic risk score explained 1.2% of the variation in BMI in the inactive group and 0.6% in the active group [41]. These results have been replicated in a larger meta-analysis [29]. In prospective analysis, the genetic risk score tended to be associated with an increase in annual BMI in physically inactive individuals, whereas the opposite trend was observed in physically active individuals [41].

FTO is not in itself associated with physical activity [40]. However, a protective interaction between *FTO* risk allele and physical activity on body fat and body fat distribution has been reported [40]. In adults, the *FTO* risk allele was associated with an odds ratio (OR) of 1.23 for obesity in the active group, and OR 1.3 in the inactive group, representing a 27% reduced risk of obesity in physically active individuals. No interaction between the *FTO* locus and physical activity on BMI was observed in children and adolescents. The above findings have clinical implications since they support that those with the highest genetic risk for obesity benefit the most from physical exercise.

5.8.4 Effects of Combined Life-Style Modifications in Relation to Genetic Risk

The potential interaction between genetic susceptibility and life style intervention on changes in body weight has been assessed in a few RCTs. Meta-analyses of RCTs including dietary, physical activity or drug-based interventions have not identified any interaction with BMI-associated genetic loci on weight loss outcome [42, 43]. Although of limited size, these RCTs consolidate the results from the epidemiological studies and suggest that life style only to a minor degree modify genetic influence on BMI.

5.8.5 Smoking

Smoking has many negative effects, particularly on lung function and CAD. One reason to continue smoking despite the side effects may be that cessation is associated with weight gain [44]. It is therefore of interest that the impact of specific SNPs associated with BMI or body fat distribution on adiposity has been reported to be dependent on smoking [45]. Besides established pathways underlying genetic susceptibility to obesity, e.g. central regulation of food intake, the genetic loci dependent on smoking implicate additional factors such as nitric oxide synthesis in body fat regulation [45]. The variance in BMI explained by BMI associated-SNPs interacting with smoking was larger among smokers than nonsmokers. By contrast, SNPs interacting with smoking explained a greater proportion of variance in body fat distribution among nonsmokers. These results are potentially clinically important as they suggest that smoking may increase genetic susceptibility to central fat accumulation, but attenuate the genetic effects on BMI. Thus, among subjects carrying high risk alleles, smoking cessation might have a positive effect on central (abdominal) fat accumulation since the interaction between smoking and risk alleles will no longer be present.

5.8.6 *Sleep*

All living organisms have a circadian rhythm, i.e. an underlying 24 h physiological cycle for e.g. body temperature and hormones. The "chronotype" is the propensity of an individual to sleep at a particular time during a 24-hour period. The normal variation in chronotype ranges from around two hours earlier to two hours later than average. Furthermore, short and/or poor sleep is associated with obesity and T2D [46, 47]. Genes controlling circadian rhythm are important for chronotype, In addition, chronotype is likely influenced by environmental factors including light, feeding, and social behavior. BMI is genetically associated with chronotype, undersleeping (<7 h), and oversleeping (>8 h) [48]. However, in Mendelian randomization experiments there is no evidence for causality between BMI and sleep pattern. While these results are of interest, no study has to our knowledge investigated if sleep influences genetic impact on BMI.

5.9 Gene Environment Interaction and T2D

Improvement in lifestyle including healthy diet and increased physically activity can reduce the risk of progression to T2D in high-risk individuals by ~50% [7]. Knowledge of whether some individuals display a better response to intervention due to e.g. genetic profile would benefit clinical practice and primary prevention. Consequently, a number of studies have assessed the interaction between genetic risk loci for T2D and life style on incidence of T2D.

5.9.1 *Diet*

Western diet has been blamed for the recent T2D epidemic whereas the "Mediterranean" diet reduces the risk of developing the disease [49], but do they modify the genetic risk? A synergistic interaction between Western dietary pattern and a T2D genetic risk score on T2D incidence has been reported [50]. Western dietary pattern increases T2D risk among those with a higher, but not among those with lower, genetic risk. By contrast, there seems to be no interaction between a genetic risk score and Mediterranean diet on incident T2D [51]. Nor do specific macronutrients or food items interact with genetic risk for T2D [52]. Gut hormones, such as incretins, are of major importance for T2D pathophysiology and glucose control. Both genetic variants and dietary factors influence incretin release and function. A significant interaction between coffee consumption and a genetic risk score comprising T2D-associated SNPs in incretin-related genes (*GIPR, KCNQ1, TCF7L2 and WFS1*), as well as *TCF7L2* gene variants on its own, on T2D risk has been reported. Coffee protected against T2D in individuals carrying the *TCF7L2* risk allele [53]. Overall, however, dietary pattern seems to have at most a modest impact on genetic susceptibility to T2D.

5.9.2 Physical Activity

A significant interaction between physical activity and a genetic risk score for T2D, but not individual SNPs, on T2D incidence has been reported [54]. The protective effect of physical activity was weaker among those with a high genetic risk. Interestingly, the interaction was observed for SNPs implicated in regulation of insulin resistance as opposed to insulin secretion, suggesting that the former genes are easier to influence through behavioral changes. However, in even larger studies no interaction between physical activity and genetic risk score on T2D incidence was observed [51]. Furthermore, a genetic risk score was not associated with T2D incidence or regression to normoglycemia in the DPP trial encompassing life style change or pharmacological intervention with Metformin [55]. Thus it is presently unclear whether genetic risk for T2D can be modified by physical activity. Larger studies, also including more recently identified susceptibility gene loci for T2D, may clarify this.

5.10 Limitations

There are several aspects that limit the generalizability of the discussed results. First, the most clinically relevant crossover interactions are situations where someone at high genetic risk of disease in one setting may be protected if environmental exposure improves. However, genetic variants involved in such interactions are unlikely to be detected in GWAS if the environmental trigger is no accounted for, which has usually not been the case, i.e. if a gene allele increases the risk of T2D in one group and reduces the risk of the disease in another group, association between the allele and T2D is difficult to detect unless the group effect is taken into account. Thus, although hundreds of genetic loci associated with BMI, WHR and T2D have been identified, there might be other as yet unidentified loci that display stronger interaction with life style. Second, there is a large heterogeneity in the studies included in the meta-analyses referred to above, and a large bias can be expected in self-reported data. Finally, many reported studies analyze genetic risk scores, which has greater power than individual SNPs for detecting interactions with environmental factors; however, genetic risk scores provides little guidance in identifying specific culprit pathways affected by the environmental factor.

5.11 Conclusions

This chapter has focused on the genetic and environmental interactions in common forms of obesity and T2D. While genetic variations explain a substantial proportion of the heritability, known genetic risk loci can only explain a minor fraction of the inter-individual variations in the two conditions. Environmental factors have a major impact on obesity and to some degree on T2D, but only modulate the genetic influences on disease to a minor degree. As a consequence, despite the high expectations for precision-based medicine there is currently limited (if any) benefits of subdividing subjects according to genetic risk score, at least for interventions aimed against obesity and T2D.

New approaches, such as refining T2D diagnoses into different subgroups with different patient characteristics might improve the power to detect clinically relevant gene –environment interactions, e.g. physical exercise might be most important among the subgroup of T2D patients that are most insulin resistant [56]. More advanced genetic instruments may improve the predictive value and enable precision medicine but these technical approaches must always be compared with simple and cheap assessments such as asking for the body weight of the parents, T2D family history and determining simple anthropometric measures such as BMI and waist circumference. There is certainly a lot to be done in order to improve our understanding of gene-environmental interactions and their pathophysiological role in metabolic disorders.

References

1. Collaboration NCDRF. Trends in adult body-mass index in 200 countries from 1975 to 2014: a pooled analysis of 1698 population-based measurement studies with 19.2 million participants. Lancet 2016 Apr 2;387(10026):1377–1396.
2. Organization WH. Global report on diabetes WHO 2016. 2016.
3. Ebrahim S, Kinra S, Bowen L, Andersen E, Ben-Shlomo Y, Lyngdoh T, et al. The effect of rural-to-urban migration on obesity and diabetes in India: a cross-sectional study. PLoS Med. 2010 Apr 27;7(4):e1000268.
4. Goodarzi MO. Genetics of obesity: what genetic association studies have taught us about the biology of obesity and its complications. Lancet Diabetes Endocrinol. 2018 Mar;6(3):223–36.
5. Akiyama M, Okada Y, Kanai M, Takahashi A, Momozawa Y, Ikeda M, et al. Genome-wide association study identifies 112 new loci for body mass index in the Japanese population. Nat Genet. 2017 Oct;49(10):1458–67.
6. Stefan N, Haring HU, Hu FB, Schulze MB. Metabolically healthy obesity: epidemiology, mechanisms, and clinical implications. Lancet Diabetes Endocrinol. 2013 Oct;1(2):152–62.
7. Schellenberg ES, Dryden DM, Vandermeer B, Ha C, Korownyk C. Lifestyle interventions for patients with and at risk for type 2 diabetes: a systematic review and meta-analysis. Ann Intern Med. 2013 Oct 15;159(8):543–51.
8. Bouchard C, Tremblay A, Despres JP, Nadeau A, Lupien PJ, Theriault G, et al. The response to long-term overfeeding in identical twins. N Engl J Med. 1990 May 24;322(21):1477–82.
9. Mustelin L, Silventoinen K, Pietilainen K, Rissanen A, Kaprio J. Physical activity reduces the influence of genetic effects on BMI and waist circumference: a study in young adult twins. Int J Obes. 2009 Jan;33(1):29–36.

10. Maes HH, Neale MC, Eaves LJ. Genetic and environmental factors in relative body weight and human adiposity. Behav Genet. 1997 Jul;27(4):325–51.
11. Silventoinen K, Jelenkovic A, Sund R, Yokoyama Y, Hur YM, Cozen W, et al. Differences in genetic and environmental variation in adult BMI by sex, age, time period, and region: an individual-based pooled analysis of 40 twin cohorts. Am J Clin Nutr. 2017 Aug;106(2):457–66.
12. Yeo GSH. Genetics of obesity: can an old dog teach us new tricks? Diabetologia. 2017 May;60(5):778–83.
13. Frayling TM, Timpson NJ, Weedon MN, Zeggini E, Freathy RM, Lindgren CM, et al. A common variant in the FTO gene is associated with body mass index and predisposes to childhood and adult obesity. Science. 2007 May 11;316(5826):889–94.
14. Gerken T, Girard CA, Tung YC, Webby CJ, Saudek V, Hewitson KS, et al. The obesity-associated FTO gene encodes a 2-oxoglutarate-dependent nucleic acid demethylase. Science. 2007 Nov 30;318(5855):1469–72.
15. Stratigopoulos G, Burnett LC, Rausch R, Gill R, Penn DB, Skowronski AA, et al. Hypomorphism of Fto and Rpgrip1l causes obesity in mice. J Clin Invest. 2016 May 2;126(5):1897–910.
16. Claussnitzer M, Dankel SN, Kim KH, Quon G, Meuleman W, Haugen C, et al. FTO obesity variant circuitry and adipocyte Browning in humans. N Engl J Med. 2015 Sep 3;373(10):895–907.
17. Shungin D, Winkler TW, Croteau-Chonka DC, Ferreira T, Locke AE, Magi R, et al. New genetic loci link adipose and insulin biology to body fat distribution. Nature. 2015 Feb 12;518(7538):187–96.
18. Replication DIG, Meta-analysis C. Asian genetic epidemiology network type 2 diabetes C, south Asian type 2 diabetes C, Mexican American type 2 diabetes C, type 2 diabetes genetic exploration by Nex-generation sequencing in muylti-ethnic samples C, et al. genome-wide trans-ancestry meta-analysis provides insight into the genetic architecture of type 2 diabetes susceptibility. Nat Genet. 2014 Mar;46(3):234–44.
19. Langenberg C, Lotta LA. Genomic insights into the causes of type 2 diabetes. Lancet. 2018 Jun 16;391(10138):2463–74.
20. Scott RA, Scott LJ, Magi R, Marullo L, Gaulton KJ, Kaakinen M, et al. An expanded genome-wide association study of type 2 diabetes in Europeans. Diabetes. 2017 Nov;66(11):2888–902.
21. Flannick J, Johansson S, Njolstad PR. Common and rare forms of diabetes mellitus: towards a continuum of diabetes subtypes. Nat Rev Endocrinol. 2016 Jul;12(7):394–406.
22. Morris AP, Voight BF, Teslovich TM, Ferreira T, Segre AV, Steinthorsdottir V, et al. Large-scale association analysis provides insights into the genetic architecture and pathophysiology of type 2 diabetes. Nat Genet. 2012 Sep;44(9):981–90.
23. Renstrom E. Impact of transcription factor 7-like 2 (TCF7L2) on pancreatic islet function and morphology in mice and men. Diabetologia. 2012 Oct;55(10):2559–61.
24. Chen X, Ayala I, Shannon C, Fourcaudot M, Acharya NK, Jenkinson CP, et al. The diabetes gene and Wnt pathway effector TCF7L2 regulates adipocyte development and function. Diabetes. 2018 Apr;67(4):554–68.
25. Jin T. Current understanding on role of the Wnt signaling pathway effector TCF7L2 in glucose homeostasis. Endocr Rev. 2016 Jun;37(3):254–77.
26. Walter S, Mejia-Guevara I, Estrada K, Liu SY, Glymour MM. Association of a Genetic Risk Score with Body Mass Index across Different Birth Cohorts. JAMA. 2016 Jul 5;316(1):63–9.
27. Rosenquist JN, Lehrer SF, O'Malley AJ, Zaslavsky AM, Smoller JW, Christakis NA. Cohort of birth modifies the association between FTO genotype and BMI. Proc Natl Acad Sci U S A. 2015 Jan 13;112(2):354–9.
28. Stutzmann F, Tan K, Vatin V, Dina C, Jouret B, Tichet J, et al. Prevalence of melanocortin-4 receptor deficiency in Europeans and their age-dependent penetrance in multigenerational pedigrees. Diabetes. 2008 Sep;57(9):2511–8.
29. Tyrrell J, Wood AR, Ames RM, Yaghootkar H, Beaumont RN, Jones SE, et al. Gene-obesogenic environment interactions in the UK biobank study. Int J Epidemiol. 2017 Apr 1;46(2):559–75.
30. Nettleton JA, Follis JL, Ngwa JS, Smith CE, Ahmad S, Tanaka T, et al. Gene x dietary pattern interactions in obesity: analysis of up to 68 317 adults of European ancestry. Hum Mol Genet. 2015 Aug 15;24(16):4728–38.

31. Rukh G, Ericson U, Andersson-Assarsson J, Orho-Melander M, Sonestedt E. Dietary starch intake modifies the relation between copy number variation in the salivary amylase gene and BMI. Am J Clin Nutr. 2017 Jul;106(1):256–62.
32. Qi Q, Chu AY, Kang JH, Huang J, Rose LM, Jensen MK, et al. Fried food consumption, genetic risk, and body mass index: gene-diet interaction analysis in three US cohort studies. BMJ. 2014 Mar 19;348:g1610.
33. Qi Q, Kilpelainen TO, Downer MK, Tanaka T, Smith CE, Sluijs I, et al. FTO genetic variants, dietary intake and body mass index: insights from 177,330 individuals. Hum Mol Genet. 2014 Dec 20;23(25):6961–72.
34. Qi Q, Downer MK, Kilpelainen TO, Taal HR, Barton SJ, Ntalla I, et al. Dietary intake, FTO genetic variants, and adiposity: a combined analysis of over 16,000 children and adolescents. Diabetes. 2015 Jul;64(7):2467–76.
35. Vartanian LR, Schwartz MB, Brownell KD. Effects of soft drink consumption on nutrition and health: a systematic review and meta-analysis. Am J Public Health. 2007 Apr;97(4):667–75.
36. Qi Q, Chu AY, Kang JH, Jensen MK, Curhan GC, Pasquale LR, et al. Sugar-sweetened beverages and genetic risk of obesity. N Engl J Med. 2012 Oct 11;367(15):1387–96.
37. Rask-Andersen M, Karlsson T, Ek WE, Johansson A. Gene-environment interaction study for BMI reveals interactions between genetic factors and physical activity, alcohol consumption and socioeconomic status. PLoS Genet. 2017 Sep;13(9):e1006977.
38. Young AI, Wauthier F, Donnelly P. Multiple novel gene-by-environment interactions modify the effect of FTO variants on body mass index. Nat Commun. 2016 Sep 6;7:12724.
39. de Lauzon-Guillain B, Clifton EA, Day FR, Clement K, Brage S, Forouhi NG, et al. Mediation and modification of genetic susceptibility to obesity by eating behaviors. Am J Clin Nutr. 2017 Oct;106(4):996–1004.
40. Kilpelainen TO, Qi L, Brage S, Sharp SJ, Sonestedt E, Demerath E, et al. Physical activity attenuates the influence of FTO variants on obesity risk: a meta-analysis of 218,166 adults and 19,268 children. PLoS Med. 2011 Nov;8(11):e1001116.
41. Li S, Zhao JH, Luan J, Ekelund U, Luben RN, Khaw KT, et al. Physical activity attenuates the genetic predisposition to obesity in 20,000 men and women from EPIC-Norfolk prospective population study. PLoS Med. 2010 Aug 31;7(8):e1000332.
42. Papandonatos GD, Pan Q, Pajewski NM, Delahanty LM, Peter I, Erar B, et al. Genetic predisposition to weight loss and regain with lifestyle intervention: analyses from the diabetes prevention program and the look AHEAD randomized controlled trials. Diabetes. 2015 Dec;64(12):4312–21.
43. Livingstone KM, Celis-Morales C, Papandonatos GD, Erar B, Florez JC, Jablonski KA, et al. FTO genotype and weight loss: systematic review and meta-analysis of 9563 individual participant data from eight randomised controlled trials. BMJ. 2016 Sep 20;i4707:354.
44. Aubin HJ, Farley A, Lycett D, Lahmek P, Aveyard P. Weight gain in smokers after quitting cigarettes: meta-analysis. BMJ. 2012 Jul 10;345:e4439.
45. Justice AE, Winkler TW, Feitosa MF, Graff M, Fisher VA, Young K, et al. Genome-wide meta-analysis of 241,258 adults accounting for smoking behaviour identifies novel loci for obesity traits. Nat Commun. 2017 Apr 26;8:14977.
46. Cappuccio FP, Taggart FM, Kandala NB, Currie A, Peile E, Stranges S, et al. Meta-analysis of short sleep duration and obesity in children and adults. Sleep. 2008 May;31(5):619–26.
47. Schmid SM, Hallschmid M, Schultes B. The metabolic burden of sleep loss. Lancet Diabetes Endocrinol. 2015 Jan;3(1):52–62.
48. Jones SE, Tyrrell J, Wood AR, Beaumont RN, Ruth KS, Tuke MA, et al. Genome-wide association analyses in 128,266 individuals identifies new morningness and sleep duration loci. PLoS Genet. 2016 Aug;12(8):e1006125.
49. Salas-Salvado J, Bullo M, Babio N, Martinez-Gonzalez MA, Ibarrola-Jurado N, Basora J, et al. Reduction in the incidence of type 2 diabetes with the mediterranean diet: results of the PREDIMED-Reus nutrition intervention randomized trial. Diabetes Care. 2011 Jan;34(1):14–9.
50. Qi L, Cornelis MC, Zhang C, van Dam RM. Hu FB. Genetic predisposition, Western dietary pattern, and the risk of type 2 diabetes in men. Am J Clin Nutr. 2009 May;89(5):1453–8.

51. Langenberg C, Sharp SJ, Franks PW, Scott RA, Deloukas P, Forouhi NG, et al. Gene-lifestyle interaction and type 2 diabetes: the EPIC interact case-cohort study. PLoS Med. 2014 May;11(5):e1001647.
52. Li SX, Imamura F, Ye Z, Schulze MB, Zheng J, Ardanaz E, et al. Interaction between genes and macronutrient intake on the risk of developing type 2 diabetes: systematic review and findings from European prospective investigation into Cancer (EPIC)-InterAct. Am J Clin Nutr. 2017 Jul;106(1):263–75.
53. InterAct C. Investigation of gene-diet interactions in the incretin system and risk of type 2 diabetes: the EPIC-InterAct study. Diabetologia. 2016 Dec;59(12):2613–21.
54. Klimentidis YC, Chen Z, Arora A, Hsu CH. Association of physical activity with lower type 2 diabetes incidence is weaker among individuals at high genetic risk. Diabetologia. 2014 Dec;57(12):2530–4.
55. Hivert MF, Jablonski KA, Perreault L, Saxena R, McAteer JB, Franks PW, et al. Updated genetic score based on 34 confirmed type 2 diabetes loci is associated with diabetes incidence and regression to normoglycemia in the diabetes prevention program. Diabetes. 2011 Apr;60(4):1340–8.
56. Ahlqvist E, Storm P, Karajamaki A, Martinell M, Dorkhan M, Carlsson A, et al. Novel subgroups of adult-onset diabetes and their association with outcomes: a data-driven cluster analysis of six variables. Lancet Diabetes Endocrinol. 2018 May;6(5):361–9.

Chapter 6
Gene-Environment Interaction and Cancer

Vittoria D'Esposito, Maria Rosaria Ambrosio, Giuseppe Perruolo, Michele Libutti, and Pietro Formisano

Keywords Epigenetic · Cancer · Nutrition · Endocrine disruptors · Circadian rhythms

6.1 Cancer as Epigenetic Disease

For the past 30 years cancer has been thought to arise from a single cell. According to this view, a series of genetic alterations is responsible for continued clonal selection and tumor cell heterogeneity. This clonal genetic model gives rise to tumor proliferation, invasion, metastasis and drug resistance and has been supported by the discovery of dominantly acting oncogenes and recessively acting tumor-suppressor genes. The large number of genes discovered so far has led to the view that cancer is a heterogeneous group of diseases with diverse etiology and pathogenesis. However, classic genetic alone cannot explain the diversity of phenotypes within populations. Pathological epigenetic changes are increasingly being considered as alternatives to mutations and chromosomal alterations in disrupting gene function, and great advances have been made in characterizing epigenetic altera-

V. D'Esposito
Unità di Ricerca "Genomica del Diabete", Istituto di Endocrinologia ed Oncologia Sperimentale del Consiglio Nazionale delle Ricerche, Napoli, Italy

M. R. Ambrosio
Dipartimento di Scienze Mediche Traslazionali, Università degli Studi di Napoli "Federico II", Napoli, Italy

G. Perruolo
Azienda Ospedaliera Universitaria "Federico II", Napoli, Italy

M. Libutti
Oncologia, Azienda Sanitaria Locale Napoli 3, Napoli, Italy

P. Formisano (✉)
Unità di Ricerca "Genomica del Diabete", Istituto di Endocrinologia ed Oncologia Sperimentale del Consiglio Nazionale delle Ricerche, Napoli, Italy

Dipartimento di Scienze Mediche Traslazionali, Università degli Studi di Napoli "Federico II", Napoli, Italy
e-mail: fpietro@unina.it

R. Teperino (ed.), *Beyond Our Genes*,
https://doi.org/10.1007/978-3-030-35213-4_6

tions in cancer [24, 25]. The neoplastic tissue encompasses many cell types (Cancer cells, Tumor Infiltrating Lymphocytes, Macrophages, Cancer Associated Fibroblasts, Stromal Cells, Stem Cells) in a system rich of molecular crosstalks mediated by cytokines, growth factors, vesicles and lipids. Depending on the distance from vascular tree, cells undergo hypoxia, acidosis, redox changes and oxygen and nutrients shortage too, together with epigenetic and mutational changes. In order to upregulate the membrane lipids and DNA production, metabolic adaptive changes in many pathways implicated in energy production (aerobic hyperglycolisis, TCA reprogramming, OxPhos downregulation, alterations in glutamine metabolism) occur both in cancer and in neighboring cells. Products of such reactions (alfa2-ketoglutarate, lactate, citrate, pyruvate) are inducers of angiogenetic and pro-inflammatory factors; they can lead to epigenetic changes but can also be modulated by epigenetic processes [54].

Epigenetic regulation of gene expression occurs at the level of DNA, histones and RNA and include global alterations, such as hypomethylation of DNA and hypoacetylation of chromatin, as well as gene-specific hypomethylation and hypermethylation.

Methylation of DNA is generally regarded as one of the most important epigenetic modifications; it is a process largely controlled by enzymes known as DNA methyltransferases (DNMTs) and involves the addition of a methyl moiety to the cytosine residues of CpG dinucleotides [43]. The genome is sparsely populated with CpG sites in intergenic regions and repetitive sequences, and many of these sites are methylated. In cancer, hypomethylation of these regions often takes place; hence, the chromatin becomes less densely packaged and the DNA can be transcribed [9] leading to chromosomal instability as well as gene-specific oncogene activation, such as R-ras in gastric cancer, and cyclin D2 (CCND2) and maspin in pancreatic cancer [25]. Moreover, residues of CpG dinucleotides are clustered in what is termed CpG islands, often within the regulatory regions of gene promoters. Approximately 60% of human genes contain a high density of CpG islands that are usually unmethylated; DNA hypermethylation of a gene promoter usually leads to gene silencing [19, 83]. Site-specific CpG promoter hypermethylation is also a common epigenetic feature of cancer [42]. Promoter hypermethylation is at least as common as the mutation-dependent disruption of classic tumor-suppressor genes in human cancer. Nearly 50% of genes - that cause familial forms of cancer when mutated in the germline - are known to undergo methylation-associated silencing in various sporadic forms of cancer. These include retinoblastoma 1 (RB1), p16 (also known as cyclin-dependent kinase inhibitor 2A -CDKN2A), von Hippel–Lindau tumor suppressor (VHL) and MutL protein homologue 1 (MLH1) [43].

Modifications of histones involve changes within the basic structure of the chromatin unit - known as the nucleosome - and occur both during embryonic development and throughout life. Remodeling of histones can be a result of modifications via methylation, ubiquitylation, phosphorylation, biotinylation, sumoylation, ADP ribosylation and acetylation processes. Such modifications define the chromosomal structure and gene expression state [83]. The combination of specific histone modifications are responsible for changes in gene expression levels. Commonly, histone methylation leads to a condensed chromatin structure and suppressed gene expression. Just as lysine and arginine residues on histones can be methylated, lysine can also be acetylated. The methylation of the histone is catalyzed by histone methyltransferase (HMT); such event could be reversed by specific histone demethylase and the differential histone demethylase activity is associated with aberrant histone methylation [42], (Bishop K.S. and Ferguson L.R., Nutrition 2015). Histone acetylation usually leads to an open chromatin structure promoting gene expression. It also is a dynamic process regulated by histone acetyltransferases (HAT) and histone deacetylases (HDAC) that, respectively, catalyze the addition or the removal of acetyl group to lysine residues of histones. Such event, removing lysine positive charge, reduces attraction to the negatively charged DNA strand and produces a loosened chromatin structure, that allows various transcription factors to access the DNA and promotes the transcriptional activation of genes. Both altered histone methylation and acetylation have been shown in several types of cancers thus promoting gene repression and silencing [9, 19, 42].

Non-coding RNAs- consisting of microRNAs (miRNAs), piwi-interacting RNAs (piRNAs), small-interfering RNAs (siRNAs) and small nucleolar RNAs - exhibit the potential to modulate gene expression by a number of mechanisms, including heterochromatin formation and inhibition of translation. Non-coding RNAs can induce DNA methylation or histone modifications that result in silenced or enhanced gene expression [83]. In turn, epigenetic effects are known to regulate expression of specific miRNAs. Dysregulation of miRNAs is associated with the development of a number of cancers, particularly metastatic cancers; for example, partial methylation of the promoters of miR-29a and miR-1256 in prostate cancer cells increases the expression of cancer progression associated genes; statistically significant dysregulation of miR-10b and miR-145 has been found in breast cancer versus normal mammary tissue [9].

In this scenario, besides tumor initiation and progression driven by alterations in gene expression as a result of specific mutations - activating in oncogenes and prometastatic genes or inactivating in tumor suppressor genes - cancer needs to be considered as an epigenetic disease. Epigenetic – even more than genetic - events are susceptible to environmental and lifestyle factors, including diet, endocrine disruptors and circadian rhythm.

6.2 Environmental Hits on Cancer Progression

6.2.1 Nutrition and Metabolism

Nutrition is thought to be the most influential of all the external environmental factors due to its ability to affect transcriptional activity and expression of certain genes. The occurrence of cancer is dependent on the interplay between genome and epigenome, which together interact with nutrition [9] thus linking diet, in both its quantity and quality, to cancer incidence and prognosis [19]. A variety of natural compounds from different sources have been shown to directly regulate metabolism related to epigenetics. The intake of certain bioactive food components can modulate cancer risk, tumor development and cancer progression. Obviously, the quantity of food components, the frequency of intake, and the duration of intake by individuals also play a significant role in cancer development. Well-characterized bioactive food components include polyphenols and isothiocyanates [42, 83].

Poliphenols exert chemo-preventive effects in part by modulating various components of the epigenetic machinery in humans [83]. They consist predominantly of flavonoids, stilbenes, phenolic acids, benzoquinones, acetophenones, lignins and xanthones and can range from simple molecules to highly complex compounds; more than 8000 distinct dietary polyphenols exist. The chemo-preventive potential of dietary polyphenols can be traced to their ability to inhibit DNMTs – thus reactivating methylation-silenced genes - as well as their ability to act as histone modifiers. Both of these properties of dietary polyphenols can significantly change the epigenome of cancer cells and are viewed as attractive possibilities for anticancer therapeutics. Isoflavones represent the largest class of polyphenolic compounds; among them, the most studied is the estrogen-like genistein. Several mechanisms have been found to explain the anticarcinogenic properties of genistein including its ability to regulate gene transcription by affecting histone acetylation and/or DNA methylation. Indeed, genistein is able to inhibit DNMT1, 3a and 3b and to increase acetylation by enhancing HAT or inhibiting HDAC activities. It has been demonstrated that genistein treatment induces the expression of the tumor suppressor genes p16 and p21 by altering histone and promoter methylation in prostate cells, and partially reverse DNA hypermethylation thus reactivating p16, retinoic acid receptor b (RARβ) and O6-methylguanine methyltransferase (MGMT) in esophageal squamous and prostate cell carcinoma. In addition, genistein is able to induce demethylation of GSTP1 (Glutathione S-Transferase Pi 1), RARβ and CCND2 promoter genes in breast cancer cells, GSTP1 and EPHB2 (EPH Receptor B2) promoter genes in prostate cancer cells, and BTG3 (BTG Anti-Proliferation Factor 3) gene in prostate and renal cells. Genistein, as well as other isoflavones, has also been found to regulate miRNA expression in several cancer cell lines. For example, genistein inhibits the expression of miR-27a in human uveal melanoma cells, which is believed to be associated with its growth inhibitory actions; similarly, genistein regulation of miRNA-29a and miRNA-1256 in prostate cancer has been reported [19, 34, 75]. Other examples of common polyphenols include EpiGalloCatechin-3-Gallate (EGCG; found in green tea), resveratrol (found in peanuts, mulberries, cranberries and blueberries and abundant in the skin of grapes) and curcumin (found in curry). A positive correlation between the consumption of EGCG and the inhibition of oral, breast, prostate, gastric, ovarian, skin, colorectal and pancreatic cancers has been reported; such anticancer effects occur through several different mechanisms - much of which can be altered by epigenetic mechanisms - including induction of apoptosis and cell cycle arrest, inhibition of oxidative stress and angiogenesis, regulation of signal transduction and reduction of cancer cell proliferation. Specifically, EGCG exhibits a highly efficient activity in targeting DNMTs. Indeed, EGCG treatment (1) reduces DNMT activity via hypomethylation and re-expression of p16, RAR β, MGMT and MLH1 in human esophageal, colon and prostate cancer cells, (2) suppresses promoter methylation of p15 and p16 in breast cancer and promyelocytic leukemia cells, (3) decreases hTERT promoter methylation in breast cancer cells. Similar effects have been reported in oral carcinoma, lung cancer and melanoma cells. Notably, EGCG is able to reactivate estrogen receptor-α (ERα) in breast cancer cells via the decreased binding of the transcriptional repressor complex, Rb/p130–E2F4/5– HDAC1–SUV39H1–DNMT1 in the regulatory regions of the ERα gene promoter. EGCG also induces a number of histone modifications that, in turn, can induce transcriptional activation of tumor suppressor genes. Recent studies also suggest that changes in miRNA expression as well as histone modifications may be mediated by EGCG [34, 75, 83]. Antioxidant, anti-inflammatory and anti-cancer properties of resveratrol occur through various molecular and biochemical pathways. It exerts antiproliferative properties impacting on signaling pathways that control cell division, cell growth, apoptosis, angiogenesis and tumor metastasis. Such properties have been reported in liver, skin, breast, prostate, lung and colon cancer cells. Although resveratrol is not as potent as EGCG in DNMT inhibition, it can prevent the silencing of certain tumor suppressor genes (i.e. BRCA1). In addition, one of the most important properties of resveratol is its strong ability to activate SIRT1, a known HDAC inhibitor [19, 34, 83]. Curcumin is a bioactive dietary component showing anti-inflammatory, anti-oxidant, anti-angiogenic and anti-cancer properties. It inhibits DNMT activity and also functions as HDAC/HAT inhibitor. While the inhibition of both HATs and HDACs may seem contradictory, recent investigations provide evidence that HAT inhibitors have a potential role in cancer therapies and that inhibition of both HATs and HDACs together may provide a potent strategy for cancer treatment [34]. Of note, polyphenols, as well as various other compounds associated with the 'epigenetic diet', have also been shown to influence glycolysis and the energy metabolism of cancer cells. It is known that the loss of p53 function, often associated with carcinogenesis, has been linked to the upregulation of glucose transporters, facilitating the increased uptake of glucose, characteristic of cancer cells. Moreover, recent studies have shown that mutations in the KRAS oncogene enhance the rate of glycolysis and subsequent uptake of glucose by upregulation of glucose transporter 1 (GLUT1). Interestingly, flavonoids exist in the diet as conjugates with glucose and that they utilize glucose transporters, acting as competitive inhibitors to the uptake of glucose; EGCG or green tea polyphenols were shown to modulate the expression of key genes involved in the metabolism of glucose and fats; EGCG reduced the accumulation of body fat and significantly downregulated the expression of hepatic glucokinase, a key enzyme of liver glycolysis [19].

Isothiocyanates are dietary compounds present in cruciferous vegetables (broccoli, cabbage and kale) and include sulforaphane. They elicit both proapoptotic and antiproliferative properties. Isothiocyanates are known to affect the epigenome. Specifically, sulforaphane has many effects, including HDAC inhibition: it is thought to be involved in the regulation of cancer-related genes in colorectal and prostate cancer cells as well as in peripheral blood mononucleocytes. Sulforaphane induced the hyperacetylation of TERT promoter and allowed the binding of repressor proteins. This finding is significant since hTERT is overexpressed in approximately 90% of cancers. Lastly, sulforaphane can inhibit DNMTs in breast, prostate and colon cancer cells, thus increasing the expression of genes involved in regulating cell cycle arrest and decreasing cell proliferation [9, 19, 34, 75].

Other dietary components affecting the epigenome Selenium is a nutrient found in nuts, chicken, game meat and beef. It is an essential element with antioxidant, proapoptotic, DNA repair and anticancer properties and its deficiencies have been linked to various human diseases including cancer. However, epigenetic effects of selenium have not been clearly defined. It has been linked to DNA methylation in cellular and animal models: it has been shown to cause global hypomethylation and promoter methylation of p16 and p53 tumor suppressor genes. In addition, decreased histone deacetylase activity, causing increased levels of histone acetylation, has been described. Although these studies are intriguing, further studies involving the epigenetic influence of selenium are needed to fully appreciate the impact of selenium on the epigenome [34, 75]. Garlic cloves contain several compounds including: vitamins A, B-complex, C, E, fiber, free amino acids, sulfur/organosulfur compounds and proteins. Garlic acts to inhibit cell cycle progression, induce apoptosis, inhibit angiogenesis and modifies histones. Studies conducted revealed that garlic inhibits histone deacetylase and enhances Sp3 binding on the P21/WAF1 promoter, which results in elevated p21 protein expression and cell cycle arrest. In addition, investigations demonstrated the induction of histone acetylation in cancer cells treated with garlic compounds [34]. Folate is a cofactor acting as a carrier of the methyl group and is involved in the generation of S-adenosyl methionine (SAM) that, in turn, becomes the primary donor of methyl groups for gene hypermethylation. Furthermore, dietary factors (e.g., vitamin B6, vitamin B12, and Zn) that feed into the folate cycle have an effect on methyl group availability. A number of fortified foods contain folic acid. Folates, present in high concentrations in green leafy vegetables, maintain DNA stability through their ability to donate one-carbon units for cellular metabolism. Mammals cannot synthesize folate de novo; therefore, they get it either from natural foods (green leafy vegetables), supplemented foods, or from microbial breakdown during digestion. Folate deficiencies lead to hypomethylated genomic DNA, which is associated with tumorigenesis and are reported to contribute to the development of several cancers including breast, cervix, ovary, brain, lung, liver and colorectal [83, 34]. Other bioactive components are apigenin (parsley), baicalein (Indian trumpet), cyanidins (grapes), rosmarinic acid (rosemary) and silymarin (milk thistle) [42, 83]).

Notably, while most natural dietary products have shown beneficial effects on the epigenome, not all dietary components share this characteristic. In fact, alcohol consumption is associated with harmful epigenetic modifications as well as the development/progression of several human cancers [34]. Other nutrients, such as glucose and lipid-derived free fatty acids, can affect epigenetic machinery by promoting or inhibiting enzymes involved in DNA methylation and histone modifications [44]. The high metabolic demands of cancer cells allow them to utilize such nutrients with an altered metabolic program able to support their high proliferative rates and adapt to hostile tumor microenvironment. Numerous metabolic genes have been identified as driver genes mutated in some cancers. Such "metabolic rewiring" of cancer cells is considered as hallmarks of cancer and could impact on their epigenetic machinery. Indeed, it could affect the availability of cofactors required for epigenetic modification enzymes and generate oncometabolites acting as agonists and/or antagonists for epigenetic modification enzymes, finally impacting on the epigenetic landscape. On the other hand, epigenetic dysfunction modifies metabolism by directly affecting the expression of metabolic enzymes and altering the signal transduction cascades involved in the control of cell metabolism [87]. Therefore, the epigenetic-metabolomic interplay has a critical role in tumorigenesis by coordinately sustaining cell proliferation, metastasis and pluripotency [44]. It is intuitive that such effects could be amplified when metabolic disorders (i.e. diabetes, obesity, etc.) occur. Both hyperglycemia and hyperlipidemia are related to cancer incidence/outcome and a number of review articles proposed hypotheses concerning the mechanisms linking obesity and diabetes to neoplasia [45, 71]. Among them, the excessing nutrients have been considered attractive candidates to explain the relationship between metabolic imbalance and cancer also for a variety of evidence, both in humans and animal models, that support the association between changes in nutritional status and epigenetic modifications [36]. Indeed, many studies have clearly demonstrated the influence of hyperglycemia on abnormal epigenetic mechanisms leading to altered post-transcriptional histone modifications, skewed action of DNMTs and altered levels of numerous miRNAs [85]. Nutrition is also thought to be a factor involved in inflammation and a modulator of risk toward some cancers. Prolonged low grade chronic inflammation may lead (together with oxidative stress) to epigenetic changes and are increasingly recognized as a contributor to cancer. Several studies have noted inflammatory signals as a new epigenetic mechanism that silence specific genes causing inflammation-induced cellular changes, thus highlighting the complexity of linkages between dietary components and epigenetic mechanisms, including how these may affect the inflammation phenotype and the development of cancer [44]. Therefore, an abundance of nutrients, like glucose and fatty acids, may contribute to metabolic imbalance-cancer onset/progression association, either directly acting on cancer cells, either indirectly contributing to epigenetic modifications. In this regard, low molecular weight byproducts generated from the gut microbes can also influence the epigenome through DNA modifications as well as chromatin remodeling [50]. Microbial-induced epigenetic modifications have been recently suggested as a result of gut or alveolar microbe mechanisms, by which the microbiome translates environmental signals to a fine-tuning of the host DNA expression and subsequently its corresponding functions (a modern version of Lamarckism) [1, 6].

Thus, cancer is a metabolically driven process with dynamic nutrient-responsive alterations within the human genome. The epigenetic machinery acting on the human genome is heavily susceptible to alterations of metabolism and nutrition. This emerging knowledge strongly supports the current interest in nutri-epigenetics or nutri-epigenomics [44].

6.2.2 Endocrine Disruptors

Cancer is one of the major disease caused by environmental and exogenous components [15]. It is estimated that about 90% of cancers are due to the environmental contaminants. Among these substances, several agents are named "endocrine-disrupting chemicals" (EDCs) [67]. An EDC is defined by the US Environmental Protection Agency as 'an exogenous agent that interferes with synthesis, secretion, transport, metabolism, binding, action, or elimination of natural blood-borne hormones that are present in the body and are responsible for homeostasis, reproduction, and developmental process'. The sources of EDCs are diverse as too are their structure and action, making prediction of their endocrine-disrupting properties a challenge. Furthermore, EDCs are also a highly heterogenous set of compounds ranging from synthetic chemicals used in plastics, pesticides and pharmaceutical agents to natural chemicals found in human and animal food such as phytoestrogens [46]. The contact of EDCs with humans has become inevitable today [67]. EDCs act primarily through nuclear hormone receptors, such as estrogen and progesterone receptors. However, it has now been clearly demonstrated that the mechanisms of EDCs are much broader than originally recognized. For instance, recent reports have indicated that EDCs alter the epigenomic landscape in cancers, including breast, prostate, testes, endometrium, ovary and thyroid [31, 41, 46].

It appears that mis-timed exposure of tissues to hormonally active agents can interfere with the subtle processes of gene silencing, and that disruption of these processes might be one factor that predisposes towards cancer [41]. Environmental perturbations can lead to aberrant epigenetic modifications that persist into later life, inducing disease states, including cancer, with the capacity of being transgenerationally inherited [26]. In vitro, animal and human investigations have linked EDCs with changes in DNA methylation, histone modifications and miRNAs [39]. However, to date, the precise mechanism by which EDCs modify the epigenome, contributing to gene-environment interaction and cancer, remains largely unknown.

Nowadays, there are hundreds or more environmental chemicals with EDC activity; among these, some classes are most common and most commonly studied, and include plasticizers (bisphenol A and phthalates), polychlorinated biphenyls, polybrominated diethyl ethers, dioxins, pesticides (methoxychlor, chlorpyrifos, dichlorodiphenyltrichloroethane), fungicides (vinclozolin), and herbicides [31, 67].

Bisphenol A (BPA) is the most widely produced chemical worldwide [26] with 15 billion pounds produced in 2013. It is used in a very wide array of manufacturing, food packaging, toys and other applications (i.e. resins for the lining of many canned foods and beverages). BPA is an ubiquitous lipophilic compound able, for example, to migrate from can coatings into foods and beverages during storage, depending on heating process, contact with oil or acetic acid, such that virtually everyone is exposed continuously [31, 81]. In 2008, BPA was found in the urine of 92.6% of American men and women [88]. Some studies have linked early life exposure to BPA with the development of breast, prostate and, more recently, hepatic cancer. BPA, however, does not seem to be directly involved in the promotion of thyroid proliferative lesions, as well as in the increase of uterine tumorigenesis [26]. BPA acts as an estrogen mimetic and can interact with the ligand-binding domain of ERα, increasing cellular proliferation and inducing a gene expression profile that clusters with poor breast cancer prognosis. BPA also interacts with the orphan nuclear receptor, estrogen-related receptor γ (ERRγ), and may also signal via the G-protein-coupled receptor GPER/GPR30 [46]. Several studies have identified specific genes epigenetically altered in response to BPA that may increase breast epithelial proliferation and tumor development. Treatment of primary human breast epithelial cells with BPA increased DNA methylation levels in the promoter region of LAMP3 (lysosomal-associated membrane protein 3), whose DNA methylation is a characteristic of ER-positive tumors. Furthermore, in vitro treatment of MCF-7, ER-positive breast cancer cell lines, with BPA resulted in increased expression of EZH2 (Enhancer Of Zeste 2 Polycomb Repressive Complex 2 Subunit), a histone methyltransferase enzyme, which has been previously implicated in breast tumorigenesis. This determined an increase in histone H3 trimethylation which was also detected in the mammary glands of mice exposed to BPA in utero. Spheres of human mammary epithelial cells displayed increased size and proliferation in response to BPA treatment as well as increased methylation levels of key genes associated with tumor development, namely BRCA1, CCNA1 (Cyclin A1), CDKN2A, THBS1 (Thrombospondin 1), TNFRSF10C and TNFRSF10D (TNF Receptor Superfamily Member 10c and 10d). Additional cell line studies have demonstrated aberrant methylation of the genome resulting a unique miRNA signature in BPA-treated cells distinct from miRNAs induced by estradiol [23, 26, 46]. Rats and mice exposed prenatally to BPA showed accelerated growth of the mammary gland coupled with hyperplastic epithelial ducts, more rapid development of mature structures and increased proliferation of stromal cells at an earlier age [31, 46]. Moreover, oral prenatal exposure to BPA increased mammary cancer susceptibility in offspring and shifted the window of susceptibility for dimethylbenzanthracene (DMBA)-induced tumorigenesis in the rat mammary gland [88]. Additional studies in rats determined that a brief perinatal exposure to low-dose BPA (10 μg/kg body weight) epigenetically reprogrammed the prostate tissues and markedly increased the incidence and severity of precancerous lesions upon treatment with estradiol as adults [31]. Neonatal exposure of rats to environmentally relevant doses of BPA induced hypermethylation of Pde4d promoter, resulting in its overexpression in prostate. These aberrant changes were associated with an increase in the risk of developing prostate lesions in the adult rats [26, 88]. Furthermore, upon BPA exposure, early and persistent overexpression of prostate DNA DNMT3a/b and methyl-CpG binding domain proteins with demethylase activity (Mbd2/4) were noted; it may mechanistically underlie early-life reprogramming and allow dynamic changes in response to secondary estrogenic exposures throughout life. Together, these results highlight the complexity of developmental reprogramming events initiated by BPA exposure that may predispose to carcinogenesis with aging [31]. Interestingly, nutrition factors, such as high-fat diets, are able to increase the susceptibility to BPA via epigenetic mechanisms. For instance, rats fed with high-fat butter diet (HFB) + low dose of BPA (25 μg/kg body weight) during acclimation and gestation periods showed: a) a doubling of terminal end buds (TEBs) in postnatal day-21 mammary glands (these are presumed targets of carcinogenesis); b) ERα-associated cell proliferation in the epithelial cells of these TEBs; c) a significant loss of 5′ hydroxymethylation of cytosine (hmC) and 5′ methylation of cytosine (mC) in the epithelial cells of the TEBs and d) a dramatic loss of the histone H4 lysine 20 monomethylation (H4K20me1) mark in these cells. These early cellular and epigenetic changes correlate tightly with the dramatic increase in DMBA-induced tumor incidence from 45% in the HFB diet group to 90% in the HFB + BPA group. This work supports the hypothesis that 'gestation is a critical window for dietary fatty acids–BPA interaction that reprograms the mammary epigenome, resulting in aberrant gene expression and increased breast cancer risk in adulthood'[46].

Polychlorinated biphenyls (PCBs) are a class of industrial chemicals with paired phenolic rings and variable degrees of chlorination. The commercial production of polybrominated diphenyl ethers (PBDEs) began in the late 1970s, just about the time that PCB production was banned. They were used as flame retardants in upholstered products, mattresses, and clothing [31]. PCBs and PBDEs seem to be associated with the risk of breast and prostate cancers, however human, animal and in vitro studies are controversial [31, 41, 46]. PCBs interfere with estrogen metabolism, increase the amount of bioavailable estradiol and synergistically regulate estrogen-responsive genes. In vitro studies of the estrogenic effects of PCBs have provided conflicting data, with some studies demonstrating an increase in breast cancer cellular proliferation and others showing no or opposing effects. More recent findings have revealed roles of PCBs in breast cancer cell metastasis and metabolism [46]. An in vitro analysis of the effects of selected PCBs on the human prostate cancer cell line LNCaP found several compounds that reduced cell proliferation, PSA secretion and 5α-reductase activity, whereas others (including PCB153 and 118) exhibited biphasic effects, inducing proliferation and PSA secretion at low concentrations [31]. PCBs have been identified as EDCs capable of influencing epigenetic modifying processes. It was found that PCBs decreased global genome DNA methylation either in animals either in humans. The reduction in DNA methylation levels by PCBs may well be correlated with the increased transcriptional activity of target genes in breast cancer; indeed, PCBs have also been found to reduce levels of H4K16Ac histone post-translational modifications, a hallmark of human cancer. Conversely, in utero exposure to PCBs in rats has been shown to reduce promoter methylation of the tumor suppressor gene p16 in hepatic cells [46].

p,p'-Dichlorodiphenyltrichloroethane (DDT) is a synthetic industrial and household insecticide with a long half-life, extensive use, and lipophilic nature. DDT and its metabolites, dichlorodiphenyldichloroethylene (DDE) and dichlorodiphenyldichloroethane (DDD), have been associated with endocrine-related diseases such as testicular tumors, endometrial, pancreatic and breast cancers [31]. In vitro studies have demonstrated the ability of DDT to mimic estradiol by binding to ERα and regulating breast cancer target genes. DDT and its metabolites have also been shown to regulate estrogen target genes in ERα-independent mechanisms. HER2/Neu mice with p,p′-DDE metabolite implants were found to have increased mammary tumor growth; however, it was suggested that DDE exposure alone could not be the cause of tumorigenesis but rather that it is due to hormonal contributions [46].

There are limited investigations into the epigenetic actions of DDT and its metabolites. Of note are the findings that low doses of DDT correlate with global hypomethylation and demethylation of CpG islands [8, 46]. Recent work has also emerged describing the potential for DDT to play a prominent role in miRNA regulation in breast. This research identified both common and distinct miRNA profiles induced by either estradiol or DDT in MCF-7 breast cancer cells [46].

Diethylstilbestrol (DES) is a synthetic estrogen that was used to prevent miscarriages in pregnant women between the 1940s and the 1960s. Since then, DES has been repeatedly shown to be associated with developmental programming of adult-onset chronic diseases including several reproductive tract abnormalities and increased vaginal and cervical cancer risk in women [79]. A moderate increase in breast cancer risk has been shown both in daughters of women who were treated with DES during pregnancy, as well as in their daughters. It has been demonstrated that the expression of 82 miRNAs (9.1% of the 898 miRNAs evaluated) were altered in breast epithelial cells when exposed to DES. In particular, the suppression of miR-9-3 expression was accompanied by promoter hypermethylation of its coding gene in DES-treated epithelial cells [39]. In utero exposure of mice to DES resulted in a > two-fold increase in the expression level of EZH2, a histone methyltransferase that is linked to breast cancer risk and epigenetic regulation of tumorigenesis, in adult mammary tissue compared with controls [23]. In other studies, the expression of Hoxa10 was increased in human endometrial cells after DES exposure [12], whereas in mice, in utero DES exposure resulted in hypermethylation and long-term altered expression of Hoxa10 [29, 79].

Dioxin is a general name for a family of about 30 chlorinated hydrocarbons typically used to refer to one isomer, TCDD, 2,3,7,8-tetrachloridibenzo-p-dioxin, a pollutant byproduct of combustion and manufacture of chemicals [79]. It has been detected in Agent Orange, the herbicide of choice for the United States Army in its bombings during the Vietnam War, thus exposing many troops and civilians to this endocrine disruptor. Due to its long half-life of 7–8 years and its lipophilic chemical properties, TCDD tends to accumulate in both the environment and human bodies. TCDD binds with high affinity to the aromatic hydrocarbon receptor (AhR) and is classified as an endocrine disruptor with strong anti-estrogenic properties [46]. Dioxins have been classified as human carcinogens by the International Agency for Research on Cancer [39]. Studies in rats have revealed that gestational and lactational exposure to TCDD resulted in permanent alterations in the development of the mammary epithelium of female offspring. These rats displayed delayed maturation of mammary structures and multiple immature undifferentiated structures, leading to an increased susceptibility to chemical-carcinogen-induced tumorigenesis [31, 46]. Developmental exposure to TCDD increased prostate hyperplastic lesions with aging in a wildtype mouse model, suggesting that timing of TCDD exposure may be a critical variable in determining prostatic effects, with the developing prostate being more sensitive than the adult prostate [31]. Evidence exists to indicate that TCDD has a significant effect on the epigenome in the breast, prostate and other tissues. TCDD increased association of DNMT1, histone methylation markers and methyl-binding-domain protein 2 to the promoter region of BCRA1, leading to hypermethylation and epigenetic silencing in breast cancer cell lines. In a follow-up study, Sprague–Dawley rats treated during gestation with TCDD increased the number of TEBs and reduced BRCA1 expression in mammary tissue of offspring, a process induced by occupancy of BRCA1 by DNMT1 and CpG methylation. Interestingly, these changes were partially overridden by pre-exposure to the dietary AhR antagonist resveratrol. In keratinocytes, TCDD is able to immortalize cell division through methylation-mediated repression of p53 and p16. Both these genes and their regulatory components are known to be hypermethylated and silenced in breast cancer, and these data indicate that TCDD may be one of the causes of epigenetic silencing and tumor development. TCDD significantly altered the methylome in the sperm of rats three generations after in utero exposure and increased incidence of prostate and polycystic ovarian diseases, and pubertal abnormalities. In utero treatment of developing mouse embryos with TCDD decreased expression of the imprinted genes, H19 and IGF2, through increased methylation in their promoter region. This was correlated with the increased activity of DNMTs [31, 46]. Finally, in a xenograft mouse model of hepatocellular carcinoma, dioxins up-regulated miR-191, whose inhibition decreased cell proliferation, suggesting that such miRNA may contribute to dioxin-induced carcinogenity [39].

Polycyclic aromatic hydrocarbons (PAHs) are the most widespread organic compounds found in the environment with endocrine-disrupting properties mainly produced by burning coal, gasoline, diesel and tobacco. PAHs are lipophilic compounds of high stability, acting as carcinogens with genotoxic and mutagenic properties in many diseases, such as lung cancer. The carcinogenic properties of PAHs in human breast cancer remain unclear [15, 46]. The parental PAH molecules that enter pulmonary cells are considered procarcinogens because they do not directly induce DNA damage. Rather, it is the transformation of a single PAH into its carcinogenic metabolites that contribute to cancer etiology. In general, PAHs are metabolized by CYPs and other metabolic enzymes into phenols, catechols, and quinones, resulting in the formation of diol-epoxides, radical cations, or reactive and redox-active o-quinones, which may all react with DNA to produce DNA adducts. DNA adducts can cause mismatch in DNA replication, as well as altered promoter methylation and/or promoter binding, leading to an inheritable DNA mutation or abnormal gene expression, and ultimately tumorigenesis. Although PAHs are not considered liver carcinogens, they do become metabolized to DNA-reactive metabolites in liver following oral exposure [29].

The PAH benzo[a]pyrene (BaP), mainly sourced from wood burning, but also detected in cigarettes, has been shown to have many epigenetic-disrupting properties in a number of human and animal models [46]. Human bronchial epithelial cells treated with BaP showed DNA hypermethylation that was associated with the transformation of cultured cells [15]. BaP are ligands of AhR, whose activation has a variety of other downstream effects, including the formation of DNA adducts, tumorigenesis, inflammation, cell proliferation, and loss of cell-cell adhesion. Numerous studies have shown that the AhR plays an important role in the development of lung cancer [58]. Finally, in vitro studies using human breast cancer cell lines have demonstrated a number of facets of BaP action, including DNA mismatch repair, altered cell cycle, metabolism and invasion. Epigenomic studies in breast cells reveal that the actions of BaP are widespread, altering both DNA methylation and histone patterns [15, 58].

Vinclozolin is a common fungicide used in vineyards and other agricultural settings [9]. As a pesticide it is used on fruits and vegetables, meaning that most exposure comes through the consumption of residual contamination in foods and drinking water. Vinclozolin is an antiandrogen. In male it is associated with reducing prostate weight, modulating sex steroid levels and elevating the prevalence of tumors in the prostate and testis. In the female, vinclozolin exposure in rats results in the formation of mammary tumors and in disruption of mammary gland development [46]. A number of studies have been conducted investigating the transgenerational epigenetic effects of vinclozolin. In rodent models, it has been demonstrated that prenatal exposure to this chemical can decrease the adult sperm concentration and motility, and also can increase the risk of hypercholesterolemia, kidney and prostate disease, abnormalities in immune system function, and cancer in F1 male offspring. These adverse effects persisted through F1 to F4 generations of male offspring [4]. This research was the first study which indicated the possibility of transgenerational inheritance of adult-onset disease in EDC-exposed rodents [79].

Exposure to several types of heavy metals has been directly linked to increased risk of multiple cancers, including prostate cancer. Among these molecules, there are two trace metals, arsenic and cadmium, which are classified as EDCs due to their ability to act as a ligand and/or interact with members of the steroid receptor superfamily [31]. Inorganic arsenic (iAs) is found naturally in soils and in ground and surface water. Cadmium exposures come from smoking, diet and a variety of occupations. Studies in rodent models found that brief exposure to low dose cadmium increased prostate epithelial hyperplasia and upregulated AR and ERα levels, suggesting endocrine perturbations leading to abnormal proliferation. Either iAs and cadmium transform normal human prostate epithelial cells in vitro, enabling them to recruit nearby stem cells into an oncogenic phenotype [31]. Many possible mechanisms of cadmium carcinogenesis have been suggested and, among them, induction of ROS and alteration of DNA methylation seem to play a predominant biological role. Cadmium reduces genome methylation, inhibiting DNMTs in a noncompetitive manner [8, 76]. However, other studies showed that cadmium-induced malignant transformation was associated with the overexpression of DNMT1 and DNMT3a, by increased global DNA methylation [15]. Cadmium, transactivated ERα and promotes cancer cells growth by inducing hypermethylation of Ras-association domain-containing protein 1 (RASSF1A) and p16 promoters [88]. In rat-liver epithelial cell lines treated with chronic low arsenic doses, cellular transformation was associated with depressed SAM levels, global DNA hypomethylation, and decreased DNA methyltrasferase activity [89].

Notably, most EDCs are highly lipophilic and resistant to degradation in both humans and environment. This group of chemicals has been defined "Persistent Organic Pollutants" (POPs) [65] Exposure to any type of POPs has the potential for long-term disruption of metabolic, immune, and endocrine system functions. Consequently, POPs are strongly linked to obesity, type 2 diabetes and metabolic syndrome, each of which is intimately linked to cancer [65, 45]. For instance, POPs accumulate in white adipose tissue with important systemic and local consequences [31]. A wide number of tumors grows in proximity of adipocytes or metastasizes to predominantly adipocyte-dominated host environment (i.e. breast, prostate, colon) thus establishing an intimate crosstalk between cancer cells and the adipose stroma [14]. Metabolic and environmental perturbations lead to adipocyte dysfunctions which in turn enhance cancer cell proliferation, invasiveness and drug resistance [3, 17, 18, 63]. It has been shown that low dose POPs impairs adipogenesis and generates dysfunctional adipocytes [5, 80]. These altered adipocytes display a significant activation of inflammatory pathways with an increased secretion of pro-inflammatory cytokines which, in turn, sustain the low grade chronic inflammation that, through epigenetic (and not) mechanisms, contributes to cancer.

Thus, EDCs are a large class of components able to induce epigenetic changes trans-generationally hereditable, thus perturbing gene-environment interaction and contributing to cancer onset.

6.2.3 Circadian Rhythm

Circadian disruption has been classified as "potential carcinogenic to humans" by the World Health Organization's International Agency for Research on Cancer [7, 74]. This statement is mainly supported by epidemiological evidence coming from shift work [21, 68]. In this context, circadian disruption has been implicated in the development of different human cancers [53, 57, 59, 66, 77, 84].

The circadian clock is under the control of environmental factors, particularly light exposure and food intake rhythms. Many of the circadian disruptions are due to dramatic changes derived from the industrial development and from the consequent changes in lifestyle over the past few hundred years [73]. Thus, exposure to electric light, sleep/awake cycles and feeding timing have been hypothesized as having an impact on cancer onset and/or progression.

A hallmark of industrialization is the increasing use of electricity to light the night, both indoor and outdoor. The impact of electric lighting may occur at different levels, and particularly on melatonin production [51]. Specific retinal ganglion cells contain melanopsin, the primary photoreceptor to transduce light information to the circadian system [55]. Melatonin is mostly secreted at nighttime and exposure to light reduces melatonin secretion in a dose-dependent manner [86]. Therefore, nighttime light exposure reduces the rhythmic secretion of melatonin from the pineal gland [11]. Interestingly, melatonin may exert anti-cancer functions. For instance, it reduces mammary tumor growth in experimental animals [37]. Exposure of rats to light at night disrupted melatonin profiles, as compared with those exposed to no light. This correlated with increased tumor growth and resistance to tamoxifen, which could be reversed by melatonin supplementation [20]. In other mouse model studies, mice with arrhythmic light-dark cycle or suprachiasmatic nucleus (SCN) ablation exhibited accelerated transplanted tumor growth rate compared with control mice [27, 28].

In humans, reduction of nocturnal melatonin levels may increase the effects of estrogen and thereby contribute to breast cancer risk. Epidemiological studies in nurses also reveal that shift work induces a significant increase in circulating estradiol levels, which could further disrupt mammary estrogen signalling and thereby promote cancer [10]. Several reports have also described an inverse relationship between sleep duration and cancer risk [82]. Moreover, a reduced risk of breast cancer has been reported in blind women [33]. Conversely, cancer patients with irregular circadian rhythm had poorer prognosis compared with those with regular rhythms [60, 78].

At molecular level, the anti-cancer effects of melatonin might be attributed to its anti-estrogen action, by inhibiting both aromatase activity and estrogen signaling [2, 32, 61]. Moreover, melatonin promotes genomic stability which may facilitate its anti-metastatic activity [37, 72].

In addition to the lighting cues, circadian rhythm is also dictated by feeding habits, which may specifically impinge on central as well as peripheral biological clocks [69]. Very surprising evidence comes from a recent study [47] indicating that adherence to diurnal eating patterns and, specifically, a long interval between last meal and sleep are associated with a lower cancer risk (at least for breast and prostate cancer).

Thus, both light and food timing might affect cancer related phenotypes.

Regulation of Clock genes and cancer

The circadian clock, which exists in almost all eukaryotic organisms, consists of both central (located in the suprachiasmatic nucleus, SCN) and peripheral (in most body cells) biological oscillators [62]. At the molecular level, BMAL1 and CLOCK genes can be considered the master regulators of circadian rhythm. The CLOCK:BMAL1 complex binds to E-boxes of target genes, including CRY1, CRY2, PER1, PER2, and PER3 and increases their expression levels. In turn, the protein products of these genes oppose the activity of the CLOCK:BMAL1 complex and consequently form a negative feedback loop that suppresses their own expression [64, 52].

Furthermore, a rather precise molecular oscillator established when the function of canonical feedback loop is modulated by additional circadian genes, such as Dec1, Dec2, Rev-Erbα, Rorα, CK1ε and Npas2 [38, 64].

In mammals, approximately 10% of all genes are controlled by the circadian clock, such as p21, CCNA, CCNE, c-myc and WEE-1, known as clock controlled genes (CCGs) which regulate cell metabolism, proliferation, differentiation, DNA damage repair, apoptosis and autophagy [38, 70].

Disruption of the expression of clock genes has been observed in cancer patients [53, 56, 57]. Disrupted circadian rhythms may then lead to up- or down-regulation of genes and proteins, which may favor cancer-related phenotypes. Dysfunction of the clock machinery and cellular oscillators is involved in tumorigenesis. For example, PER1 and PER2 are known tumor suppressor genes, and their knockdown results in the doubling of tumor number and cancer growth; in contrast, overexpression of these genes decreases tumor number and cancer growth [40]. In addition, transcriptional silencing of the BMAL1 gene through hypermethylation of its promoter CpG island has been observed in hematologic malignancies [77].

It has been clearly documented that disruption of circadian rhythms leads to epigenetic modifications. Studies on shift workers, for example have demonstrated changes in the DNA methylation of their genes [35].

Long interspersed element-1 (L1) is a protein complex that promotes genomic instability through DNA double-strand breaks and insertional mutagenesis. Up-regulation of L1 has been reported in many human malignancies. Melatonin receptor 1 acts as an inhibitor of L1 mobilization by down-regulating L1 mRNA and the open reading frame 1 (ORF1) protein. Hence, exposure to environmental light regulates the expression of L1 through the regulation of melatonin production. This association indicates that suppression of melatonin production due to forced exposure to light increases L1-induced genomic instability and consequently promotes carcinogenesis [22]. Furthermore, it has been shown that some long non-coding RNAs (lncRNAs) directly and indirectly alter melatonin synthesis. It has also been shown that the abundance of these lncRNAs changes in a circadian manner. These findings altogether indicate that circadian disruption may also alter melatonin expression and consequently promote carcinogenesis by changing the abundance of certain lncRNAs [16].

CLOCK and TIMELESS genes were found overexpressed, while PER1, PER2, PER3, CRY2 were found to be significantly down-expressed in breast cancer specimen compared with non tumor samples [49]. Hypermethylation of PER1, PER2, CRY1 and BMAL1 promoters has also been reported in 37 of 53 breast cancer cell lines [48]. This observation provides evidence for the underlying epigenetic mechanism of clock gene deregulation and its carcinogenic effects [48].

Coordinated co-expression of clock genes, indicative of a functional circadian clock, is maintained in ER-positive, HER2-negative, low grade and non-metastasizing tumors but is compromised in more aggressive carcinomas [13]. Thus, besides dramatic metabolic alterations, cancer cells display severe changes in the clock phenotype with likely consequences in tumor progression and treatment response. Interestingly, reciprocal connections have been identified between "metabolic" genes (for example in the glycolytic pathways) and clock genes in primary tumors, but they may be lost in the metastatic cells [30].

6.3 Conclusions

1. Common environmental and dietary chemicals act on the human genome, either directly or indirectly, to alter gene expression or structure.
2. Imbalanced nutrition, environmental pollution and circadian rhythm disruption largely play their role by impinging on epigenetic mechanisms.
3. Cancer onset and progression is a combination of genetic and epigenetic hits.

4. Epigenetic modifications may occur early during development (or even in the previous generations) and contribute to cancer cell phenotypes.
5. Some diet-regulated genes (because of genetic or epigenetic regulation) are likely to play a role in the onset, incidence, progression, and/or severity of cancer.
6. An incorrect diet can be a risk factor for cancer.
7. Endocrine disrupting chemicals may have either direct (on cancer cells) or indirect (on metabolic imbalance) effects on cancer onset and progression.
8. Disruption of circadian rhythms (light, sleep/awake cycles, feeding) may favor the onset and the progression of some forms of cancer.
9. Dietary and lifestyle intervention can be used to prevent, mitigate, or cure chronic diseases.
10. Understanding the detailed epigenetic mechanisms will lead to the identification of novel personalized therapeutic strategies.

References

1. Alenghat T, Artis D. Epigenomic regulation of host-microbiota interactions. Trends Immunol. 2014 Nov;35(11):518–25.
2. Alvarez-García V, González A, Martínez-Campa C, Alonso-González C, Cos S. Melatonin modulates aromatase activity and expression in endothelial cells. Oncol Rep. 2013 May;29(5):2058–64.
3. Ambrosio MR, D'Esposito V, Costa V, Liguoro D, Collina F, Cantile M, Prevete N, Passaro C, Mosca G, De Laurentiis M, Di Bonito M, Botti G, Franco R, Beguinot F, Ciccodicola A, Formisano P. Glucose impairs tamoxifen responsiveness modulating connective tissue growth factor in breast cancer cells. Oncotarget. 2017 Nov 20;8(65):109000–17.
4. Anway MD, Cupp AS, Uzumcu M, Skinner MK. Epigenetic transgenerational actions of endocrine disruptors and male fertility. Science. 2005 Jun 3;308(5727):1466–9.
5. Ariemma F, D'Esposito V, Liguoro D, Oriente F, Cabaro S, Liotti A, Cimmino I, Longo M, Beguinot F, Formisano P, Valentino R. Low-dose bisphenol-a impairs Adipogenesis and generates dysfunctional 3T3-L1 adipocytes. PLoS One. 2016 Mar 4;11(3):e0150762.
6. Asgari S. Epigenetic modifications underlying symbiont-host interactions. Adv Genet. 2014;86:253–76.
7. Baan R, Grosse Y, Straif K, Secretan B, El Ghissassi F, Bouvard V, Benbrahim-Tallaa L, Guha N, Freeman C, Galichet L. Cogliano V; WHO International Agency for Research on Cancer monograph working group. A review of human carcinogens--part F: chemical agents and related occupations. Lancet Oncol. 2009 Dec;10(12):1143–4.
8. Baccarelli A, Bollati V. Epigenetics and environmental chemicals. Curr Opin Pediatr. 2009;21(2):243–51.. Review
9. Bishop KS, Ferguson LR. The interaction between epigenetics, nutrition and the development of cancer. Nutrients. 2015 Jan 30;7(2):922–47.
10. Bracci M, Manzella N, Copertaro A, Staffolani S, Strafella E, Barbaresi M, Copertaro B, Rapisarda V, Valentino M, Santarelli L. Rotating-shift nurses after a day off: peripheral clock gene expression, urinary melatonin, and serum 17-β-estradiol levels. Scand J Work Environ Health. 2014 May 1;40(3):295–304.
11. Brainard GC, Richardson BA, King TS, Reiter RJ. The influence of different light spectra on the suppression of pineal melatonin content in the Syrian hamster. Brain Res. 1984 Mar 5;294(2):333–9.

12. Bromer JG, Wu J, Zhou Y, Taylor HS. Hypermethylation of homeobox A10 by in utero diethylstilbestrol exposure: an epigenetic mechanism for altered developmental programming. Endocrinology. 2009 Jul;150(7):3376–82.
13. Cadenas C, van de Sandt L, Edlund K, Lohr M, Hellwig B, Marchan R, Schmidt M, Rahnenführer J, Oster H, Hengstler JG. Loss of circadian clock gene expression is associated with tumor progression in breast cancer. Cell Cycle. 2014;13(20):3282–91.
14. Calle EE, Kaaks R. Overweight, obesity and cancer: epidemiological evidence and proposed mechanisms. Nat Rev Cancer. 2004 Aug;4(8):579–91.
15. Cao Y. Environmental pollution and DNA methylation: carcinogenesis, clinical significance, and practical applications. Front Med. 2015 Sep;9(3):261–74.
16. Coon SL, Munson PJ, Cherukuri PF, Sugden D, Rath MF, Møller M, Clokie SJ, Fu C, Olanich ME, Rangel Z, Werner T;. NISC comparative sequencing program, Mullikin JC, Klein DC. Circadian changes in long noncoding RNAs in the pineal gland. Proc Natl Acad Sci U S A 2012;109(33):13319–13324.
17. D'Esposito V, Passaretti F, Hammarstedt A, Liguoro D, Terracciano D, Molea G, Canta L, Miele C, Smith U, Beguinot F, Formisano P. Adipocyte-released insulin-like growth factor-1 is regulated by glucose and fatty acids and controls breast cancer cell growth *in vitro*. Diabetologia. 2012;55:2811–22.
18. D'Esposito V, Liguoro D, Ambrosio MR, Collina F, Cantile M, Spinelli R, Raciti GA, Miele C, Valentino R, Campiglia P, De Laurentiis M, Di Bonito M, Botti G, Franco R, Beguinot F, Formisano P. Adipose microenvironment promotes triple negative breast cancer cell invasiveness and dissemination by producing CCL5. Oncotarget. 2016 Apr 26;7(17):24495–509.
19. Daniel M, Tollefsbol TO. Epigenetic linkage of aging, cancer and nutrition. J Exp Biol. 2015 Jan 1;218(Pt 1):59–70.
20. Dauchy RT, Xiang S, Mao L, Brimer S, Wren MA, Yuan L, Anbalagan M, Hauch A, Frasch T, Rowan BG, Blask DE, Hill SM. Circadian and melatonin disruption by exposure to light at night drives intrinsic resistance to tamoxifen therapy in breast cancer. Cancer Res. 2014 Aug 1;74(15):4099–110.
21. Davis S, Mirick DK, Stevens RG. Night shift work, light at night, and risk of breast cancer. J Natl Cancer Inst. 2001 Oct 17;93(20):1557–62.
22. deHaro D, Kines KJ, Sokolowski M, Dauchy RT, Streva VA, Hill SM, Hanifin JP, Brainard GC, Blask DE, Belancio VP. Regulation of L1 expression and retrotransposition by melatonin and its receptor: implications for cancer risk associated with light exposure at night. Nucleic Acids Res. 2014 Jul;42(12):7694–707.
23. Doherty LF, Bromer JG, Zhou Y, Aldad TS, Taylor HS. In utero exposure to diethylstilbestrol (DES) or bisphenol-A (BPA) increases EZH2 expression in the mammary gland: an epigenetic mechanism linking endocrine disruptors to breast cancer. Horm Cancer. 2010 Jun;1(3):146–55.
24. Esteller M. Epigenetics in cancer. N Engl J Med. 2008 Mar 13;358(11):1148–59.
25. Feinberg AP, Ohlsson R, Henikoff S. The epigenetic progenitor origin of human cancer. Nat Rev Genet. 2006 Jan;7(1):21–33.
26. Ferreira LL, Couto R, Oliveira PJ. Bisphenol a as epigenetic modulator: setting the stage for carcinogenesis? Eur J Clin Investig. 2015 Jan;45(Suppl 1):32–6.
27. Filipski E, King VM, Li X, Granda TG, Mormont MC, Liu X, et al. Host circadian clock as a control point in tumor progression. J Natl Cancer Inst. 2002;94:690–7.
28. Filipski E, Delaunay F, King VM, Wu MW, Claustrat B, Grechez-Cassiau A, et al. Effects of chronic jet lag on tumor progression in mice. Cancer Res. 2004;64:7879–85.
29. Fleisch AF, Wright RO, Baccarelli AA. Environmental epigenetics: a role in endocrine disease? J Mol Endocrinol. 2012 Aug 30;49(2):R61–7.
30. Fuhr L, El-Athman R, Scrima R, Cela O, Carbone A, Knoop H, Li Y, Hoffmann K, Laukkanen MO, Corcione F, Steuer R, Meyer TF, Mazzoccoli G, Capitanio N, Relógio A. The circadian clock regulates metabolic phenotype rewiring via HKDC1 and modulates tumor progression and drug response in colorectal Cancer. EBioMedicine. 2018 Jul;33:105–21.

31. Gore AC, Chappell VA, Fenton SE, Flaws JA, Nadal A, Prins GS, Toppari J, Zoeller RT. EDC-2: the Endocrine Society's second scientific statement on endocrine-disrupting chemicals. Endocr Rev. 2015 Dec;36(6):E1–E150.
32. Grant SG, Melan MA, Latimer JJ, Witt-Enderby PA. Melatonin and breast cancer: cellular mechanisms, clinical studies and future perspectives. Expert Rev Mol Med. 2009 Feb 5;11:e5.
33. Hahn RA. Profound bilateral blindness and the incidence of breast cancer. Epidemiology. 1991 May;2(3):208–10.
34. Hardy TM, Tollefsbol TO. Epigenetic diet: impact on the epigenome and cancer. Epigenomics. 2011 Aug;3(4):503–18.
35. Haus EL, Smolensky MH. Shift work and cancer risk: potential mechanistic roles of circadian disruption, light at night, and sleep deprivation. Sleep Med Rev. 2013 Aug;17(4):273–84.
36. Hernández-Aguilera A, Fernández-Arroyo S, Cuyàs E, Luciano-Mateo F, Cabre N, Camps J, Lopez-Miranda J, Menendez JA, Joven J. Epigenetics and nutrition-related epidemics of metabolic diseases: current perspectives and challenges. Food Chem Toxicol. 2016 Oct;96:191–204.
37. Hill SM, Belancio VP, Dauchy RT, Xiang S, Brimer S, Mao L, Hauch A, Lundberg PW, Summers W, Yuan L, Frasch T, Blask DE. Melatonin: an inhibitor of breast cancer. Endocr Relat Cancer. 2015 Jun;22(3):R183–204.
38. Honma S, Kawamoto T, Takagi Y, Fujimoto K, Sato F, Noshiro M, et al. Dec1 and Dec2 are regulators of the mammalian molecular clock. Nature. 2002;419:841–4.
39. Hou L, Zhang X, Wang D, Baccarelli A. Environmental chemical exposures and human epigenetics. Int J Epidemiol. 2012 Feb;41(1):79–105.
40. Hrushesky WJ, Grutsch J, Wood P, Yang X, Oh EY, Ansell C, Kidder S, Ferrans C, Quiton DF, Reynolds J, Du-Quiton J, Levin R, Lis C, Braun D. Circadian clock manipulation for cancer prevention and control and the relief of cancer symptoms. Integr Cancer Ther. 2009 Dec;8(4):387–97.
41. Jobling S, Bjerregaard P, Blumberg B, et al. Evidence for endocrine disruption in humans and wildlife: endocrine disruptors and hormone-related cancers. In: Bergman A, Heindel JJ, Jobling S, Kidd KA, Zoeller RT, eds. *State of the Science of Endocrine Disrupting Chemicals – 2012*. Chapter 2.7. Geneva: World Health Organization; 2012:126–142.
42. Johnson C, Warmoes MO, Shen X, Locasale JW. Epigenetics and cancer metabolism. Cancer Lett. 2015 Jan 28;356(2 Pt A):309–14.
43. Jones PA, Baylin SB. The fundamental role of epigenetic events in cancer. Nat Rev Genet. 2002 Jun;3(6):415–28.
44. Joseph PV, Abey SK, Henderson WA. Emerging role of Nutri-epigenetics in inflammation and Cancer. Oncol Nurs Forum. 2016 Nov 1;43(6):784–8.
45. Klil-Drori AJ, Azoulay L, Pollak MN. Cancer, obesity, diabetes, and antidiabetic drugs: is the fog clearing? Nat Rev Clin Oncol. 2017 Feb;14(2):85–99.
46. Knower KC, To SQ, Leung YK, Ho SM, Clyne CD. Endocrine disruption of the epigenome: a breast cancer link. Endocr Relat Cancer. 2014 Mar 12;21(2):T33–55.
47. Kogevinas M, Espinosa A, Castelló A, Gómez-Acebo I, Guevara M, Martin V, Amiano P, Alguacil J, Peiro R, Moreno V, Costas L, Fernández-Tardón G, Jimenez JJ, Marcos-Gragera R, Perez-Gomez B, Llorca J, Moreno-Iribas C, Fernández-Villa T, Oribe M, Aragones N, Papantoniou K, Pollán M, Castano-Vinyals G, Romaguera D. Effect of mistimed eating patterns on breast and prostate cancer risk (MCC-Spain study). Int J Cancer. 2018 Jul;17
48. Kuo SJ, Chen ST, Yeh KT, Hou MF, Chang YS, Hsu NC, Chang JG. Disturbance of circadian gene expression in breast cancer. Virchows Arch. 2009 Apr;454(4):467–74.
49. Lesicka M, Jabłońska E, Wieczorek E, Seroczyńska B, Siekierzycka A, Skokowski J, Kalinowski L, Wąsowicz W, Reszka E. Altered circadian genes expression in breast cancer tissue according to the clinical characteristics. PLoS One. 2018 Jun 29;13(6):e0199622.
50. Lewis KA, Tollefsbol, TO. The influence of an epigenetics diet on the cancer epigenome. Epigenomics. 2017 Sep;9(9):1153–5.
51. Lewy AJ, Wehr TA, Goodwin FK, Newsome DA, Markey SP. Light suppresses melatonin secretion in humans. Science. 1980 Dec 12;210(4475):1267–9.

52. Liang X, FitzGerald GA. Timing the microbes: the circadian rhythm of the gut microbiome. J Biol Rhythm. 2017;32:505–15.
53. Litlekalsoy J, Rostad K, Kalland KH, Hostmark JG, Laerum OD. Expression of circadian clock genes and proteins in urothelial cancer is related to cancer-associated genes. BMC Cancer. 2016 Jul 27;16:549.
54. Lopez J, Percharde M, Coley HM, Webb A, Crook T. The context and potential of epigenetics in oncology. Br J Cancer. 2009 Feb 24;100(4):571–7.
55. Lucas RJ, Lall GS, Allen AE, Brown TM. How rod, cone, and melanopsin photoreceptors come together to enlighten the mammalian circadian clock. Prog Brain Res. 2012;199:1–18.
56. Mazzoccoli G, Vinciguerra M, Papa G, Piepoli A. Circadian clock circuitry in colorectal cancer. World J Gastroenterol. 2014 Apr 21;20(15):4197–207.
57. Mazzoccoli G, Colangelo T, Panza A, Rubino R, Tiberio C, Palumbo O, Carella M, Trombetta D, Gentile A, Tavano F, Valvano MR, Storlazzi CT, Macchia G, De Cata A, Bisceglia G, Capocefalo D, Colantuoni V, Sabatino L, Piepoli A, Mazza T. Analysis of clock gene-miRNA correlation networks reveals candidate drivers in colorectal cancer. Oncotarget. 2016 Jul 19;7(29):45444–61.
58. Moorthy B, Chu C, Carlin DJ. Polycyclic aromatic hydrocarbons: from metabolism to lung cancer. Toxicol Sci. 2015 May;145(1):5–15.
59. Mteyrek A, Filipski E, Guettier C, Okyar A, Lévi F. Clock gene Per2 as a controller of liver carcinogenesis. Oncotarget. 2016 Dec 27;7(52):85832–47.
60. Nagata C, Tamura T, Wada K, Konishi K, Goto Y, Nagao Y, et al. Sleep duration, nightshift work, and the timing of meals and urinary levels of 8-isoprostane and 6-sulfatoxymelatonin in Japanese women. Chronobiol Int. 2017;34:1187–96.
61. Nooshinfar E, Safaroghli-Azar A, Bashash D, Akbari ME. Melatonin, an inhibitory agent in breast cancer. Breast Cancer. 2017 Jan;24(1):42–51.
62. Panda S, Antoch MP, Miller BH, Su AI, Schook AB, Straume M, et al. Coordinated transcription of key pathways in the mouse by the circadian clock. Cell. 2002;109:307–20.
63. Park J, Morley TS, Kim M, Clegg DJ, Scherer PE. Obesity and cancer- mechanisms underlying tumour progression and recurrence. Nat Rev Endocrinol. 2014;10(8):455–65.
64. Preitner N, Damiola F, Lopez-Molina L, Zakany J, Duboule D, Albrecht U, et al. The orphan nuclear receptor REV-ERBalpha controls circadian transcription within the positive limb of the mammalian circadian oscillator. Cell. 2002;110:251–60.
65. Reaves DK, Ginsburg E, Bang JJ, Fleming JM. Persistent organic pollutants and obesity: are they potential mechanisms for breast cancer promotion? Endocr Relat Cancer. 2015 Apr;22(2):R69–86.
66. Reszka E, Przybek M. Circadian genes in breast cancer. Adv Clin Chem. 2016;75:53–70.
67. Samtani R, Sharma N, Garg D. Effects of endocrine-disrupting chemicals and epigenetic modifications in ovarian cancer: a review. Reprod Sci. 2018 Jan;25(1):7–18.
68. Schernhammer ES, Laden F, Speizer FE, Willett WC, Hunter DJ, Kawachi I, Colditz GA. Rotating night shifts and risk of breast cancer in women participating in the nurses' health study. J Natl Cancer Inst. 2001 Oct 17;93(20):1563–8.
69. Schibler U, Sassone-Corsi P. A web of circadian pacemakers. Cell. 2002 Dec 27;111(7):919–22.
70. Shearman LP, Sriram S, Weaver DR, Maywood ES, Chaves I, Zheng B, et al. Interacting molecular loops in the mammalian circadian clock. Science. (New York, NY. 2000;288:1013–9.
71. Shi Y, Hu FB. The global implications of diabetes and cancer. Lancet. 2014 Jun7;383(9933):1947–8.
72. Stevens RG, Brainard GC, Blask DE, Lockley SW, Motta ME. Adverse health effects of nighttime lighting: comments on American Medical Association policy statement. Am J Prev Med. 2013 Sep;45(3):343–6.
73. Stevens RG, Brainard GC, Blask DE, Lockley SW, Motta ME. Breast cancer and circadian disruption from electric lighting in the modern world. CA Cancer J Clin. 2014 May-Jun;64(3):207–18.

74. Straif K, Baan R, Grosse Y, Secretan B, El Ghissassi F, Bouvard V, Altieri A, Benbrahim-Tallaa L. Cogliano V; WHO international agency for research on cancer monograph working group. Carcinogenicity of shift-work, painting, and fire-fighting. Lancet Oncol. 2007 Dec;8(12):1065–6.
75. Supic G, Jagodic M, Magic Z. Epigenetics: a new link between nutrition and cancer. Nutr Cancer. 2013;65(6):781–92.
76. Takiguchi M, Achanzar WE, Qu W, Li G, Waalkes MP. Effects of cadmium on DNA-(Cytosine-5) methyltransferase activity and DNA methylation status during cadmium-induced cellular transformation. Exp Cell Res. 2003;286:355–65.
77. Taniguchi H, Fernández AF, Setién F, Ropero S, Ballestar E, Villanueva A, Yamamoto H, Imai K, Shinomura Y, Esteller M. Epigenetic inactivation of the circadian clock gene BMAL1 in hematologic malignancies. Cancer Res. 2009 Nov 1;69(21):8447–54.
78. Truong T, Liquet B, Menegaux F, Plancoulaine S, Laurent-Puig P, Mulot C, et al. Breast cancer risk, nightwork, and circadian clock gene polymorphisms. Endocr Relat Cancer. 2014;21:629–38.
79. Vaiserman A. Early-life exposure to endocrine disrupting chemicals and later-life health outcomes: an epigenetic bridge? Aging Dis. 2014 Jan 28;5(6):419–29.
80. Valentino R, D'Esposito V, Passaretti F, Liotti A, Cabaro S, Longo M, Perruolo G, Oriente F, Beguinot F, Formisano P. Bisphenol-a impairs insulin action and up-regulates inflammatory pathways in human subcutaneous adipocytes and 3T3-L1 cells. PLoS One. 2013 Dec 9;8(12):e82099.
81. Valentino R, D'Esposito V, Ariemma F, Cimmino I, Beguinot F, Formisano P. Bisphenol a environmental exposure and the detrimental effects on human metabolic health: is it necessary to revise the risk assessment in vulnerable population? J Endocrinol Investig. 2016 Mar;39(3):259–63.
82. Verkasalo PK, Lillberg K, Stevens RG, Hublin C, Partinen M, Koskenvuo M, Kaprio J. Sleep duration and breast cancer: a prospective cohort study. Cancer Res. 2005 Oct 15;65(20):9595–600.
83. Verma M. Cancer control and prevention by nutrition and epigenetic approaches. Antioxid Redox Signal. 2012 Jul 15;17(2):355–64.
84. Wang Q, Ao Y, Yang K, Tang H, Chen D. Circadian clock gene Per2 plays an important role in cell proliferation, apoptosis and cell cycle progression in human oral squamous cell carcinoma. Oncol Rep. 2016 Jun;35(6):3387–94.
85. Wegner M, Neddermann D, Piorunska-Stolzmann M. Jagodzinski PP. Role of epigenetic mechanisms in the development of chronic complications of diabetes. Diabetes Res Clin Pract. 2014 Aug;105(2):164–75.
86. West KE, Jablonski MR, Warfield B, Cecil KS, James M, Ayers MA, Maida J, Bowen C, Sliney DH, Rollag MD, Hanifin JP, Brainard GC. Blue light from light-emitting diodes elicits a dose-dependent suppression of melatonin in humans. J Appl Physiol (1985). 2011 Mar;110(3):619–626.
87. Wong CC, Qian Y, Yu J. Interplay between epigenetics and metabolism in oncogenesis: mechanisms and therapeutic approaches. Oncogene. 2017 Jun 15;36(24):3359–74.
88. Zhang X, Ho SM. Epigenetics meets endocrinology. J Mol Endocrinol. 2011 Feb;46(1):R11–32.
89. Zhao CQ, Young MR, Diwan BA, Coogan TP, Waalkes MP. Association of arsenic-induced malignant transformation with DNA hypomethylation and aberrant gene expression. Proc Natl Acad Sci U S A. 1997;94:10907–12.

Chapter 7
The Role of Gene-Environment Interaction in Mental Health and Susceptibility to the Development of Psychiatric Disorders

Elham Assary, John Vincent, Sandra Machlitt-Northen, Rob Keers, and Michael Pluess

Keywords Psychiatric genetics · Gene-environment interaction · Mental health · Environmental sensitivity · Differential susceptibility

7.1 Introduction: Gene-Environment Interactions (GxE) and Psychiatric Disorders

Mental health disorders account for around a third of all disabilities worldwide [96], placing an enormous burden on affected individuals, and on society more generally [33]. Psychological research suggests that environmental influences such as childhood maltreatment and stressful life events increase the risk for a range of mental health disorders [11, 93, 124], while findings from behavioural genetic studies (i.e., twin studies) highlight a substantial role of genetic factors, with heritability estimates ranging from 37% for major depression [121], to 65–80% for schizophrenia [79, 120], and 60–85% for bipolar disorder [18, 85, 118]. Behavioural genetic studies suggest that genetic and environmental influences are both important in the development of psychiatric conditions. However, these factors do not work in isolation. Rather, mental health and wellbeing is determined by a complex interplay between the two, including gene-environment interaction (GxE) and gene-environment correlation (rGE) [38, 69, 70]. GxE occurs when an individual's

E. Assary · J. Vincent · S. Machlitt-Northen · R. Keers · M. Pluess (✉)
Department of Biological and Experimental Psychology, School of Biological and Chemical Sciences, Queen Mary University of London, London, UK
e-mail: e.assary@qmul.ac.uk; j.p.vincent@qmul.ac.uk; s.machlitt-northen@qmul.ac.uk; r.keers@qmul.ac.uk; m.pluess@qmul.ac.uk

R. Teperino (ed.), *Beyond Our Genes*,
https://doi.org/10.1007/978-3-030-35213-4_7

genotype influences their *response* (biological/psychological) to an environmental factor [66]. For instance, those with a high genetic risk of major depression have been shown to be more negatively affected by stressful life events [71]. On the other hand, rGE occurs when genetic factors influence *exposure* to a given environment in the form of *active*, *passive* and *evocative* rGE. An *active* rGE occurs when the genotype influences the probability of exposure to an environmental factor [110]. In other words, individuals select, modify and create their own environments in a way that reflects their genetically influenced propensities, including personality traits [101]. A *passive* rGE may occur because parents not only provide their children with genes, but also shape the environment in which they are raised [77]. For example, parental depressive symptoms are associated with reduced positivity towards their offspring, which in turn partially mediates intergenerational transmission of depression from parents to their offspring [108]. An *evocative* rGE can occur when a genetic predisposition gives rise to a behaviour that evokes a particular response from other individuals [22]. For instance, children who express a low level of autonomy may evoke more controlling behaviour from their mothers, while those with high level of autonomy may experience significantly lower maternal control [74].

While behavioural genetics studies have been fundamental in evidencing and promoting GxE and rGE models of mental health and disease, they are limited in their remit, given that they only provide statistical estimates of the contribution that genes and environments make to a disorder, and therefore provide no information as to the specific genes or genetic variants involved. Advances in genetic technology in the past two decades have facilitated examination of GxE models at the molecular level, via variations in DNA sequences. Early molecular genetic investigations of GxE relied on the candidate-gene approach, testing a small number of genetic variants hypothesised to be biologically relevant to psychiatric disorders. The excitement surrounding the first of these studies, published in 2002, initiated a decade long boom in GxE research that saw over one hundred candidate-gene GxE studies of mental health disorders published [42]. However, growing concerns regarding the methodological limitations of candidate-gene GxE studies, coupled with the increasing availability of genome-wide data, have ushered in a new era of genome-wide GxE research, allowing the exploration of genotype by environment interplay across the whole human genome. It is worth noting that much of the research using either methodology has concentrated disproportionately on GxE, rather than rGE models, an important omission, the implications of which will be discussed later on in the chapter.

The aim of the current chapter is to present a concise but comprehensive review of the current state of GxE research in the field of psychiatric genetics, specifically focusing on molecular genetic studies. The chapter is organised in three main parts, starting with a review of GxE research in mental health disorders, followed by a critical evaluation of the research, taking into account theoretical GxE concepts, as well as methodological limitations of the studies in the field. The final section will include suggestions and future directions for GxE research in the field.

7.2 Review of Gene-Environment Interaction Research in Psychiatric Disorders

7.2.1 Candidate GxE Research

Caspi et al. [27] conducted the first pioneering GxE study of mental health, by examining the interaction between monoamine-oxidase A (*MAOA*) gene and childhood maltreatment in the development of antisocial behaviour in adulthood. Childhood maltreatment is a well-documented risk factor for antisocial behaviour. However, the occurrence of maltreatment is not necessarily resulting in this behaviour in all people. That is, while some individuals exposed to maltreatment exhibit antisocial behaviour, others appear to be resilient to these effects. Caspi et al. [27] reasoned that differences in an individual's genotype may account for some of these differences. The gene encoding monoamine oxidase A *(MAOA)* was chosen based on evidence showing that a common mutation that affects the expression of the gene, was associated with aggressive behaviour in previous animal and linkage studies [21, 26]. Consistent with their hypothesis, they found that male carriers of the genotype conferring low levels of *MAOA* gene expression showed higher levels of antisocial behaviour in adulthood, but only if they had experienced maltreatment in childhood. Caspi et al. [27]'s novel approach, to examine the interactive effects of genetic and environmental risks in the aetiology of a behavioural trait, paved the way for a plethora of candidate GxE studies focusing on a wide variety of further traits and psychiatric disorders including major depression, attention deficit hyperactivity disorder, schizophrenia, and bipolar disorder. These studies tested the moderating effects of the *MAOA* variant, but also variants in other candidate genes including the serotonin transporter *(SLC6A4)*, dopamine receptor D4 *(DRD4)* and D2 *(DRD2)*, catechol-O-methyl transferase *(COMT)*, and brain-derived neurotrophic factor *(BDNF)*.

Caspi et al. [29] were also the first group to examine the moderating effects of the serotonin-transporter-linked polymorphic region *(5-HTTLPR)* on depression within a GxE framework. In this seminal study, Caspi et al. [27] hypothesized that the 's' variant of the *5-HTTLPR,* which confers reduced expression of the serotonin transporter, may be implicated in depression, by moderating the serotonergic response to stress. They found in their longitudinal sample of 1037 individuals, that those with a history of stressful life events were at higher risk of depression and suicidality, but only if they were also carriers of the s-allele. Over fifty studies have since aimed to replicate these findings, some more successful (e.g. [30]) than others (e.g. [51, 122]). The failure to replicate results of GxE studies involving the serotonin transporter and other candidate genes has generated a fierce debate about the robustness of these findings. Subsequent meta-analytic studies have done little to settle this debate, with some finding support for [65, 113, 127] and against the proposed GxE interaction effects [36, 90, 109]. Risch et al. [109] and Munafò et al. [90] concluded that the significant interaction effects likely reflect false positive findings, due to the small sample sizes of the majority of these studies, rendering them statistically underpowered to detect the small moderating effects of single genes in the prediction of psychiatric outcomes. Karg et al. [65], on the other hand, suggested that previous meta-analyses by Risch et al. [109] and Munafò et al. [90] which only included studies of depression and stressful life events in adulthood, preferentially excluded higher quality studies with larger interaction effects and with a focus on childhood maltreatment. Karg et al. [65]'s more comprehensive meta-analysis found a significant interaction effect for childhood maltreatment, suggesting that the lack of a robust association with stressful life events was due to greater variation in how and when stressful life events have been conceptualized and measured across different studies. In the most recent meta-analysis study of 31 published and unpublished studies of *5-HTTLPR* and stressful life events with sample sizes of 300 or more, Culverhouse et al. [36] did not find support for a significant interaction effect. The field, therefore, remains divided regarding GxE findings involving *5-HTTLPR* and depression.

Caspi et al. [27]'s initial findings on *MAOA* moderating the effects of childhood maltreatment on antisocial behaviour, however, appear to be more robust, with replication attempts mostly successful [45, 48, 73], though not all (e.g. [54]). Overall, meta-analytic studies of *MAOA* have consistently found support for a moderating effect of *MAOA* genotype on the pathway between childhood maltreatment/adversity and antisocial behaviour [23, 73, 123]. The results of a meta-analysis by Byrd and Manuck [23] also indicated that some observed inconsistencies in results may reflect gender differences, whereby most studies with females have either not found a significant interaction effect, or that high-activity, rather than low-activity, *MAOA* genotype is found to infer an increased risk of antisocial behaviour.

GxE research in schizophrenia have typically sought to investigate the role of *COMT* in relation to environmental factors that have consistently been linked to schizophrenia, such as obstetric complications, regular cannabis use, and stress [132]. *COMT* is a protein coding gene on chromosome 22, the enzyme product of which is involved in degradation of catecholamine neurotransmitters such as dopamine, epinephrine and norepinephrine [53]. One of the most studied functional polymorphisms on this gene is a substitution of valine with methionine (Val158Met, rs4680), with homozygous Val genotype showing higher levels of dopamine degradation. GxE studies of psychosis report that cannabis use in adolescents increases the risk for schizophreniform disorder in adulthood [28] and exacerbates psychotic symptoms (e.g. hallucinations) in patients prone to psychotic illness [58], but only in individuals carrying the *COMT* Val-allele. However, the reported interactions have failed to replicate in high-quality studies of healthy controls and patients [64, 137, 138]. Furthermore, some studies have reported the opposite effect, with Met as the risk allele [117, 119], or found that the effects of *COMT* genotype differs as a function of the interaction with other genes [99]. With regards to stress, it appears also to be the Met rather than the Val allele that has been identified as the risk allele, whereby patients with the Met allele showing increased levels of psychotic experiences in response to stress [34, 133]. These inconsistent results have been suggested to reflect variations in study design and measurement [87], with samples ranging from 35 individuals to over 2000, and including patients with a diagnosis, patients at risk, healthy controls, or a combination of these groups, using different symptoms of clinically diagnosed schizophrenia or only psychotic experiences.

7.2.2 Genome-Wide GxE Research

Technological advances and a substantial reduction in the cost of genotyping has led to the increasing use of genome-wide association studies (GWAS) in psychiatric disorders. In contrast to the candidate gene approach, which assesses variation in a small number of loci, genome-wide studies attempt to capture variation across the entire genome by genotyping up to one million single nucleotide polymorphisms (SNPs). Each SNP is then examined for associations with the outcome of interest, with resulting P values corrected for multiple testing [135]. Many of the early GWAS studies failed to detect significant associations between genetic variants and psychiatric disorders that survived the substantial correction for multiple testing. However, later data pooling and formation of consortia resulted in more replicable findings. For instance, while the initial studies of schizophrenia were only able to identify a single locus with a small effect on the disorder explaining less than 1% of the variance [106, 114], the most recent study with a much larger sample identified 108 loci explaining up to 7% [112].

Despite these encouraging GWAS findings, the total identified genetic influence is often very small compared to the more substantial heritability estimates found in twin studies (e.g., 80% for schizophrenia and 37% for major depression). A potential explanation for this missing heritability [82] is the influence of GxE interactions that are not detected when examining main effects of genes [126]. Subsequent GWAS-based research has considered GxE in two ways: A) by applying genome-wide gene-environment interaction study (GWEIS) designs, or B), by using polygenic score gene-by environment interaction (PGSxE) designs.

The GWEIS approach is similar to the candidate GxE approach, except that GxE effects are examined for all available variants across the genome, rather than for a single candidate [5]. Only a handful of GWEISs have been conducted to date, all of which have focused on depressive symptoms [43, 61, 97]. Dunn et al. [43] conducted one the largest GWEIS studies, investigating social support and Stressful Life Events (SLEs) as environmental factors implicated in depressive symptoms in a sample of over 10,000 women of African American and Hispanic/Latina descent. They found a significant interaction between a genetic variant located near the *CEP350* gene and SLEs, predicting increased depressive symptoms in African Americans. The findings, however, could not be replicated in an independent sample. Similarly, while the other two studies found a significant interaction between SLEs and bone morphogenetic protein 2 (*BMP2*) gene [61] and rs10510057 polymorphism [97] in the prediction of depressive symptoms, the results have not yet been examined in an independent sample in the former, and were not robust to statistical adjustment for significance threshold in the latter.

The lack of replication may reflect the main limitation of the genome-wide approaches and highlights the requirement for very large sample sizes, and the need for more stringent statistical thresholds [3, 5]. More stringent statistical criteria are required in GWIES to counter the potential problem of increased rates of false positive findings given the typical testing of over one million SNPs for their interaction with an environmental factor. The significance threshold for genome-wide findings are therefore commonly adjusted to $p < 5 \times 10^{-8}$ [98]. While the multiple testing correction addresses the type I error rates, it increases the possibility of false negative results, especially in cases where the sample is underpowered to detect the very small effect sizes at this high significance threshold level.

In the polygenic score approach to GxE, the *aggregate* effects of all available variants on a given outcome are summarised into a single polygenic score (PGS). This is derived by adding up the number of trait-associated alleles in each individual, weighted by effect size from the GWAS discovery sample [41].The PGS is subsequently tested as a moderator of the environmental variable of interest in a standard GxE model. Not only does this approach alleviate problems with multiple testing, it is also more compatible with the current understanding of the aetiology of complex traits being due to the combined effects of many variants, each exerting a small effect [36]. However, information on the specific contribution of individual gene and gene variants is no longer available with this approach.

Only a handful of studies have so far used a PGSxE approach in relation to mental health, either studying schizophrenia as a function of the interaction between PGS and childhood maltreatment [125] or cannabis use [49], or depressive symptoms/major depressive disorder (MDD) as a function of the interaction of PGS with childhood maltreatment or SLEs [39, 89, 92, 100].

Regarding Schizophrenia, French et al. [49] investigated the interaction between cannabis use, PGS of schizophrenia and cortical thickness. Findings suggested that in males, higher PGS scores were associated with lower cortical thickness in the context of cannabis use during early adolescence. This was not observed in males with a low PGS for schizophrenia. In the only other PGSxE study on schizophrenia, Trotta et al. [125] found that higher PGS for schizophrenia and childhood adversity predicted psychosis status but there were no significant interaction effects, probably due to the relatively small sample size ($N = 200$).

With regards to depression, two studies to date have examined whether polygenic scores of MDD moderate the effect of maltreatment on MDD case/control status [89, 100]. Both studies found a significant interaction between the PGS and childhood maltreatment, though the pattern of this interaction differed across studies. Peyrot et al. [100] found that the effects of maltreatment on MDD were greater for those with a *higher* PGS for MDD, whereas Mullins et al. [89] found that the effects were greater for those with a *lower* PGS. It is unclear why the findings were contradictory, though differences in study design, such as how maltreatment was measured (self-report vs. clinical interviews) may be an important contributing factor. Three other studies have examined interactions between adult SLEs and PGS of depression on depressive symptoms [39, 92], and MDD [89]. While Mullins et al. [89] and Musliner et al. [92] did not find a significant interaction between PGS and SLEs on depression, the longitudinal analysis in Domingue et al. [39]'s study yielded a significant interaction effect between death of a spouse and PGS for depression. Specifically, they investigated whether several polygenic scores (including MDD, depressive symptoms and subjective wellbeing) moderated the effects of the death of a spouse on changes in depressive symptoms, in over 8000 married older adults. The authors reported that those who experienced the death of a spouse showed a significant increase in depressive symptoms, but returned to baseline levels 2 years after the bereavement. The increase in depression was significantly greater in those with a higher PGS for MDD/depressive symptoms, but significantly lower for those with a higher wellbeing PGS. Those with a higher wellbeing PGS also experienced a more rapid return to baseline levels. The detection of GxE effects in this study was likely facilitated by the longitudinal assessment of the outcome and consideration of within-person changes, rather than merely looking at cross-sectional GxE effects, an important consideration for future GxE studies.

Overall, research findings of both candidate and genome-wide studies over the past two decades indicate that both genetic and environmental influences play an important role in explaining individual differences in susceptibility to mental health disorders. Despite the fact that GxE research has been crucial for our understanding of the interplay between genetic and environmental factors in risk for psychiatric disorders, it is important to evaluate the findings in the context of several limitations, a brief discussion of which is included in the following sections.

7.3 Critical Evaluation of GxE Research in Psychiatric Disorders

7.3.1 Theoretical Models of Gene-Environment Interaction

Traditionally, the guiding theoretical framework for the majority of GxE studies has been the *Diathesis-Stress* model [88]. According to this model, the adverse effects of environmental influences lead to psychopathology predominately in the presence of specific inherent factors that renders individuals more vulnerable. Such individual-specific factors include those that pertain to biological differences between individuals, including genetic factors. Therefore, according to the *Diathesis-Stress* model, some individuals are more vulnerable to adverse environmental influences, as a function of their specific biological (e.g. genetic) and/or psychological (temperament) make up, but this vulnerability does not necessarily predict psychopathology in the absence of adversity. Individuals who do not succumb to the effects of environmental stressors, either as function of not having these genetic vulnerabilities or due to the presence of other protective factors, are considered resilient. The *Diathesis-Stress* concept of GxE therefore adopts a genetic *vulnerability* perspective, whereby an environmental stressor increases risk for psychopathology only in those groups of individuals who carry a vulnerability/risk genotype.

However, the *Diathesis-Stress* conceptualization of GxE has recently been challenged by *Environmental Sensitivity,* models including Sensory Processing Sensitivity: Aron and Aron [4]; Differential susceptibility Theory: Belsky and Pluess [14]; Biological Sensitvity to Context: Boyce and Ellis [20]**.** Each of these models suggest that some individuals are *generally more sensitive* to both negative *and* positive environmental influences as a function of their genotypes. This means that genetically more sensitive individuals are more likely to develop psychopathology in response to the adverse effects of stressors, but also disproportionately likely to benefit from positive and supportive environmental influences. Higher sensitivity therefore functions in a *"for better and for worse"* manner [12, 13], rather than a risk-only model as *Diathesis-Stress* suggests. Related to the notion of *Differential Susceptibility* is the *Vantage Sensitivity* [103] model, which refers to differences in sensitivity to exclusively positive aspects of the environment (not making predictions about individual differences in the response to adversity). That is, certain genotypes may influence who benefits from protective/positive environmental factors such as greater social support or psychological interventions.

The *Differential Susceptibility* hypothesis is an evolutionary-inspired developmental model, which suggests that natural selection led to the emergence of both high and low susceptibility types, each characterised by specific advantages and disadvantages. While higher susceptibility may increase reproductive fitness through enhanced adaptation to the environment, this is counterbalanced by the risk of decreased fitness due to heightened vulnerability to adversity (e.g. see: [37, 47, 115]). In other words, both high and low susceptibility have evolutionary pros and cons, depending on the quality of the environment. This explains why some of the genetic variants associated with vulnerability to adversity are found at a high frequency in the general population despite their apparent deleterious effects [15].

More recent GxE studies of mental health have adopted a perspective of *Environmental Sensitivity*, providing support for the cross-over interaction effects of genetic factors traditionally understood as vulnerability, indicating these may reflect general sensitivity to the environment. Meta-analyses of studies with serotonin transporter gene variants [129] and dopamine-related genes [7] have found GxE interaction patterns consistent with *Environmental Sensitivity* theory. For example, the *5-HTTLPR* s-allele has been associated with higher neuroticism in the context of negative life events, but also found to be associated with lower levels of neuroticism in the context of positive life events [104]. Elsewhere, the same genotype has been found to moderate, *for better and for worse*, the impact of parenting practices on children's positive affect [57], and of perceived racial discrimination and child maltreatment on conduct problems and antisocial behaviour [31]. In other studies with *DRD4* as marker of sensitivity, higher genetic sensitivity (*DRD4* 7-repeat genotype) was associated with higher inattention in the context of insensitive early maternal care, but also with lower levels of inattention in the context of more sensitive maternal care [17], with development of social competence in interaction with quality of child-care [16], and prosocial behaviour [75] and children's externalizing behaviour [9] in interaction with parenting practices. A meta-analysis of experimental studies by van Ijzendoorn and Bakermans-Kranenburg [130], which included 22 experimental GxE studies (N = 3000), with *DRD4* and *5-HTTLPR* as the sensitivity genes and a range of developmental outcomes including externalizing problems, internalizing behaviours, and cognitive development, further supports the hypothesised increased sensitivity to positive experiences.

The various frameworks for GxE are not necessarily competitive or contrary in explaining the interaction between genotype and environment on outcomes. Indeed, it is likely that the gross genetic risk for psychiatric disorders includes a mix of genetic variants that operate in various ways. That is, genetic risk for a psychiatric disorder may include variants that have a direct effect on psychiatric disorders, variants that increase risk by increasing sensitivity to adversity (i.e., *Diathesis-Stress*), variants that increase risk by decreasing sensitivity to protective factors such as social support (i.e., *Vantage Resistance*), as well as variants that increase risk in adverse environments but decrease risk in protective environments (i.e., *Differential Susceptibility*). While there is no research to date that examined an integrated GxE model which takes all the different GxE patterns into account, current empirical evidence supports the existence and relevance of all three GxE interaction models for explaining the link between environmental factors, genotype and mental health outcomes. Hence, considering the various ways (i.e., patterns) in which genetic factors interact with environmental influences to bring about mental health problems or protect against them is an important point for further investigation.

7.3.2 Methodological Limitations of Candidate and Genome-Wide GxE Studies

Despite the important contribution of GxE studies to the field of psychiatry, it is important to acknowledge several limitations. First, the main requirement of the candidate gene approach is the selection of candidate genes based on strong biological hypotheses. However, knowledge regarding the specific biological mechanisms underlying complex psychiatric disorders remains surprisingly limited and, therefore, the risk of selecting the wrong candidates is high. Furthermore, a publication bias for significant novel results over null or negative results has been reported [19, 32], which may make candidate genes appear more robust than they actually are. In addition, small sample sizes in candidate gene studies and the lack of sufficient statistical power to detect small interaction effects [42, 91] further increases the probability of false positive results [42, 91]. The low power is especially an issue considering that the genetic architecture of common psychiatric traits is most probably highly complex and including many variants of very small effect [36, 40]. Consequently, the reliable detection of a genuine effect of a single variant requires larger samples. Relatedly, and of particular concern, is the general difficulty in replicating candidate gene findings, perhaps as a function of the limitations noted herein. Together, these limitations require more cautious interpretation of the role of these specific candidate genes in psychiatric disorders.

Second, despite genome-wide GxE approaches addressing some of the main limitations of the candidate gene approach (i.e., the requirement for an a-priori biological hypothesis), the hypothesis-free approach of GWAS increases the possibility of false negative results due to stringent correction for multiple testing, and more robust findings requires very large sample sizes (over one million) that are yet to be assembled in psychiatric genetics research. The PGS approach does not necessitate the same stringent criteria for multiple testing correction, because the SNPs are not considered for their singular contribution to the trait. However, the often low power at the initial stage of PGS construction, impacts the down-stream processes when summary statistics are used to arrive at the final score. Importantly, although this approach is conceptually more in line with the highly complex genetic architecture of psychiatric disorders, none of the studies conducted so far have followed up on the biological mechanisms that implicate these variants in response to environmental influences and development of disorder. An important limitation that must be addressed in future research. Finally, it must be acknowledged that genome-wide approaches are still in the early stage, and until further research can address the current inconsistencies in the results, the findings remain exploratory.

Third, the majority of GxE studies have been conducted from a *Diathesis-Stress* perspective, an approach that requires re-evaluation in light of recent research suggesting that many of the common genetic variants in these studies seem to reflect generally heightened susceptibility to various environmental influences, rather than vulnerability for the specific disorders being studied. Incorporating a perspective of *Environmental Sensitivity* is therefore crucial when interpreting the role of genetic factors in the development of mental health disorders [102].

Finally, both genome wide and candidate GxE studies only explain a very small proportion of the variance in psychopathology. Therefore, although GxE research has had important implications for theoretical models of disease, knowledge of an individual's genotype does not (yet) provide greater accuracy in predicting who might develop mental health problems in response to life stressors and trauma (e.g. [89, 92]). Caution must be taken therefore when considering the relevance of GxE in mental health outcomes, though that is not to say that future research won't prove more successful.

7.3.3 Gene-Environment Correlations

One of the biggest challenges for gene-environment interplay research is the disentangling of rGE and GxE [105]. rGE can potentially mask the identification of GxE (or vice versa) due to the difficulty in assessing whether the gene modifies environmental effects (GxE), or whether the genetic risk is more prevalent in particular environments (rGE) [111]. For instance, does genetic risk for depression make an individual more susceptible to the adverse effects of negative life events (GxE) or does a genetic predisposition for depression simply increase the individual's likelihood of selecting unfavourable environments (rGE)? Lau and Eley [76] investigated this by examining the development of depressive symptoms in adolescents. They demonstrated that adolescents with depressive genotypes may be subject to the experience of increased social adversity due to rGE, but are at the same time also more susceptible to the development of depressive symptoms in response to these factors due to GxE. The findings suggest that psychiatric conditions may be shaped by the combined effects of different types of gene-environment interplay [76, 111]. One reason that it is difficult to differentiate GxE from rGE, is that a large portion of gene-environment interplay studies investigated rGE and GxE independently [136], and often tried to exclude rGE by only selecting environmental factors which the individual and their family cannot have influence over [116]. It is however challenging to exclude rGE effects, since family environments are also heritable [136] and rGE and GxE often co-exist for complex mental health conditions, such as depression [44]. Furthermore, focusing on either rGE or GxE alone can sometimes lead to conflicting findings. For example, Narusyte et al. [94] suggest that maternal criticism could be the result of evocative rGE emanating externalising behaviour from their offspring. In contrast, Cadoret et al. [24] proposed in an earlier adoption study that adverse home environments may increase conduct disorder problems and aggression in children with a genetic risk for antisocial behaviour, which is consistent with GxE.

Identifying which form of gene-environment interplay contributes to a particular disorder or behaviour is absolutely crucial in order to select suitable intervention efforts [62]. For example, Leve et al. [78] explain in a child gene-environment interplay study that interventions often focus on improving parenting processes, with the aim to reduce proximal or immediate risks in families or school setting. But such interventions may not work equally well for all individuals, due to genetic differences in *Vantage Sensitivity* [78]. Looking back at our example of antisocial behaviour and externalisation symptoms in children and adolescents, this would mean that improving adverse home environments alone will not necessarily be a successful intervention for these children and adolescents, due to the combined effects of evocative rGE and GxE.

7.3.4 GxE From a Life Course Perspective

The impact of environmental influences likely changes throughout the course of life, with greater impact for adverse events being observed in earlier stages of development. Indeed, several studies and meta-analyses of candidate genes such as *5-HTTLPR*, *BDNF*, *DRD4* and *FKBP5* suggest that early environments are of greater importance when assessing GxE for psychiatric disorders [1, 7, 50, 129].

While the effects of early environments on development of mental health problems later in life is widely studied, only a handful of studies have examined the possible interactions across various stages of development, and how effects may vary as a function of genotype. For example, in a multi-wave longitudinal study, using a PGS of *Environmental Sensitivity* derived from nine candidate gene variants, and a measure of childhood and adult material environments, Keers and Pluess [68] found evidence for environment-by-environment interaction (ExE) and gene-by-environment-by-environment (GxExE) interactions predicting psychological distress in adulthood, but no GxE effects. The findings suggest that poor childhood material environment exacerbated the negative impact of poor adulthood environment on psychological distress. Importantly, this interaction was moderated by an individual's genetic sensitivity, whereby more sensitive individuals showed greater levels of distress in the context of poor childhood and poor adulthood environmental quality, but lower levels of distress if the childhood material environmental quality was high. In a recent GxE life-course study using the same data as Keers & Pluess [68], Assary et al. [6] used a genome-wide polygenic score of sensitivity and quality of psychosocial environment to predict psychological distress from ages 7–50 years old. The results showed a significant effect of time on the GxE interactions, whereby the interaction between poor psychosocial environmental quality and genetic sensitivity predicted more psychological distress in childhood, but less in adulthood. The results highlight the dynamic nature of GxE in mental health outcomes across the life span and that developmental changes across life require closer inspection when investigating GxE.

7.3.5 Specificity of GxE Effects in Predicting Psychiatric Disorders

Considering the high comorbidity of psychiatric disorders [72], it is not surprising to find that many of the previously noted candidate genes have been examined and associated with multiple disorders. *BDNF*, for instance, has been found to be related to depression and bipolar disorder, as well as to schizophrenia [2, 59, 60]. The involvement of these candidate genes in multiple disorders may indicate a shared genetic aetiology. Indeed, GWAS of multiple disorders have found shared genetic influences between commonly comorbid/symptomatically overlapping disorders, for example schizophrenia, major depression, bipolar disorder, autism and ADHD [35]. The shared genetic aetiology and involvement of the same candidate genes in multiple disorders may implicate these genes as potentiating general vulnerability to a wide range of psychopathologies, and that the specific course that such an initial general genetic vulnerability takes may depend on other factors such as specific environmental influences. Relatedly, GxE research tends to focus largely on the same environmental factors (i.e., stressful life events, childhood maltreatment) and the same diagnostically distinct disorders (e.g., schizophrenia, major depression, autism). Therefore, while GxE findings have demonstrated that genetic differences play an important role in influencing the trajectory of environmental adversity from severe risk for disorders in some individuals to minor risk in others, these studies are yet to specify the specific environmental influences that implicate a specific disorder.

7.4 Future Directions

To improve on some of the limitations of the research reviewed so far, future studies may benefit from incorporating some of the ideas presented in the following sections. These include studying intermediate phenotypes, using experimental study designs and taking a life course approach to study GxE interaction effects in the prediction and development of psychiatric disorders.

7.4.1 *Intermediate Phenotypes*

Due to the complex genetics underlying psychiatric disorders, studying intermediate phenotypes, i.e., *endophenotypes*, which are constructs that exist along the pathway between a genotype and a phenotype, is suggested to represent an important strategy for investigating pathology and uncovering new susceptibility genes [25, 52, 86]. Gottesman and Gould [52] suggested that endophenotypes could be used to dissect the genetics of particular psychiatric disorders into smaller and more manageable components, likely having simpler genetic architectures than the complex disorder of interest.

Cognitive biases, such as attention bias and interpretation bias, have been examined as potential cognitive endophenotypes for anxiety and depression, respectively. In particular, attentional bias towards threat has been consistently linked to anxiety [10, 83]), whilst interpretation biases have been said to play a possible causal role in the onset and maintenance of depression [63]. Experimental manipulation of attentional bias, using computer based attentional bias modification (ABM) techniques, has therefore shown promise for research into the mechanisms of psychopathology [55, 56, 81].

Cognitive theories of depression and anxiety also suggest that biased cognitions may increase risk for psychopathology by increasing individuals' emotional reactivity to both negative and positive influences. However, to date, only a small number of prospective studies regarding cognitive biases and emotional reactivity have been conducted, all of which suggest that attentional bias towards threat predict reactivity to real world [80, 107, 128, 134], and laboratory induced stressors [95]. Given that the pathways from genotype to phenotype in psychiatric disorders are relatively indirect, discerning endophenotypes that exist closer to the link with genetic variation may be an excellent way of improving attempts to associate genes with disorders.

7.4.2 *Experimental Studies*

Experimental study designs such as randomised controlled trials (RCT) represent a powerful method of testing GxE. Assessing the moderating effect of genetic variants in a manipulated, or experimental environment, known as gene-by-experimental environment interaction (GxeE) [8, 130, 131], accounts for the confounding effects of rGE, as participants cannot select environments that correlate with their genotype. Furthermore, this approach also reduces measurement error, requires fewer participants, and has greater statistical power than correlational studies [84]. In one such candidate gene study testing the *for-better-or-for-worse* hypothesis of *Differential Susceptibility*, allelic variation of *5-HTTLPR* was examined as a possible predictor of sensitivity to a computer based ABM task designed to induce either a bias towards negative or positive affective pictures [46]. Findings provided evidence for *5-HTTLPR* as sensitivity rather than vulnerability factor. Specifically, the authors showed that s-allele carriers of the 5-HTTLPR developed stronger biases towards negative and positive affective pictures when compared to individuals homozygous for the l-allele. This suggests that in line with the differential susceptibility hypothesis, s-allele carriers may be more sensitive to both negative *and* positive environmental influences and therefore acquire either negative or positive biases depending on the nature of the environmental context.

This type of approach has been extended to genome-wide GxE studies, for example, by investigating individual differences in response to cognitive behavioural therapy (CBT). In a recent study using genome-wide PGS of Environmental Sensitivity, Keers et al. [67] compared differences in the response to CBT treatment as a function of genetic sensitivity and found that a PGS that increases sensitivity to the environment also moderated the response to treatment. The more genetically sensitive individual showed more discriminant response to treatment type, with intensive treatments in the genetically sensitive individuals leading to larger reductions in anxiety symptoms. The logical next step in experimental GxE studies, following on from candidate gene and whole-genome studies conducted thus far, would be to extend the G component through the inclusion of genetic pathways, whilst the inclusion of methylation and gene expression data would serve to extend the E component of GxE tests [8].

7.4.3 Alternative Theoretical Models of Gene-Environment Interaction

As noted earlier, future research may benefit from considering alternative GxE models of individual differences in response to environmental influences. Investigating GxE from the perspective of *Environmental Sensitivity,* rather than exclusively *Diathesis-Stress,* represents a promising approach which will not only improve our understanding of genetic susceptibility to environmental influences but also address issues regarding what works for whom in intervention settings (i.e., *Vantage Sensitivity*).

Furthermore, future research should also consider the different way genes and environment influence each other when investigating mental health outcomes. A joint analysis of rGE and GxE through multi-disciplinary approaches may lead to further recognition of co-occurrence and help ensuring that the outcomes of one do not bias the effects of the other. This would help us to provide more effective targeted medical interventions and allow for better prevention of mental health problems during crucial development windows in order to significantly improve the quality of life of at-risk individuals.

7.5 Conclusion

Despite limitations in current GxE studies of psychiatric disorders, the research has been highly valuable in terms of taking a first step in identifying genetic variants implicated in the risk for psychiatric disorders, and in explaining individual differences in response to both adverse and beneficial environmental influences. Genome-wide GxE studies of psychiatric disorders, although few, will likely uncover more of the aetiology of psychiatric disorders as sample sizes become larger, alternative theoretical models of diseases are considered, and more efficient statistical methodologies are applied. However, the challenge that remains is to combine these new approaches with detailed and accurate longitudinal measurements of the environment as well as outcomes across the life span in order to get a clearer understanding of the development of psychiatric disorders and better prediction of individual trajectories in response to adversity and support.

References

1. Aguilera M, Arias B, Wichers M, Barrantes-Vidal N, Moya J, Villa H, et al. Early adversity and 5-HTT/BDNF genes: new evidence of gene–environment interactions on depressive symptoms in a general population. Psychol Med. 2009;39(09):1425–32.
2. Alemany S, Arias B, Aguilera M, Villa H, Moya J, Ibanez MI, et al. Childhood abuse, the BDNF-Val66Met polymorphism and adult psychotic-like experiences. Br J Psychiatry. 2011;199(1):38–42. https://doi.org/10.1192/bjp.bp.110.083808.
3. Almli LM, Duncan R, Feng H, Ghosh D, Binder EB, Bradley B, et al. Correcting systematic inflation in genetic association tests that consider interaction effects: application to a genome-wide association study of posttraumatic stress disorder. JAMA Psychiat. 2014;71(12):1392–9.
4. Aron EN, Aron A. Sensory-processing sensitivity and its relation to introversion and emotionality. J Pers Soc Psychol. 1997;73(2):345–68.
5. Aschard H, Lutz S, Maus B, Duell EJ, Fingerlin TE, Chatterjee N, et al. Challenges and opportunities in genome-wide environmental interaction (GWEI) studies. Hum Genet. 2012;131(10):1591–613.
6. Assary E, Keers R, Krapohl E, Pluess M. Polygenic score of environmental senstivity moderates the effects of psychosocial environment on psychological distress across life-span. in preparation.
7. Bakermans-Kranenburg MJ, van Ijzendoorn MH. Differential susceptibility to rearing environment depending on dopamine-related genes: new evidence and a meta-analysis. Dev Psychopathol. 2011;23(01):39–52.
8. Bakermans-Kranenburg MJ, van IJzendoorn MH. The hidden efficacy of interventions: gene × environment experiments from a differential susceptibility perspective. Annu Rev Psychol. 2015;66:381–409.
9. Bakermans-Kranenburg MJ, Van IMH, Pijlman FT, Mesman J, Juffer F. Experimental evidence for differential susceptibility: dopamine D4 receptor polymorphism (DRD4 VNTR) moderates intervention effects on toddlers' externalizing behavior in a randomized controlled trial. Dev Psychol. 2008;44(1):293–300. https://doi.org/10.1037/0012-1649.44.1.293.
10. Bar-Haim Y, Lamy D, Pergamin L, Bakermans-Kranenburg MJ, Van Ijzendoorn MH. Threat-related attentional bias in anxious and nonanxious individuals: a meta-analytic study. Psychol Bull. 2007;133(1):1–24.

11. Battle CL, Shea MT, Johnson DM, Yen S, Zlotnick C, Zanarini MC, et al. Childhood maltreatment associated with adult personality disorders: findings from the collaborative longitudinal personality disorders study. J Personal Disord. 2004;18(2):193–211. https://doi.org/10.1521/pedi.18.2.193.32777.
12. Belsky J, Pluess M. Beyond diathesis stress: differential susceptibility to environmental influences. Psychol Bull. 2009;135(6):885–908. https://doi.org/10.1037/a0017376.
13. Belsky J, Pluess M. Beyond risk, resilience, and dysregulation: phenotypic plasticity and human development. Dev Psychopathol. 2013a;25(4 Pt 2):1243–61. https://doi.org/10.1017/S095457941300059X.
14. Belsky J, Pluess M. Genetic moderation of early child-care effects on social functioning across childhood: a developmental analysis. Child Dev. 2013b;84(4):1209–25. https://doi.org/10.1111/cdev.12058.
15. Belsky J, Bakermans-Kranenburg MJ, van IJzendoorn MH. For better and for worse: differential susceptibility to environmental influences. Curr Dir Psychol Sci. 2007a;16(6):300–4. https://doi.org/10.1111/j.1467-8721.2007.00525.x.
16. Belsky J, Fearon RMP, Bell B. Parenting, attention and externalizing problems: testing mediation longitudinally, repeatedly and reciprocally. J Child Psychol Psychiatry. 2007b;48(12):1233–42.
17. Berry D, Deater-Deckard K, McCartney K, Wang Z, Petrill SA. Gene-environment interaction between dopamine receptor D4 7-repeat polymorphism and early maternal sensitivity predicts inattention trajectories across middle childhood. Dev Psychopathol. 2013;25(2):291–306. https://doi.org/10.1017/s095457941200106x.
18. Bertelsen A, Harvald B, Hauge M. A Danish twin study of manic-depressive disorders. Br J Psychiatry. 1977;130(4):330–51.
19. Bosker FJ, Hartman CA, Nolte IM, Prins BP, Terpstra P, Posthuma D, et al. Poor replication of candidate genes for major depressive disorder using genome-wide association data. Mol Psychiatry. 2011;16(5):516–32. https://doi.org/10.1038/mp.2010.38.
20. Boyce WT, Ellis BJ. Biological sensitivity to context: I. An evolutionary–developmental theory of the origins and functions of stress reactivity. Dev Psychopathol. 2005;17(02):271–301.
21. Brunner HG, Nelen M, Breakefield X, Ropers H, Van Oost B. Abnormal behavior associated with a point mutation in the structural gene for monoamine oxidase A. Science. 1993;262:578–80.
22. Burt A. A mechanistic explanation of popularity: genes, rule breaking, and evocative gene-environment correlations. J Pers Soc Psychol. 2009;96(4):783–94. https://doi.org/10.1037/a0013702.
23. Byrd AL, Manuck SB. MAOA, childhood maltreatment, and antisocial behavior: meta-analysis of a gene-environment interaction. Biol Psychiatry. 2014;75(1):9–17. https://doi.org/10.1016/j.biopsych.2013.05.004.
24. Cadoret RJ, Yates WR, Troughton E, Woodworth G, Stewart MA. Genetic-environmental interaction in the genesis of aggressivity and conduct disorders. Arch Gen Psychiatry. 1995;52(11):916–24.
25. Cannon TD, Keller MC. Endophenotypes in the genetic analyses of mental disorders. Annu Rev Clin Psychol. 2006;2:267–90.
26. Cases O, Seif I, Grimsby J, Gaspar P, Chen K, Pournin S, et al. Aggressive behavior and altered amounts of brain serotonin and norepinephrine in mice lacking MAOA. Science. 1995;268(5218):1763–6. https://doi.org/10.1126/science.7792602.
27. Caspi A, McClay J, Moffitt TE, Mill J, Martin J, Craig IW, et al. Role of genotype in the cycle of violence in maltreated children. Science. 2002;297(5582):851–4.
28. Caspi A, Sugden K, Moffitt TE, Taylor A, Craig IW, Harrington H, et al. Influence of life stress on depression: moderation by a polymorphism in the 5-HTT gene. Science. 2003;301(5631):386–9.
29. Caspi A, Moffitt TE, Cannon M, McClay J, Murray R, Harrington H, et al. Moderation of the effect of adolescent-onset cannabis use on adult psychosis by a functional polymorphism in the catechol-O-methyltransferase gene: longitudinal evidence of a gene X environment interaction. Biol Psychiatry. 2005;57(10):1117–27.

30. Cervilla JA, Molina E, Rivera M, Torres-Gonzalez F, Bellon JA, Moreno B, et al. The risk for depression conferred by stressful life events is modified by variation at the serotonin transporter 5HTTLPR genotype: evidence from the Spanish PREDICT-gene cohort. Mol Psychiatry. 2007;12(8):748–55.
31. Cicchetti D, Rogosch FA, Thibodeau EL. The effects of child maltreatment on early signs of antisocial behavior: genetic moderation by tryptophan hydroxylase, serotonin transporter, and monoamine oxidase a genes. Dev Psychopathol. 2012;24(3):907–28.
32. Collins PY, Patel V, Joestl SS, March D, Insel TR, Daar AS, et al. Grand challenges in global mental health. Nature. 2011;475(7354):27–30. https://doi.org/10.1038/475027a.
33. Collins AL, Kim Y, Sklar P, O'Donovan MC, Sullivan PF. Hypothesis-driven candidate genes for schizophrenia compared to genome-wide association results. Psychol Med. 2012;42(3):607–16. https://doi.org/10.1017/s0033291711001607.
34. Collip D, van Winkel R, Peerbooms O, Lataster T, Thewissen V, Lardinois M, et al. COMT Val158Met-stress interaction in psychosis: role of background psychosis risk. CNS Neurosci Ther. 2011;17(6):612–9. https://doi.org/10.1111/j.1755-5949.2010.00213.x.
35. Cross-Disorder Group of the Psychiatric Genomics, C. (2013). Genetic relationship between five psychiatric disorders estimated from genome-wide SNPs. Nat Genet, 45(9), 984–994. doi:https://doi.org/10.1038/ng.2711.http://www.nature.com/ng/journal/v45/n9/abs/ng.2711.html – supplementary-information.
36. Culverhouse RC, Saccone NL, Horton AC, Ma Y, Anstey KJ, Banaschewski T, et al. Collaborative meta-analysis finds no evidence of a strong interaction between stress and 5-HTTLPR genotype contributing to the development of depression. Mol Psychiatry. 2017;23:133–42. https://doi.org/10.1038/mp.2017.44.
37. DeWitt TJ, Sih A, Wilson DS. Costs and limits of phenotypic plasticity. Trends Ecol Evol. 1998;13(2):77–81. https://doi.org/10.1016/s0169-5347(97)01274-3.
38. Dick DM, Rose RJ, Viken RJ, Kaprio J, Koskenvuo M. Exploring gene-environment interactions: socioregional moderation of alcohol use. J Abnorm Psychol. 2001;110(4):625–32.
39. Domingue BW, Liu H, Okbay A, Belsky DW. Genetic heterogeneity in depressive symptoms following the death of a spouse: polygenic score analysis of the US Health and Retirement Study. Am J Psychiatry. 2017; https://doi.org/10.1176/appi.ajp.2017.16111209.
40. Donnelly P. Progress and challenges in genome-wide association studies in humans. Nature. 2008;456(7223):728–31.
41. Dudbridge F. Power and predictive accuracy of polygenic risk scores. PLoS Genet. 2013;9(3):e1003348. https://doi.org/10.1371/journal.pgen.1003348.
42. Duncan LE, Keller MC. A critical review of the first 10 years of candidate gene-by-environment interaction research in psychiatry. Am J Psychiatr. 2011;168(10):1041–9. https://doi.org/10.1176/appi.ajp.2011.11020191.
43. Dunn EC, Wiste A, Radmanesh F, Almli LM, Gogarten SM, Sofer T, et al. Genome-Wide Association Study (gwas) and Genome-Wide by Environment Interaction Study (gweis) of depressive symptoms in african american and hispanic/latina women. Depress Anxiety. 2016;33(4):265–80.
44. Eaves L, Silberg J, Erkanli A. Resolving multiple epigenetic pathways to adolescent depression. J Child Psychol Psychiatry. 2003;44(7):1006–14.
45. Foley DL, Eaves LJ, Wormley B, Silberg JL, Maes HH, Kuhn J, Riley B. Childhood adversity, monoamine oxidase a genotype, and risk for conduct disorder. Arch Gen Psychiatry. 2004;61(7):738–44.
46. Fox E, Zougkou K, Ridgewell A, Garner K. The serotonin transporter gene alters sensitivity to attention bias modification: evidence for a plasticity gene. Biol Psychiatry. 2011;70(11):1049–54.
47. Frankenhuis WE, Panchanathan K. Individual differences in developmental plasticity may result from stochastic sampling. Perspect Psychol Sci. 2011;6(4):336–47. https://doi.org/10.1177/1745691611412602.
48. Frazzetto G, Di Lorenzo G, Carola V, Proietti L, Sokolowska E, Siracusano A, et al. Early trauma and increased risk for physical aggression during adulthood: the moderating role of MAOA genotype. PLoS One. 2007;2(5):e486.

49. French L, Gray C, Leonard G, Perron M, Pike GB, Richer L, et al. Early cannabis use, polygenic risk score for schizophrenia and brain maturation in adolescence. JAMA Psychiat. 2015;72(10):1002–11.
50. Gillespie NA, Whitfield JB, Williams B, Heath AC, Martin NG. The relationship between stressful life events, the serotonin transporter (5-HTTLPR) genotype and major depression. Psychol Med. 2005;35(01):101–11.
51. Gillespie CF, Phifer J, Bradley B, Ressler KJ. Risk and resilience: genetic and environmental influences on development of the stress response. Depress Anxiety. 2009;26(11):984–92.
52. Gottesman II, Gould TD. The endophenotype concept in psychiatry: etymology and strategic intentions. Am J Psychiatr. 2003;160(4):636–45.
53. Grossman MH, Emanuel BS, Budarf ML. Chromosomal mapping of the human catechol-O-methyltransferase gene to 22q11.1→q11.2. Genomics. 1992;12(4):822–5. https://doi.org/10.1016/0888-7543(92)90316-K.
54. Haberstick BC, Lessem JM, Hopfer CJ, Smolen A, Ehringer MA, Timberlake D, Hewitt JK. Monoamine oxidase a (MAOA) and antisocial behaviors in the presence of childhood and adolescent maltreatment. Am J Med Genet B Neuropsychiatr Genet. 2005;135B(1):59–64. https://doi.org/10.1002/ajmg.b.30176.
55. Hakamata Y, Lissek S, Bar-Haim Y, Britton JC, Fox NA, Leibenluft E, et al. Attention bias modification treatment: a meta-analysis toward the establishment of novel treatment for anxiety. Biol Psychiatry. 2010;68(11):982–90.
56. Hallion LS, Ruscio AM. A meta-analysis of the effect of cognitive bias modification on anxiety and depression, vol. 137: American Psychological Association; 2011. p. 940–58.
57. Hankin BL, Nederhof E, Oppenheimer CW, Jenness J, Young JF, Abela JRZ, et al. Differential susceptibility in youth: evidence that 5-HTTLPR x positive parenting is associated with positive affect 'for better and worse'. Transl Psychiatry. 2011;1:e44. https://doi.org/10.1038/tp.2011.44.
58. Henquet C, Rosa A, Krabbendam L, Papiol S, Fananas L, Drukker M, et al. An experimental study of catechol-O-Methyltransferase Val158Met moderation of [Delta]-9-Tetrahydrocannabinol-induced effects on psychosis and cognition. Neuropsychopharmacology. 2006;31(12):2748–57.
59. Hosang GM, Uher R, Keers R, Cohen-Woods S, Craig I, Korszun A, et al. Stressful life events and the brain-derived neurotrophic factor gene in bipolar disorder. J Affect Disord. 2010;125(1–3):345–9. https://doi.org/10.1016/j.jad.2010.01.071.
60. Hosang GM, Shiles C, Tansey KE, McGuffin P, Uher R. Interaction between stress and the BDNF Val66Met polymorphism in depression: a systematic review and meta-analysis. BMC Med. 2014;12:7. https://doi.org/10.1186/1741-7015-12-7.
61. Ikeda M, Shimasaki A, Takahashi A, Kondo K, Saito T, Kawase K, et al. Genome-wide environment interaction between depressive state and stressful life events. J Clin Psychiatry. 2016;77(1):e29–30.
62. Jaffee SR, Price TS. Gene-environment correlations: a review of the evidence and implications for prevention of mental illness. Mol Psychiatry. 2007;12(5):432–42. https://doi.org/10.1038/sj.mp.4001950.
63. Joormann J, Gotlib IH. Emotion regulation in depression: relation to cognitive inhibition. Cognit Emot. 2010;24(2):281–98.
64. Kantrowitz JT, Nolan KA, Sen S, Simen AA, Lachman HM, Bowers MB Jr. Adolescent cannabis use, psychosis and catechol-O-methyltransferase genotype in African Americans and Caucasians. Psychiatry Q. 2009;80(4):213–8. https://doi.org/10.1007/s11126-009-9108-4.
65. Karg K, Sen S. Gene × environment interaction models in psychiatric genetics. Curr Top Behav Neurosci. 2012;12:441 62. https://doi.org/10.1007/7854_2011_184.
66. Karg K, Burmeister M, Shedden K, Sen S. The serotonin transporter promoter variant (5-HTTLPR), stress, and depression meta-analysis revisited: evidence of genetic moderation. Arch Gen Psychiatry. 2011;68(5):444–54.
67. Keers R, Pluess M. Childhood quality influences genetic sensitivity to environmental influences across adulthood: a life-course gene× environment interaction study. Dev Psychopathol. 2017;29(5):1921–33.

68. Keers R, Coleman JR, Lester KJ, Roberts S, Breen G, Thastum M, et al. A genome-wide test of the differential susceptibility hypothesis reveals a genetic predictor of differential response to psychological treatments for child anxiety disorders. Psychother Psychosom. 2016;85(3):146–58.
69. Kendler KS, Eaves LJ. Models for the joint effect of genotype and environment on liability to psychiatric illness. Am J Psychiatry. 1986;
70. Kendler KS, Myers J. The genetic and environmental relationship between major depression and the five-factor model of personality. Psychol Med. 2010;40(5):801–6. https://doi.org/10.1017/S0033291709991140.
71. Kendler KS, Kessler RC, Walters EE, MacLean C, Neale MC, Heath AC, Eaves LJ. Stressful life events, genetic liability, and onset of an episode of major depression in women. Am J Psychiatry. 1995;152(6):833–42. https://doi.org/10.1176/ajp.152.6.833.
72. Kessler RC, McGonagle KA, Zhao S, Nelson CB, Hughes M, Eshleman S, et al. Lifetime and 12-month prevalence of DSM-III-R psychiatric disorders in the United States. Results from the National Comorbidity Survey. Arch Gen Psychiatry. 1994;51(1):8–19.
73. Kim-Cohen J, Caspi A, Taylor A, Williams B, Newcombe R, Craig IW, Moffitt TE. MAOA, maltreatment, and gene-environment interaction predicting children's mental health: new evidence and a meta-analysis. Mol Psychiatry. 2006;11(10):903–13.
74. Klahr AM, Thomas KM, Hopwood CJ, Klump KL, Burt SA. Evocative gene-environment correlation in the mother-child relationship: a twin study of interpersonal processes. Dev Psychopathol. 2013;25(1):105–18. https://doi.org/10.1017/S0954579412000934.
75. Knafo A, Israel S, Ebstein RP. Heritability of children's prosocial behavior and differential susceptibility to parenting by variation in the dopamine receptor D4 gene. Dev Psychopathol. 2011;23(1):53–67. https://doi.org/10.1017/s0954579410000647.
76. Lau JY, Eley TC. Disentangling gene-environment correlations and interactions on adolescent depressive symptoms. J Child Psychol Psychiatry. 2008;49(2):142–50. https://doi.org/10.1111/j.1469-7610.2007.01803.x.
77. Lemery-Chalfant K, Kao K, Swann G, Goldsmith HH. Childhood temperament: passive gene-environment correlation, gene-environment interaction, and the hidden importance of the family environment. Dev Psychopathol. 2013;25(1):51–63. https://doi.org/10.1017/S0954579412000892.
78. Leve LD, Neiderhiser JM, Scaramella LV, Reiss D. The early growth and development study: using the prospective adoption design to examine genotype-environment interplay. Xin Li Xue Bao. 2008;40(10):1106–15. https://doi.org/10.3724/SP.J.1041.2008.01106.
79. Lichtenstein P, Yip BH, Bjork C, Pawitan Y, Cannon TD, Sullivan PF, Hultman CM. Common genetic determinants of schizophrenia and bipolar disorder in Swedish families: a population-based study. Lancet. 2009;373(9659):234–9. https://doi.org/10.1016/s0140-6736(09)60072-6.
80. MacLeod C, Hagan R. Individual differences in the selective processing of threatening information, and emotional responses to a stressful life event. Behav Res Ther. 1992;30(2):151–61.
81. MacLeod C, Rutherford E, Campbell L, Ebsworthy G, Holker L. Selective attention and emotional vulnerability: assessing the causal basis of their association through the experimental manipulation of attentional bias. J Abnorm Psychol. 2002;111(1):107–23.
82. Maher B. Personal genomes: the case of the missing heritability. Nature. 2008;456(7218):18–21. https://doi.org/10.1038/456018a.
83. Mathews A, MacLeod C. Cognitive vulnerability to emotional disorders. Annu Rev Clin Psychol. 2005;1:167–95.
84. McClelland GH, Judd CM. Statistical difficulties of detecting interactions and moderator effects. Psychol Bull. 1993;114(2):376–90.
85. McGuffin P, Rijsdijk F, Andrew M, Sham P, Katz R, Cardno A. The heritability of bipolar affective disorder and the genetic relationship to unipolar depression. Arch Gen Psychiatry. 2003;60(5):497–502. https://doi.org/10.1001/archpsyc.60.5.497.
86. Miller GA, Rockstroh B. Endophenotypes in psychopathology research: where do we stand? Annu Rev Clin Psychol. 2013;9:177–213.

87. Modinos G, Iyegbe C, Prata D, Rivera M, Kempton MJ, Valmaggia LR, et al. Molecular genetic gene–environment studies using candidate genes in schizophrenia: a systematic review. Schizophr Res. 2013;150(2):356–65.
88. Monroe SM, Simons AD. Diathesis-stress theories in the context of life stress research: implications for the depressive disorders. Psychol Bull. 1991;110(3):406–25. https://doi.org/10.1037/0033-2909.110.3.406.
89. Mullins N, Power R, Fisher H, Hanscombe K, Euesden J, Iniesta R, et al. Polygenic interactions with environmental adversity in the aetiology of major depressive disorder. Psychol Med. 2016;46(04):759–70.
90. Munafo MR, Flint J. Replication and heterogeneity in gene x environment interaction studies. Int J Neuropsychopharmacol. 2009;12(6):727–9. https://doi.org/10.1017/s1461145709000479.
91. Munafò MR, Durrant C, Lewis G, Flint J. Gene × environment interactions at the serotonin transporter locus. Biol Psychiatry. 2009;65(3):211–9. https://doi.org/10.1016/j.biopsych.2008.06.009.
92. Musliner KL, Seifuddin F, Judy JA, Pirooznia M, Goes FS, Zandi PP. Polygenic risk, stressful life events and depressive symptoms in older adults: a polygenic score analysis. Psychol Med. 2015;45(08):1709–20.
93. Nanni V, Uher R, Danese A. Childhood maltreatment predicts unfavorable course of illness and treatment outcome in depression: a meta-analysis. Am J Psychiatry. 2012;169(2):141–51. https://doi.org/10.1176/appi.ajp.2011.11020335.
94. Narusyte J, Neiderhiser JM, Andershed AK, D'Onofrio BM, Reiss D, Spotts E, et al. Parental criticism and externalizing behavior problems in adolescents: the role of environment and genotype-environment correlation. J Abnorm Psychol. 2011;120(2):365–76. https://doi.org/10.1037/a0021815.
95. Nay WT, Thorpe GL, Roberson-Nay R, Hecker JE, Sigmon ST. Attentional bias to threat and emotional response to biological challenge. J Anxiety Disord. 2004;18(5):609–27.
96. Organization, W. H. (2008). The global burden of disease: 2004 update. (9241563710). World Health Organization.
97. Otowa T, Kawamura Y, Tsutsumi A, Kawakami N, Kan C, Shimada T, et al. The first pilot genome-wide gene-environment study of depression in the Japanese population. PLoS One. 2016;11(8):e0160823.
98. Pe'er I, Yelensky R, Altshuler D, Daly MJ. Estimation of the multiple testing burden for genomewide association studies of nearly all common variants. Genet Epidemiol. 2008;32(4):381–5.
99. Peerbooms O, Rutten BPF, Collip D, Lardinois M, Lataster T, Thewissen V, et al. Evidence that interactive effects of COMT and MTHFR moderate psychotic response to environmental stress. Acta Psychiatr Scand. 2012;125(3):247–56. https://doi.org/10.1111/j.1600-0447.2011.01806.x.
100. Peyrot WJ, Milaneschi Y, Abdellaoui A, Sullivan PF, Hottenga JJ, Boomsma DI, Penninx BW. Effect of polygenic risk scores on depression in childhood trauma. Br J Psychiatry. 2014;205(2):113–9.
101. Plomin R, DeFries JC, Knopik VS, Neiderhiser JM. Top 10 replicated findings from Behavioral genetics. Perspect Psychol Sci. 2016;11(1):3–23. https://doi.org/10.1177/1745691615617439.
102. Pluess M, Belsky J. Conceptual issues in psychiatric gene-environment interaction research. Am J Psychiatr. 2012;169(2):222–3. https://doi.org/10.1176/appi.ajp.2011.11111614.
103. Pluess M, Belsky J. Vantage sensitivity: individual differences in response to positive experiences. Psychol Bull. 2013;139(4):901–16. https://doi.org/10.1037/a0030196.
104. Pluess M, Belsky J, Way BM, Taylor SE. 5-HTTLPR moderates effects of life events on neuroticism: differential susceptibility to environmental influences. Prog Neuro-Psychopharmacol Biol Psychiatry. 2010;34:1070–4.
105. Purcell S. Variance components models for gene-environment interaction in twin analysis. Twin Res. 2002;5(6):554–71. https://doi.org/10.1375/136905202762342026.

106. Purcell SM, Wray NR, Stone JL, Visscher PM, O'donovan MC, Sullivan PF, et al. Common polygenic variation contributes to risk of schizophrenia and bipolar disorder. Nature. 2009;460(7256):748–52.
107. Pury CL. Information-processing predictors of emotional response to stress. Cognit Emot. 2002;16(5):667–83.
108. Rice F, Lewis G, Harold GT, Thapar A. Examining the role of passive gene-environment correlation in childhood depression using a novel genetically sensitive design. Dev Psychopathol. 2013;25(1):37–50. https://doi.org/10.1017/S0954579412000880.
109. Risch N, Herrell R, Lehner T, Liang KY, Eaves L, Hoh J, et al. Interaction between the serotonin transporter gene (5-HTTLPR), stressful life events, and risk of depression: a meta-analysis. JAMA. 2009;301(23):2462–71.
110. Rutter M, Silberg J. Gene-environment interplay in relation to emotional and behavioral disturbance. Annu Rev Psychol. 2002;53:463–90. https://doi.org/10.1146/annurev.psych.53.100901.135223.
111. Rutter M, Moffitt TE, Caspi A. Gene-environment interplay and psychopathology: multiple varieties but real effects. J Child Psychol Psychiatry. 2006;47(3–4):226–61.
112. Schizophrenia Working Group of the Psychiatric Genomics, C. (2014). Biological insights from 108 schizophrenia-associated genetic loci. Nature, 511(7510), 421–427. doi:https://doi.org/10.1038/nature13595http://www.nature.com/nature/journal/v511/n7510/abs/nature13595.html - supplementary-information.
113. Sharpley CF, Palanisamy SK, Glyde NS, Dillingham PW, Agnew LL. An update on the interaction between the serotonin transporter promoter variant (5-HTTLPR), stress and depression, plus an exploration of non-confirming findings. Behav Brain Res. 2014;273:89–105.
114. Shi J, Levinson DF, Duan J, Sanders AR, Zheng Y, Pe'Er I, et al. Common variants on chromosome 6p22. 1 are associated with schizophrenia. Nature. 2009;460(7256):753–7.
115. Sih A. Effects of early stress on behavioral syndromes: an integrated adaptive perspective. Neurosci Biobehav Rev. 2011;35(7):1452–65. https://doi.org/10.1016/j.neubiorev.2011.03.015.
116. Silberg J, Rutter M, Neale M, Eaves L. Genetic moderation of environmental risk for depression and anxiety in adolescent girls. Br J Psychiatry. 2001;179:116–21.
117. Simons CJP, Wichers M, Derom C, Thiery E, Myin-Germeys I, Krabbendam L, Van Os J. Subtle gene–environment interactions driving paranoia in daily life. Genes Brain Behav. 2009;8(1):5–12. https://doi.org/10.1111/j.1601-183X.2008.00434.x.
118. Smoller JW, Finn CT. Family, twin, and adoption studies of bipolar disorder. Am J Med Genet C Semin Med Genet. 2003;123c(1):48–58. https://doi.org/10.1002/ajmg.c.20013.
119. Stefanis NC, Henquet C, Avramopoulos D, Smyrnis N, Evdokimidis I, Myin-Germeys I, et al. COMT Val158Met moderation of stress-induced psychosis. Psychol Med. 2007;37(11):1651–6.
120. Sullivan PF, Neale MC, Kendler KS. Genetic epidemiology of major depression: review and meta-analysis. Am J Psychiatry. 2000;157(10):1552–62.
121. Sullivan PF, Kendler KS, Neale MC. Schizophrenia as a complex trait: evidence from a meta-analysis of twin studies. Arch Gen Psychiatry. 2003;60(12):1187–92. https://doi.org/10.1001/archpsyc.60.12.1187.
122. Surtees PG, Wainwright NW, Willis-Owen SA, Luben R, Day NE, Flint J. Social adversity, the serotonin transporter (5-HTTLPR) polymorphism and major depressive disorder. Biol Psychiatry. 2006;59(3):224–9.
123. Taylor A, Kim-Cohen J. Meta-analysis of gene-environment interactions in developmental psychopathology. Dev Psychopathol. 2007;19(4):1029–37. https://doi.org/10.1017/S095457940700051X.
124. Teicher MH, Samson JA. Childhood maltreatment and psychopathology: a case for ecophenotypic variants as clinically and neurobiologically distinct subtypes. Am J Psychiatry. 2013;170(10):1114–33. https://doi.org/10.1176/appi.ajp.2013.12070957.
125. Trotta A, Iyegbe C, Di Forti M, Sham PC, Campbell DD, Cherny SS, et al. Interplay between schizophrenia polygenic risk score and childhood adversity in first-presentation psychotic disorder: a pilot study. PLoS One. 2016;11(9):e0163319.

126. Uher R. Gene–environment interactions in severe mental illness. Front Psych. 2014;5:48.
127. Uher R, McGuffin P. The moderation by the serotonin transporter gene of environmental adversity in the etiology of depression: 2009 update. Mol Psychiatry. 2010;15(1):18–22.
128. Van Den Hout M, Tenney N, Huygens K, Merckelbach H, Kindt M. Responding to subliminal threat cues is related to trait anxiety and emotional vulnerability: a successful replication of MacLeod and Hagan (1992). Behav Res Ther. 1995;33(4):451–4.
129. van Ijzendoorn MH, Bakermans-Kranenburg MJ. Genetic differential susceptibility on trial: meta-analytic support from randomized controlled experiments. Dev Psychopathol. 2015;27(01):151–62.
130. Van Ijzendoorn MH, Bakermans-Kranenburg MJ, Belsky J, Beach S, Brody G, Dodge KA, et al. Gene-by-environment experiments: a new approach to finding the missing heritability. Nat Rev Genet. 2011;12(12):881.
131. Van IJzendoorn M, Belsky J, Bakermans-Kranenburg M. Serotonin transporter genotype 5HTTLPR as a marker of differential susceptibility? A meta-analysis of child and adolescent gene-by-environment studies. Transl Psychiatry. 2012;2(8):e147.
132. van Os J, Kenis G, Rutten BPF. The environment and schizophrenia. Nature. 2010;468(7321):203–12.
133. van Winkel R, Henquet C, Rosa A, Papiol S, Fananas L, De Hert M, et al. Evidence that the COMT(Val158Met) polymorphism moderates sensitivity to stress in psychosis: an experience-sampling study. Am J Med Genet B Neuropsychiatr Genet. 2008;147B(1):10–7. https://doi.org/10.1002/ajmg.b.30559.
134. Verhaak CM, Smeenk JM, van Minnen A, Kraaimaat FW. Neuroticism, preattentive and attentional biases towards threat, and anxiety before and after a severe stressor: a prospective study. Personal Individ Differ. 2004;36(4):767–78.
135. Visscher PM, Brown MA, McCarthy MI, Yang J. Five years of GWAS discovery. Am J Hum Genet. 2012;90(1):7–24.
136. Wilkinson PO, Trzaskowski M, Haworth CM, Eley TC. The role of gene-environment correlations and interactions in middle childhood depressive symptoms. Dev Psychopathol. 2013;25(1):93–104. https://doi.org/10.1017/S0954579412000922.
137. Zammit S, Spurlock G, Williams H, Norton N, Williams N, O'Donovan MC, Owen MJ. Genotype effects of CHRNA7, CNR1 and COMT in schizophrenia: interactions with tobacco and cannabis use. Br J Psychiatry. 2007;191(5):402–7. https://doi.org/10.1192/bjp.bp.107.036129.
138. Zammit S, Owen MJ, Evans J, Heron J, Lewis G. Cannabis, COMT and psychotic experiences. Br J Psychiatry. 2011;199(5):380–5. https://doi.org/10.1192/bjp.bp.111.091421.

Chapter 8
Gene/Environment Interaction and Autoimmune Disease

Tamia A. Harris-Tryon and Shai Bel

Keywords Autoimmune disease · Inflammatory bowel diseases · Crohn's disease · Ulcerative colitis · Microbiome · Microbiota · Environmental factors · Psoriasis · Rheumatoid arthritis · Gene/environment interaction · Cigarette smoking

8.1 Introduction

8.1.1 What Are Autoimmune Diseases

Autoimmune diseases are complex illnesses in which the body's own immune system attacks and destroys the body's own healthy tissues. Many tissues and organs can be affected by autoimmune diseases such as the skin, joints, intestines, endocrine glands (thyroid, pancreas, etc.) and blood vessels [1]. Over 80 different diseases have been recognized as autoimmune diseases and as a group they affect more than 8% of the world population [2]. Symptoms of autoimmune diseases can range from fatigue and malaise to life threatening organ failure. Although decades of research was dedicated to understanding the cause and course of these diseases, we still do not fully understand why they develop or how to cure them [3].

At the core of all autoimmune diseases is an improper response of the immune system against the body's healthy tissues. The human immune system is an immensely powerful cellular weapon, designed to attack invaders it deems as foreign, or non-self (not an integral part of the body). Many cellular regulatory processes are in place to

T. A. Harris-Tryon, M.D., Ph.D. (✉)
Department of Dermatology, The University of Texas Southwestern Medical Center, Dallas, TX, USA
e-mail: Tamia.Harris-Tryon@UTSouthwestern.edu

S. Bel (✉)
Azrieli Faculty of Medicine, Bar-Ilan University, Safed, Israel
e-mail: shai.bel@biu.ac.il

R. Teperino (ed.), *Beyond Our Genes*,
https://doi.org/10.1007/978-3-030-35213-4_8

prevent the immune system from recognizing its own body as an invader. For example, T and B cells that are found to be reactive to self-antigens are destroyed or are put in a state of anergy (immunologically non-functional) before becoming active. This recognition of the body as "self" is called self-tolerance, and malfunction of this process is at the heart of autoimmune diseases [1].

Autoimmune diseases are both chronic in nature, and currently incurable. As such, they pose a major burden on healthcare systems while causing significant individual suffering. Current treatments focus on relieving symptoms, and since autoimmune diseases are caused by faulty immune system function, these treatments attempt to suppress immune function in the patients. This leaves the patients susceptible to infections and cancer development. As with other incurable diseases, a major focus of research has been on identifying the causes of autoimmune diseases with the goal of preventing them. So far, several inheritable genetic factors were identified which explain some autoimmune diseases, while several environmental factors were found to explain others. Still, the cause for most autoimmune disease cases are unknown, leading to the notion that interactions of genetics and environmental factors are responsible for disease onset [3].

8.1.2 Genetic Factors Associated with Autoimmune Diseases

The observation that most autoimmune diseases feature familial clustering has led to the notion that genetic risk factors might be involved in development of autoimmune diseases. Advances in DNA sequencing technologies in the past century has provided much needed insight into these genetic risk factors. Indeed, genome wide association studies (GWAS) have identified that certain genetic variants are shared across multiple autoimmune diseases, suggesting that certain shared pathways are dysregulated in these diseases. The most common genetic risk factors are variants in the HLA locus, which enables recognition of many different foreign antigens, but can also impact recognition of self-antigens [1]. Also worth noting are variants in STAT4 and IL-23R, which have a central role in regulating the adaptive immune response, and are shared across many different autoimmune diseases [4, 5].

Although GWAS studied were a source for much hope, they still cannot explain most autoimmune diseases cases. Reflective of this is the finding that concordance rates of autoimmune diseases in monozygotic twins ranges between 12% for certain autoimmune diseases and 67% for others. Thus, attention has shifted to uncover environmental factors that might explain the gap between identification of disease-associated genetic variants and disease occurrence [3].

8.1.3 Environmental Factors Associated with Autoimmune Diseases

Numerous studies have established a connection between exposure to environmental factors and the development of autoimmune diseases. Indeed, the observation that geographical location and individual lifestyle choices can affect autoimmune diseases development rates supports this. These environmental factors may be physical, such as UV radiation; chemical, such as exposure to pesticides or tobacco smoke; or biological, such as infection by pathogenic microbes [2]. Another biological factor that has recently come into the spotlight is the human microbiome [6].

8.1.4 The Microbiome

Humans, like most other animals, are colonized by a multitude of microorganisms. These include archaea, bacteria, fungi, viruses and protozoa, which are collectively termed the microbiota. The microbiota of a human adult accounts for 1–2 kg of total body weight and are spread out across many tissues such as skin, urogenital tract, and respiratory tract, with the largest community of microbes residing in the gastrointestinal tract. The genomic data the microbiota contains, and thus their ability to produce proteins which affect human physiology, is collectively termed the microbiome [6].

Most members of the human microbiota require very specific growth conditions. Thus, our understanding of the function of the microbiota was limited to the species that we could culture in the lab. Advances in sequencing technology now allows us to identify most members of the human microbiota, and to infer their potential for influencing human physiology from their genome [6]. Additionally, use of germ-free mice (sterile mice reared in isolators which allow mono-colonization with a single species of bacteria or whole microbiota transfer from a human donor) has allowed us to move from correlative studies to more mechanistic studies. Now, we can identify each member of the microbiota and the proteins they produce, colonize germ-free mice carrying the genetic background of choice with these microbes, and discover how genetics and environmental factors interact in the development of many diseases [6].

In the past 15 years or so, the microbiota was found to influence almost every facet of human physiology. For example, the microbiota has been found to contribute to obesity and metabolic syndrome [6]. Recent studies have linked microbiota composition and metabolic activity to neurodegenerative diseases like Alzheimer's and Parkinson's disease [7] and other neurological conditions such as autism [8] and psychosis [9]. Most relevant, the microbiota was found to have a major role in shaping the immune system. It is now thought that the microbiota "educates" the immune system, helping it distinguish symbiotic microbes from pathogenic ones. In the context of autoimmune diseases, the microbiota was found to trigger immune regulatory factors such as maturation of T regulatory cells and secretion of anti-inflammatory cytokines. This is achieved by metabolizing nutrients into short-chain fatty acids (SCFA) which regulate the innate and adaptive immune response, and activation of pattern recognition receptors such as toll-like receptors and Nod-like receptors, amongst others [6]. The factors controlling microbiota composition are varied and are still being extensively studied. While genetics have a role in shaping microbiota composition, the most influential factors seem to be environmental such as diet, use of medicine and geography [10].

8.2 Gene/Environment Interaction and Autoimmune Disease

8.2.1 Inflammatory Bowel Diseases

Inflammatory bowel diseases (IBD) are a group of chronic inflammatory diseases affecting the gastrointestinal tract with unknown etiology and no cure, disturbing the lives of tens of millions across the world [11]. The two major types of IBD are Crohn's disease and ulcerative colitis. Crohn's disease is characterized by transmural inflammation that can manifest anywhere along the gastrointestinal tract, from the mouth to the anus, though most cases are confined to the last segment of the small intestine, the ileum. Ulcerative colitis on the other hand is confined to the terminal segment of the gastrointestinal tract, the colon, and is characterized by ulceration of the colonic mucosa. Both diseases appear in episodes of flares followed by a remission period. These diseases have a massive effect on morbidity and mortality and left untreated could develop to bowel cancer. Current treatment options include agents that suppress the immune system such as steroids and antibodies targeting cytokines, while last resort treatments are resection of inflamed sections of the intestine [12].

The pathogenesis of IBD includes some features of autoimmune disease, such as the presence of autoantibodies, but also features of immune-mediated diseases, such as dysregulation of cellular immunity and exaggerated response to luminal content. This exaggerated response of the immune system does not seem to target a certain member of the microbiota, which is evident by the fact that antibiotic treatment holds little therapeutic effect in IBD. While IBD are not characterized by autoreactive T cells, which is one of the postulates of autoimmune diseases, transfer of T cells in animal models can still transmit IBD-like conditions between hosts, which is another postulate. Thus, even though not a "classic" example of autoimmune disease, IBD are still considered as such, a classification which affects treatment routes [13, 14].

The etiology of IBD are currently unclear, though research points to an interaction of host genetics and environmental factors. GWAS studies have identified about 200 IBD susceptibility genes. The most prevalent mutations discovered to be associated with IBD are in genes involved in innate and adaptive immune responses, intestinal barrier function and autophagy. However, these can explain only a small percentage of IBD cases. Additionally, study of monozygotic twins has shown only low concordance rates for development of IBD [15]. Strikingly, studies following immigration of Asian population to America and Europe has found a sharp increase in the incidence of IBD in first and second-generation immigrants [16–18]. This, and the rise of disease prevalence in industrialized countries in the twentieth century, has led to the hypothesis that environmental factors might also be involved in development of IBD [19].

8.2.2 Gene-Cigarette Smoke Interaction in IBD

One of the most studied and reliably reproducible environmental factors affecting IBD risk is cigarette smoking. While smoking cigarettes significantly elevates the risk of developing Crohn's disease, it has a surprising protective effect on the development of ulcerative colitis, with non-smokers having a four-fold higher risk of developing ulcerative colitis compared to smokers [20]. Currently, the exact mechanism conferring disease susceptibility or resistance from cigarette smoking is not clear.

Several IBD risk associated genes have been shown to interact with cigarette smoking in affecting the risk of IBD development. An increased risk for development of Crohn's disease has been found in cigarette smokers harboring a mutation in the CYP2A6 gene, which encodes an enzyme involved in metabolism of nicotine. The same study also demonstrated that smoking cessation is associated with an increased risk of developing ulcerative colitis, but only in patients carrying a mutation in the GSTP1 gene which encodes a glutathione transferase protein [21]. A different study has used the method of logic regression to show that cigarette smoker carrying a mutation in the gene CALM3, a calcium-binding protein that affects different kinases, were three times less likely to develop ulcerative colitis compared to non-smokers carrying the same mutation [22]. Another study has identified that mutations in the IL-23R gene, which encodes for an immune related receptor previously recognized as a risk factor for development of Crohn's disease, interact with cigarette smoking to dramatically increase disease risk [23].

So far, mechanistic studies linking genetic predisposition to cigarette smoke in IBD development have been few. A large multi-center study examining about 20,000 IBD patients has discovered 64 SNPs to be associated with altered IBD risk in cigarette smokers compared to non-smokers. Most of the identified SNPs affected genes involved in immune and barrier function. They went on to demonstrate that genetic deletion in mice of two of the identified SNPs (IL-10 and NOD2) increased the animals' susceptibility to colitis after exposure to cigarette smoke, thus validating their findings in humans [24]. A different group has built on previously reported data that mutation in an autophagy gene, ATG16L1, interacts with cigarette smoking to elevate risk of developing Crohn's disease. Looking at both Crohn's disease patients and mutant mice, they show that cigarette smoke disrupts the antimicrobial function of Paneth cells, specialized secretory cells in the intestine, but only in patients and mice carrying the mutation in ATG16L1 [25]. Mechanistic works such as these will likely expand in the future to explain how cigarette smoke interacts with host genetics and help identify new prevention and treatment strategies for IBD.

8.2.3 Gene-Microbe Interaction in IBD

The microbiota of IBD patients has been extensively studied in the past decade and it was clearly demonstrated that the microbiota of IBD patients is fundamentally different than healthy controls [26]. While massive amounts of data have been generated, it still isn't clear whether this shift in microbiota composition (termed dysbiosis) precedes manifestation of these diseases and perhaps drives them, or whether this dysbiosis is the result of the chronic inflammation. Recent works have revealed that the conditions created in the gut by the inflammatory state is actually favorable for expansion of virulent bacteria and might explain the repetition of flares and remissions observed in these diseases [27]. Currently, fecal microbiota transfer has not been shown to be efficient at treating IBD [28], though it has been shown to be feasible in animal studies [29] and in other diseases such as *Clostridium difficile* infection [30]. Yet, identification of a certain composition and structure of the gut microbiota which would allow early detection and prevention of IBD has not been revealed.

Given the genetic background of IBD (and the fact that it explains only low amounts of cases), the realization that environmental factors contribute to disease risk [31], and the effect host genetics has on the microbiome [32], focus has shifted to gene-microbiota interaction to try and explain most cases of IBD [33]. Experiments in genetically altered mice have shown that mutations in certain IBD risk genes can lead to dysbiosis of the gut microbiota of these animals [34, 35]. However, transferring this microbiota to germ-free mice does not lead to IBD-like pathology, as happens with mice carrying non-IBD related mutations (T-bet$^{-/-}$, Rag2$^{-/-}$, TLR5$^{-/-}$ and NLRP6$^{-/-}$) [36–38]. Since mice with mutations in the most prevalent IBD associated gene (NOD2, ATG16L1 and IRGM) have been shown to have impaired pathogen clearance, it is possible that host mutations which drive dysbiosis might also make the host more susceptible to infection by these dysbiotic microbes.

While no specific pathogenic bacteria have been associated with all IBD cases, one such microbe, adherent-invasive *Escherichia coli* (AIEC), was found to be associated with many cases of IBD [39]. Interestingly, these associations seem to be dependent upon host genetics, thus forming a gene-microbe interaction. AIEC have the ability to attach to, and invade intestinal epithelial cells, triggering disease. AIEC attaches to intestinal epithelial cells by binding to host protein CEACAM6, which is not expressed in healthy tissue, but only in inflamed epithelial cells [40]. CD patients with ileal disease were found to abnormally have AIEC attached to their intestinal epithelium, and AIEC numbers seems to correlate with disease severity [41]. A Study has found that epithelial cells lacking IBD risk genes NOD2, ATG16L1 and IRGM were not able to clear the invading AIEC [42]. Additionally, while mice are not susceptible to AIEC colonization of the intestine, mice lacking the NOD2 gene display high numbers of infiltrating AIEC in intestinal lymph nodes

[43]. Thus, it seems that in individuals carrying genetic risk factors these bacteria take advantage of host susceptibility to invade, and might be a factor contributing to disease development [44].

As in the above example, it is possible that certain microbes take advantage of host genetic susceptibility to invade and trigger disease which might progress to an IBD. For example, mice carrying a mutation in the Crohn's disease risk gene *ATG16L1* were shown to be highly susceptible to infection by an invasive foodborne pathogen [45, 46]. Another possibility is that infection of an individual carrying mutations in IBD risk genes will enhance disease susceptibility. This was shown with mice lacking ATG16L1 that were infected with the foodborne pathogen norovirus. When these mice were infected with the virus, they displayed reduced antimicrobial protein secretion by intestinal goblet cells which left them susceptible to development of colitis in a chemical model [47].

To summarize, mutations in IBD risk genes in the host can affect microbiota composition and susceptibility to infection by various pathogens. This can then leave the genetically predisposed host in risk of developing a chronic inflammation in the bowel as is seen in IBD.

8.2.4 Psoriasis

Psoriasis is an immune-mediated systemic disease that manifests on the skin as well defined thick red plaques with overlying silver scale [48–50]. Psoriasis affects an estimated 2–3% of the population of the western world making it a common autoimmune disease [48, 51, 52]. In the majority of cases, psoriasis is limited to the skin, but anywhere between 2–10% of psoriasis patients also develop associated inflammation and destructions of the joints, termed psoriatic arthritis [48, 51, 53].

Though originally thought of as a merely cosmetic affliction with limited systemic health implications, it is now known that psoriasis is a systemic condition, with a range of comorbidities [54]. Psoriasis patients develop Crohn's disease at a higher rate than the general population and have a higher incidence of psychiatric disorders and uveitis [51, 55, 56]. In the last ten years it has also been established by large population studies that psoriasis is associated with metabolic syndrome and is an independent risk factor for cardiovascular disease [57–59].

As is the case for IBD, the classification of psoriasis as an autoimmune condition is controversial. There is little understanding of what triggers initial diseases presentation and no clearly identified self-antigen. It is also suspected that the breakdown in immune tolerance in psoriasis likely happens in the innate immune system [60]. Despite the limits in our understanding of the pathophysiology of psoriasis, there is a consensus that immune dysregulation characterized by aberrant activation of Th1 and Th17 immunity and high levels of TNF-α mediate the skin findings in psoriasis [60]. The key role of T-cells and their cytokine profile in the pathophysiology of psoriasis is highlighted by the advent of anti-TNFα, anti-IL-23, and anti-IL-17 monoclonal antibody therapeutics, which are effective in treating psoriasis and leading to disease control [48, 61–63].

8.2.5 *Genetics*

Psoriasis has been shown to have a strong genetic component, with a higher concordance of psoriasis between monozygotic twins compared to the concordance between dizygotic twins. Cases of psoriasis also cluster within families [64]. Numerous GWAS studies have been completed on families with a predisposition to psoriasis [65–67]. In gene-linkage studies the MHC class I region has the strongest association with psoriasis; with the HLA-Cw∗0602 allele being implicated in more than half of patients with plaque-type psoriasis [60, 64, 68, 69]. In keeping with the known role that T-cell subsets play in the pathophysiology of psoriasis, genes that are known to regulate T-cell function, such as *IL23R*, *IL12B*, *IL23A*, *TRAF3IP2*, *RUNX3*, *TAGAP*, and *STAT3* [5, 70], have all been implicated in GWAS studies. In addition, recent work has identified susceptibility loci that link genetic alterations in the innate immune system with psoriasis. Genes identified in these studies play a role in macrophage activation, NF-κB signaling, and interferon-mediated antiviral responses [70]. Several of the identified loci have also been associated with Crohn's disease, ankylosing spondylitis and celiac disease, strengthening the hypothesis that these autoimmune conditions might have a shared etiology [5].

8.2.6 *Environment*

It is clear that genetics alone does not explain the presentation of psoriasis. Even though disease onset and severity are similar for monozygotic twins, concordance between monozygotic twins is not 100% [71]. Furthermore, the identified MHC class I susceptibility loci are seen in only half of patients with psoriasis [64]. Taken together, these findings highlight that environmental factors also play a role in the pathogenesis of psoriasis.

8.2.6.1 Ultra-violet light

Ultra-violet (UV) light is thought to modify psoriatic disease. Though, there is little correlation between latitude and the prevalence of psoriasis in a given region, ultra-violet light exposure is known to improve psoriasis and both psoriasis and psoriatic arthritis improve in the summer months [49, 71].

8.2.6.2 Smoking

Several studies have also shown a strong association between psoriasis and people who currently or previously smoked cigarettes. Cigarette smoking increases the risk of psoriasis in a dose-dependent manner and smoking has been shown to effect response rates to therapy [72, 73]. Amplifying this observed link have been studies that have specifically looked at the combined effect of carrying a genetic susceptibility allele and smoking. Two separate studies looked at smokers that carry the HLA-Cw6 allele and genetic variants at the *CSMD1* and the *TNIP/ANXA6* loci. In both studies, people that smoke and carry these genetic variations showed an increased risk of developing psoriasis compared to non-smokers. For HLA-Cw6, the psoriasis risk was greater than ten-fold for smokers that carried the allele compared to non-smokers without the allele [74, 75].

8.2.6.3 Infection and the Microbiome

The microbiota has also been shown to be modulated in psoriasis. In pediatric populations the onset of a specific clinical presentation of psoriasis called guttate psoriasis, with small rain-drop sized psoriatic plaques, has been specifically associated with *Streptococcus pyogenes* (*S. pyogenes)* infections of the skin and of the pharynx [76, 77]. *S. pyogenes* infection has also been shown to exacerbate the more common variant of psoriasis, plaque psoriasis [78]. The T-cells of patients with psoriasis and *S. pyogenes* pharyngeal infections recognize M-proteins expressed by *S. pyogenes.* These M-protein specific T-cells show increased expression of skin homing molecules and cross-reactivity with skin antigens. Additionally, in a single prospective study, tonsillectomy led to a reduction in the circulating M-protein specific skin homing T-cells and decreased the severity of psoriasis [79, 80]. In sum, these finding suggest that *S. pyogenes* infection may play a role in the both disease initiation and propagation of psoriasis [81].

With the advent of next generation sequencing, recent studies have looked more broadly at the links between the microbiome and psoriasis. The impetus for these studies was the observation that Crohn's disease patients have a 7-times higher risk of developing psoriasis than the general population [54]. As evidence mounts that imbalances in the gut microbiome may trigger immune-mediated inflammatory disorders such as Crohn's diseases, investigators began to postulate that imbalances in the gut microbiome might also be associated with psoriasis and psoriatic arthritis [82]. An initial study used PCR to ask specifically if patients with psoriasis showed a depletion of *Faecalibacterium prausnitzii* and an increase of *Escherichia coli*- gut microbiome shifts that had been previously reported in patients with inflammatory bowel disease [83]. This report did show similar shifts in patients with psoriasis, though almost half of the psoriasis patients that were studied also had IBD [84]. A second study, which looked more globally at bacterial communities in psoriasis using 16*S* sequencing, revealed a decrease in the diversity of the gut microbiome in patients with psoriasis and psoriatic arthritis compared to health controls, with specific reductions in the taxa *Akkermansia*, *Ruminococcus*, and *Pseudobutyrivibrio* [85]. A more recent study attempted to characterize the skin microbiome in 52 patients with active psoriasis with comparison to the data of 300 healthy individuals in the NIH human microbiome studies. This study found the opposite of the first, with increased diversity of the gut microbiome in patients with psoriasis compared to controls. Increases in *Akkermansia, Ruminococcus*, and *Faecalibacterium* and decreases in *Bacteroides* were also observed [86]. The differences in these study findings likely reflects the current lack of standardization in the field, as Scher et al. completed 16*S* sequencing with sampling of the V1-V2 region compared to the use of V3-V4 sequencing used by Codoñer et al. Though the shifts are different, the studies conducted on the gut microbiome do show that patients with psoriasis have a different microbiome than patients without psoriasis. These findings are significant, as in mouse models of psoriasis mice that lack a microbiota showed a lower degree of local and systemic Th17 inflammation [87].

The skin is also home to a diverse population of bacteria [88, 89] that might function in the pathogenesis of psoriasis. The composition of the skin microbiome is site specific, with oily, wet, and dry locations of the body creating unique ecological niches with distinct microbial communities [89–91]. In a highly standardized study sampling six body sites in both psoriasis patients and controls, 16*S* sequencing showed higher diversity in the psoriasis-associated skin microbiome; with enrichment of *Staphylococcus aureus* and decreased abundance of *Staphylococcus epidermidis* and *Propionibacterium acnes* [92]. Smaller studies also support an increased abundance of *Staphylococcus aureus* and a decrease in the abundance of *Propionibacterium acnes* [93, 94]. It is still unclear whether shifts in the skin microbiome drive disease progression in psoriasis. There is some evidence that the colonization of neonatal mice with *Staphylococcus aureus* skews the immune system towards Th17 immunity, which would support the hypothesis that the increase in *Staphylococcus aureus* seen in these studies could drive the inflammation observed in psoriasis [92]. However, it is still possible that the changes observed in the skin microbiome of psoriatic plaques are secondary to the dry inflamed environment, rich in antimicrobial proteins, created by the underlying disease [50]. Additional, work will need to be completed to define an etiological function for the skin microbiome in psoriasis.

8.2.7 Rheumatoid Arthritis

Rheumatoid arthritis (RA) is a systemic autoimmune condition affecting 0.5–1% of the world population [95]. RA often presents clinically with symmetrical inflammation of the small joints of the hands, progressing with time to a destructive and debilitating systemic arthritis. In RA, a breakdown in immune tolerance in the adaptive immune system leads to the develop of characteristic auto antibodies, rheumatoid factor (RF) and anti-citrullinated protein antibodies (ACPA). RF and ACPA are associated with joint inflammation and also cause inflammation of the heart and lungs, further contributing to the increased mortality seen in this condition [96].

RA is more common in women and in people with a genetic susceptibility to the disease [97]. The allele of greatest significance is HLA-DRB1, which encodes an MHC class II molecule. HLA-DRB1 mutations are seen in more than half of patients with rheumatoid arthritis and have been specifically linked to patients who develop the auto antibodies RF and ACPA. Several different alleles of HLA-DRB1 have been linked to RA. Though these alleles vary, they all share an amino acid sequence in a single hypervariable domain. The hypervariable domains are the regions of the MHC class II molecules involved in antigen recognition. Thus, in the current paradigm, HLA-DRB1 alleles associated with RA are capable of generating an immune response to unique self-antigens and initiating the breakdown in adaptive immunity seen in this condition [96, 98–100]. Mutations in *PTPN22*, *CTLA4*, *TRAF1/C5* region, and *c-REL* have also been associated with the development of rheumatoid arthritis [101–104].

8.2.8 Environment

8.2.8.1 Smoking

The low 15–30% concordance rate in RA between monozygotic twins indicates that environmental triggers also play a role in the pathogenesis of RA [105]. As has been observed in other autoimmune diseases, smoking greatly increases the risk of developing rheumatoid arthritis. In the case of rheumatoid arthritis, the mechanism linking smoking to disease is more established. Patients with rheumatoid arthritis that harbor HLR-DR4 susceptibility alleles and smoke, develop a serological subtype of RA characterized by RF and ACPA [106]. ACPA, anti-citrullinated protein antibodies, are thought to develop in response to host proteins that undergo the post-translational modification process known as citrullination. Citrullination is catalyzed by peptidylarginine deiminase enzymes, which convert arginine within host proteins to citrullines. These citrullinated proteins act as neo-antigens. People with HLA-DR4 susceptibility alleles have immune systems that are primed to recognize citrullinated host proteins and thus develop an immune response that triggers the inflammation and auto-antibody profile seen in RA. Smoking has been shown to increase the activity of peptidylarginine deiminases and therefore initiates the first steps in the breakdown of the adaptive immune system [107]. These finding, linking specific alleles and smoking to the subsequent development of rheumatoid arthritis, have been shown and confirmed in numerous large population studies (reviewed by [108]).

8.2.8.2 Infection and the Microbiome

The first observations linking the gut microbiota to RA were in the 1960s with the finding that patients with arthritis showed an expansion in *Clostridium perfingens* type A [109]. Though this organism was not specifically linked to RA in future studies, the concept that the gut microbiota can modulate arthritic inflammation was subsequently confirmed in both mouse and human studies. As early as the late 1970s, it was shown that rats raised in a conventional facility, or colonized with *E.coli* and *Bacteroidies* were protected from developing inflammatory arthritis in a rat adjuvant arthritis model, when compared to germ free rats raised without the beneficial effects of the microbiota [110–113]. Furthermore, several genetic models that induce spontaneous inflammatory arthritis in mice have been shown to be microbiota dependent, solidifying the interplay between genetic susceptibility and the microbiome in the pathogenesis of inflammatory arthritis [114, 115]. In humans, 16*S* and metagenomic studies comparing the microbiomes of patients with RA to unrelated controls show decreased enrichment of the beneficial microbes *Bifidobacteria* and *Bacteroidies* [116, 117]. Two separate studies also showed an expansion in *Lactobacillus* in the microbiome of patients with RA compared to controls [118, 119]. Prior infections with Epstein bar virus and parvovirus B-19 have also been associated with RA in small studies, with higher titers of anti-viral antibodies in patients with RA compared to controls [120–122]. Taken together, these data establish a role for dysbiosis or antecedent infection in the pathogenesis of RA.

The oral microbiome has also been linked to RA. Early in disease, patients with RA have a higher incidence of periodontal disease with twice the carriage rates of *Porphyromonas gingivalis (P. gingivalis)* compared to people without RA [123–125]. Interestingly, *P. gingivalis* express the enzyme peptidylarginine deiminase. As described above for cigarette smoking, peptidylarginine deiminases catalyze the conversion of arginine residues within host proteins to citrullines, triggering the production of anti-citrullinated antibodies in genetically susceptible individuals. Thus, oral dysbiosis might be an initiating event in the pathogenesis of RA [126–128].

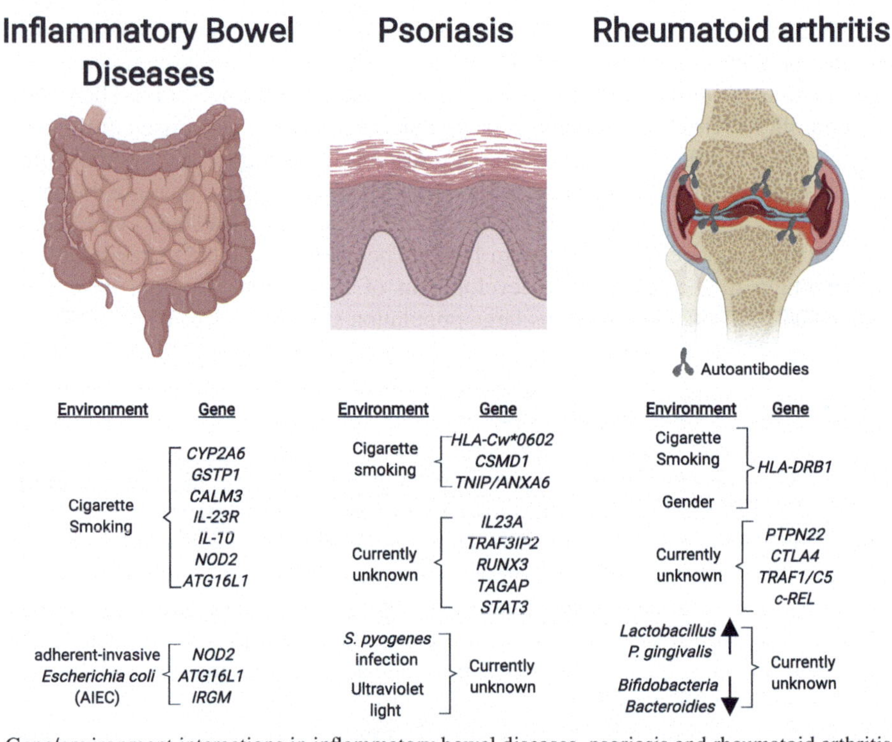

Gene/environment interactions in inflammatory bowel diseases, psoriasis and rheumatoid arthritis

References

1. Davidson A, Diamond B. Autoimmune Diseases. N Engl J Med. 2001;345:340–50.
2. Parks CG, et al. Expert panel workshop consensus statement on the role of the environment in the development of autoimmune disease. Int J Mol Sci. 2014;15:14269–97.
3. Wahren-Herlenius M, Kuchroo VK. Gene-environment interaction in induction of autoimmunity. Semin Immunol. 2011;23:65–6.
4. Genetic factors shared among diverse autoimmune disorders. 2010.
5. Cotsapas C, et al. Pervasive sharing of genetic effects in autoimmune disease. PLoS Genet. 2011;7:e1002254.
6. Shamriz O, et al. Microbiota at the crossroads of autoimmunity. Autoimmun Rev. 2016;15:859–69. https://doi.org/10.1016/j.autrev.2016.07.012.
7. Ma Q, et al. Impact of microbiota on central nervous system and neurological diseases: the gut-brain axis. J Neuroinflammation. 2019;16:53.
8. Fattorusso A, Di Genova L, Dell'Isola G, Mencaroni E, Esposito S. Autism Spectrum disorders and the gut microbiota. Nutrients. 2019;11:521.
9. Cuomo A, et al. The microbiome: a new target for research and treatment of schizophrenia and its resistant presentations? A systematic literature search and review. Front Pharmacol. 2018;9:1040.
10. Rothschild D, et al. Environment dominates over host genetics in shaping human gut microbiota. Nature. 2018;555:210–5.
11. Ng SC, et al. Worldwide incidence and prevalence of inflammatory bowel disease in the 21st century: a systematic review of population-based studies. Lancet. 2017;390:2769–78.
12. Abraham C, Cho JH. Inflammatory bowel disease. N Engl J Med. 2009;361:2066–78.
13. Wen Z, Fiocchi C. Inflammatory bowel disease: autoimmune or immune-mediated pathogenesis? Clin Dev Immunol. 2004;11:195–204.
14. Inflammatory Bowel Disease. The classic gastrointestinal autoimmune disease. Curr Probl Pediatr Adolesc Health Care. 2014;44:328–34.
15. Mirkov MU, Verstockt B, Cleynen I. Genetics of inflammatory bowel disease: beyond NOD2. Lancet Gastroenterol Hepatol. 2017;2:224–34.
16. Carr I, Mayberry JF. The effects of migration on ulcerative colitis: a three-year prospective study among Europeans and first- and second-generation south Asians in Leicester (1991-1994). Am J Gastroenterol. 1999;94:2918–22. https://doi.org/10.1016/S0002-9270(99)00494-3.
17. Pinsk V, et al. Inflammatory bowel disease in the south Asian pediatric population of British Columbia. Am J Gastroenterol. 2007;102:1077–83. https://doi.org/10.1111/j.1572-0241.2007.01124.x.
18. Tsironi E, Feakins RM, Roberts CSJ, Rampton DS. Incidence of inflammatory bowel disease is rising and abdominal tuberculosis is falling in Bangladeshis in East London, United Kingdom. Am J Gastroenterol. 2004;99:1749–55. https://doi.org/10.1111/j.1572-0241.2004.30445.x.
19. Molodecky NA, et al. Increasing incidence and prevalence of the inflammatory bowel diseases with time, based on systematic review. Gastroenterology. 2012;142:46–54.e42. https://doi.org/10.1053/j.gastro.2011.10.001.
20. Parkes GC, Whelan K, Lindsay JO. Smoking in inflammatory bowel disease: impact on disease course and insights into the aetiology of its effect. J Crohn's Colitis. 2014;8:717–25. https://doi.org/10.1016/j.crohns.2014.02.002.
21. Ananthakrishnan AN, Nguyen DD, Sauk J, Yajnik V, Xavier RJ. Genetic polymorphisms in metabolizing enzymes modifying the association between smoking and inflammatory bowel diseases. Inflamm Bowel Dis. 2014;20:783–9. https://doi.org/10.1097/MIB.0000000000000014.
22. Wang M-H, et al. Gene-gene and gene-environment interactions in ulcerative colitis. Hum Genet. 2014;133:547–58.
23. Doecke JD, et al. Smoking behaviour modifies *IL23r* -associated disease risk in patients with Crohn's disease. J Gastroenterol Hepatol. 2015;30:299–307.

24. Yadav P, et al. Genetic Factors Interact With Tobacco Smoke to Modify Risk for Inflammatory Bowel Disease in Humans and Mice. 2017; https://doi.org/10.1053/j.gastro.2017.05.010.
25. Liu T-C, Head RD, Stappenbeck TS. Interaction between smoking and ATG16L1 T300A triggers Paneth cell defects in Crohn's disease the journal of clinical investigation. J Clin Invest. 2018;128:5110–22.
26. Huttenhower C, Kostic AD, Xavier RJ. Inflammatory bowel disease as a model for translating the microbiome. Immunity. 2014;40:843–54. https://doi.org/10.1016/j.immuni.2014.05.013.
27. Litvak Y, Byndloss MX, Bäumler AJ. Colonocyte metabolism shapes the gut microbiota. Science. 2018;362(*80*):eaat9076.
28. Basso PJ, Câmara NOS, Sales-Campos H. Microbial-based therapies in the treatment of inflammatory bowel disease – an overview of human studies. Front Pharmacol. 2019;9(1571)
29. Bel S, et al. Reprogrammed and transmissible intestinal microbiota confer diminished susceptibility to induced colitis in TMF $^{-/-}$ mice. Proc Natl Acad Sci 1. 2014;6:4964–9. https://doi.org/10.1073/pnas.1319114111.
30. Youngster I, et al. Oral, capsulized, frozen fecal microbiota transplantation for relapsing Clostridium difficile infection. JAMA - J Am Med Assoc. 2014;312:1772–8. https://doi.org/10.1001/jama.2014.13875.
31. Ananthakrishnan AN, et al. Environmental triggers in IBD: a review of progress and evidence. Nat Rev Gastroenterol Hepatol. 2017;15:39–49.
32. Aschard H, et al. Genetic effects on the commensal microbiota in inflammatory bowel disease patients. PLoS Genet. 2019;15:e1008018.
33. De Souza HSP, Fiocchi C, Iliopoulos D. The IBD interactome: an integrated view of aetiology, pathogenesis and therapy. Nature Reviews Gastroenterology and Hepatology. 2017;14:739–49. https://doi.org/10.1038/nrgastro.2017.110.
34. Lavoie S, et al. The Crohn's disease polymorphism, ATG16L1 T300A, alters the gut microbiota and enhances the local Th1/Th17 response. elife. 2019:8.
35. Ramanan D, Tang MS, Bowcutt R, Loke P, Cadwell K. Bacterial sensor Nod2 prevents inflammation of the small intestine by restricting the expansion of the commensal Bacteroides vulgatus. Immunity. 2014;41:311–24.
36. Elinav E, et al. NLRP6 inflammasome regulates colonic microbial ecology and risk for colitis. Cell. 2011;145:745–57. https://doi.org/10.1016/j.cell.2011.04.022.
37. Garrett WS, et al. Communicable ulcerative colitis induced by T-bet deficiency in the innate immune system. Cell. 2007;131:33–45. https://doi.org/10.1016/j.cell.2007.08.017.
38. Ruff WE, Kriegel MA. Autoimmune host-microbiota interactions at barrier sites and beyond. Trends Mol Med. 2015;21:233–44. https://doi.org/10.1016/j.molmed.2015.02.006.
39. Darfeuille-Michaud A, et al. High prevalence of adherent-invasive Escherichia coli associated with ileal mucosa in Crohn's disease. Gastroenterology. 2004;127:412–21. https://doi.org/10.1053/j.gastro.2004.04.061.
40. Barnich N, et al. CEACAM6 acts as a receptor for adherent-invasive E. coli, supporting ileal mucosa colonization in Crohn disease. J Clin Invest. 2007;117:1566–74. https://doi.org/10.1172/JCI30504.
41. Baumgart M, et al. Culture independent analysis of ileal mucosa reveals a selective increase in invasive Escherichia coli of novel phylogeny relative to depletion of Clostridiales in Crohn's disease involving the ileum. ISME J. 2007;1:403–18. https://doi.org/10.1038/ismej.2007.52.
42. Lapaquette P, Glasser AL, Huett A, Xavier RJ, Darfeuille-Michaud A. Crohn's disease-associated adherent-invasive E. coli are selectively favoured by impaired autophagy to replicate intracellularly. Cell Microbiol. 2010;12:99–113. https://doi.org/10.1111/j.1462-5822.2009.01381.x.
43. Chassaing B, et al. Crohn disease–associated adherent-invasive E. coli bacteria target mouse and human Peyer's patches via long polar fimbriae. J Clin Invest. 2011;121:966–75.
44. Wong S-Y, Cadwell K. There was collusion: microbes in inflammatory bowel disease. PLoS Pathog. 2018;14:e1007215.

45. Bel S, Hooper LV. Secretory autophagy of lysozyme in Paneth cells. Autophagy. 2018;14:719–21.
46. Bel S, et al. Paneth cells secrete lysozyme via secretory autophagy during bacterial infection of the intestine. Science. 2017;357(*80*):eaal4677.
47. Cadwell K, et al. Virus-plus-susceptibility gene interaction determines Crohn's disease gene Atg16L1 phenotypes in intestine. Cell. 2010;141:1135–45.
48. Boehncke W-H, Schön MP. Psoriasis. Lancet. 2015;386:983–94.
49. Balato N, et al. Effect of weather and environmental factors on the clinical course of psoriasis. Occup Env Med. 2013;70:600.
50. Nestle FO, Kaplan DH, Barker J. Psoriasis. N Engl J Med. 2009;361:496–509.
51. Christophers E. Psoriasis--epidemiology and clinical spectrum. Clin Exp Dermatol. 2001;26:314–20.
52. Parisi R, et al. Global epidemiology of psoriasis: a systematic review of incidence and prevalence. J Invest Dermatol. 2013;133:377–85.
53. Ogdie A, Weiss P. The epidemiology of psoriatic arthritis. Rheum Dis Clin. 2015;41:545–68.
54. Oliveira Mde F, Rocha Bde O, Duarte GV. Psoriasis: classical and emerging comorbidities. An Bras Dermatol. 2015;90:9–20.
55. Gulliver W. Long-term prognosis in patients with psoriasis. Br J Dermatol. 2008;159(Suppl):2–9.
56. Binus AM, et al. Associated comorbidities in psoriasis and inflammatory bowel disease. J Eur Acad Dermatol Venereol. 2012;26:644–50.
57. Mehta NN, et al. Patients with severe psoriasis are at increased risk of cardiovascular mortality: cohort study using the general practice research database. Eur Heart J. 2010;31:1000–6.
58. Gelfand JM, et al. The risk of mortality in patients with psoriasis: results from a population-based study. Arch Dermatol. 2007;143:1493–9.
59. Ogdie A, et al. Risk of mortality in patients with psoriatic arthritis, rheumatoid arthritis and psoriasis: a longitudinal cohort study. Ann Rheum Dis. 2014;73:149–53.
60. Ayala-Fontanez N, Soler DC, McCormick TS. Current knowledge on psoriasis and autoimmune diseases. Psoriasis (Auckl). 2016;6:7–32.
61. Langley RG, et al. Secukinumab in plaque psoriasis--results of two phase 3 trials. N Engl J Med. 2014;371:326–38.
62. Krueger GG, et al. A human interleukin-12/23 monoclonal antibody for the treatment of psoriasis. N Engl J Med. 2007;356:580–92.
63. Leonardi CL, et al. Etanercept as monotherapy in patients with psoriasis. N Engl J Med. 2003;349:2014–22.
64. Bowcock AM, Krueger JG. Getting under the skin: the immunogenetics of psoriasis. Nat Rev Immunol. 2005;5:699–711.
65. Genetic Analysis of Psoriasis, C, et al. A genome-wide association study identifies new psoriasis susceptibility loci and an interaction between HLA-C and ERAP1. Nat Genet. 2010;42:985–90.
66. Trembath, R. C. *et al.* Identification of a major susceptibility locus on chromosome 6p and evidence for further disease loci revealed by a two stage genome-wide search in psoriasis. Hum Mol Genet. 1997;6:813–820.
67. Zhang XJ, et al. Psoriasis genome-wide association study identifies susceptibility variants within LCE gene cluster at 1q21. Nat Genet. 2009;41:205–10.
68. Nair RP, et al. Sequence and haplotype analysis supports HLA-C as the psoriasis susceptibility 1 gene. Am J Hum Genet. 2006;78:827–51.
69. Tiilikainen A, Lassus A, Karvonen J, Vartiainen P, Julin M. Psoriasis and HLA-Cw6. Br J Dermatol. 1980;102:179–84.
70. Tsoi LC, et al. Identification of 15 new psoriasis susceptibility loci highlights the role of innate immunity. Nat Genet. 2012;44:1341–8.
71. Gupta R, Debbaneh MG, Liao W. Genetic epidemiology of psoriasis. Curr Dermatol Rep. 2014;3:61–78.

72. Naldi L. Psoriasis and smoking: links and risks. *Psoriasis*. Auckland, NZ. 2016;6:65–71.
73. Naldi L, et al. Cigarette smoking, body mass index, and stressful life events as risk factors for psoriasis: results from an Italian case-control study. J Invest Dermatol. 2005;125:61–7.
74. Jin Y, et al. Combined effects of HLA-Cw6 and cigarette smoking in psoriasis vulgaris: a hospital-based case-control study in China. J Eur Acad Dermatol Venereol. 2009;23:132–7.
75. Yin XY, et al. TNIP1/ANXA6 and CSMD1 variants interacting with cigarette smoking, alcohol intake affect risk of psoriasis. J Dermatol Sci. 2013;70:94–8.
76. Whyte HJ, Baughman RD. Acute GUTTATE psoriasis and streptococcal infection. Arch Dermatol. 1964;89:350–6.
77. Telfer NR, Chalmers RJ, Whale K, Colman G. The role of streptococcal infection in the initiation of guttate psoriasis. Arch Dermatol. 1992;128:39–42.
78. Gudjonsson JE, Thorarinsson AM, Sigurgeirsson B, Kristinsson KG, Valdimarsson H. Streptococcal throat infections and exacerbation of chronic plaque psoriasis: a prospective study. Br J Dermatol. 2003;149:530–4.
79. Sigurdardottir SL, Thorleifsdottir RH, Valdimarsson H, Johnston A. The association of sore throat and psoriasis might be explained by histologically distinctive tonsils and increased expression of skin-homing molecules by tonsil T cells. Clin Exp Immunol. 2013;174:139–51.
80. Thorleifsdottir RH, et al. Improvement of psoriasis after tonsillectomy is associated with a decrease in the frequency of circulating T cells that recognize streptococcal determinants and homologous skin determinants. J Immunol. 2012;188:5160–5.
81. Valdimarsson H, Thorleifsdottir RH, Sigurdardottir SL, Gudjonsson JE, Johnston A. Psoriasis – as an autoimmune disease caused by molecular mimicry. Trends Immunol. 2009;30:494–501.
82. Salem I, Ramser A, Isham N, Ghannoum MA. The gut microbiome as a major regulator of the gut-skin Axis. Front Microbiol. 2018;9(1459)
83. Sokol H, et al. Faecalibacterium prausnitzii is an anti-inflammatory commensal bacterium identified by gut microbiota analysis of Crohn disease patients. Proc Natl Acad Sci. 2008;105:16731–6.
84. Eppinga H, et al. Similar depletion of protective Faecalibacterium prausnitzii in psoriasis and inflammatory bowel disease, but not in hidradenitis Suppurativa. J Crohns Colitis. 2016;10:1067–75.
85. Scher JU, et al. Decreased bacterial diversity characterizes the altered gut microbiota in patients with psoriatic arthritis, resembling dysbiosis in inflammatory bowel disease. Arthritis Rheumatol. 2015;67:128–39.
86. Codoñer FM, et al. Gut microbial composition in patients with psoriasis. Sci Rep. 2018;8:3812.
87. Zakostelska Z, et al. Intestinal microbiota promotes psoriasis-like skin inflammation by enhancing Th17 response. PLoS One. 2016;11:e0159539.
88. Oh J, et al. Temporal stability of the human skin microbiome. Cell. 2016;165:854–66.
89. Grice EA, et al. Topographical and temporal diversity of the human skin microbiome. Science. 2009;324(*80*):1190–2.
90. Findley K, Grice EA. The skin microbiome: a focus on pathogens and their association with skin disease. PLoS Pathog. 2014;10:e1004436.
91. Grice EA. The skin microbiome: potential for novel diagnostic and therapeutic approaches to cutaneous disease. Semin Cutan Med Surg. 2014;33:98–103.
92. Chang HW, et al. Alteration of the cutaneous microbiome in psoriasis and potential role in Th17 polarization. Microbiome. 2018;6:154.
93. Gao Z, Tseng CH, Strober BE, Pei Z, Blaser MJ. Substantial alterations of the cutaneous bacterial biota in psoriatic lesions. PLoS One. 2008;3:e2719.
94. Fahlén A, Engstrand L, Baker BS, Powles A, Fry L. Comparison of bacterial microbiota in skin biopsies from normal and psoriatic skin. Arch Dermatol Res. 2012;304:15–22.
95. Silman AJ, Pearson JE. Epidemiology and genetics of rheumatoid arthritis. Arthritis Res. 2002;4(Suppl 3):S265–72.

96. McInnes IB, Schett G. The pathogenesis of rheumatoid arthritis. N Engl J Med. 2011;365:2205–19.
97. van Vollenhoven RF. Sex differences in rheumatoid arthritis: more than meets the eye. BMC Med. 2009;7:12.
98. Gregersen PK, Silver J, Winchester RJ. The shared epitope hypothesis. An approach to understanding the molecular genetics of susceptibility to rheumatoid arthritis. Arthritis Rheum. 1987;30:1205–13.
99. Weyand CM, Goronzy JJ. Disease-associated human histocompatibility leukocyte antigen determinants in patients with seropositive rheumatoid arthritis. Functional role in antigen-specific and allogeneic T cell recognition. J Clin Invest. 1990;85:1051–7.
100. De Almeida DE, et al. Immune dysregulation by the rheumatoid arthritis shared epitope. J Immunol. 2010;185:1927–34.
101. Remmers EF, et al. *STAT4* and the risk of rheumatoid arthritis and systemic lupus erythematosus. N Engl J Med. 2007;357:977–86.
102. Plenge RM, et al. Two independent alleles at 6q23 associated with risk of rheumatoid arthritis. Nat Genet. 2007;39:1477–82.
103. Kurreeman FAS, et al. A candidate gene approach identifies the TRAF1/C5 region as a risk factor for rheumatoid arthritis. PLoS Med. 2007;4:e278.
104. Begovich AB, et al. A missense single-nucleotide polymorphism in a gene encoding a protein tyrosine phosphatase (PTPN22) is associated with rheumatoid arthritis. Am J Hum Genet. 2004;75:330–7.
105. MacGregor AJ, et al. Characterizing the quantitative genetic contribution to rheumatoid arthritis using data from twins. Arthritis Rheum. 2000;43:30–7.
106. Padyukov L, Silva C, Stolt P, Alfredsson L, Klareskog L. A gene-environment interaction between smoking and shared epitope genes in HLA-DR provides a high risk of seropositive rheumatoid arthritis. Arthritis Rheum. 2004;50:3085–92.
107. Makrygiannakis D, et al. Smoking increases peptidylarginine deiminase 2 enzyme expression in human lungs and increases citrullination in BAL cells. Ann Rheum Dis. 2008;67:1488–92.
108. Ellis JA, Kemp AS, Ponsonby A-L. Gene–environment interaction in autoimmune disease. Expert Rev Mol Med. 2014;16:e4.
109. MANSSON I, COLLDAHL H. The intestinal FLORA in patients with bronchial asthma and rheumatoid arthritis. Allergy. 1965;20:94–104.
110. Kohashi O, Kohashi Y, Takahashi T, Ozawa A, Shigematsu N. Reverse effect of gram-positive bacteria vs. gram-negative bacteria on adjuvant-induced arthritis in germfree rats. Microbiol Immunol. 1985;29:487–97.
111. Kohashi O, Kohashi Y, Takahashi T, Ozawa A, Shigematsu N. Suppressive effect of Escherichia coli on adjuvant-induced arthritis in germ-free rats. Arthritis Rheum. 1986;29:547–53.
112. Kohashi O, et al. Susceptibility to adjuvant-induced arthritis among germfree, specific-pathogen-free, and conventional rats. Infect Immun. 1979;26:791–4.
113. Rath HC, et al. Normal luminal bacteria, especially Bacteroides species, mediate chronic colitis, gastritis, and arthritis in HLA-B27/human beta2 microglobulin transgenic rats. J Clin Invest. 1996;98:945–53.
114. Wu H-J, et al. Gut-residing segmented filamentous bacteria drive autoimmune arthritis via T helper 17 cells. Immunity. 2010;32:815–27.
115. Abdollahi-Roodsaz S, et al. Stimulation of TLR2 and TLR4 differentially skews the balance of T cells in a mouse model of arthritis. J Clin Invest. 2008;118:205–16.
116. Vaahtovuo J, Munukka E, Korkeamäki M, Luukkainen R, Toivanen P. Fecal microbiota in early rheumatoid arthritis. J Rheumatol. 2008;35:1500–5.
117. Scher JU, et al. Expansion of intestinal Prevotella copri correlates with enhanced susceptibility to arthritis. elife. 2013;2:e01202.
118. Zhang X, et al. The oral and gut microbiomes are perturbed in rheumatoid arthritis and partly normalized after treatment. Nat Med. 2015;21:895–905.
119. Liu X, Zou Q, Zeng B, Fang Y, Wei H. Analysis of fecal Lactobacillus community structure in patients with early rheumatoid arthritis. Curr Microbiol. 2013;67:170–6.

120. Kozireva SV, et al. Incidence and clinical significance of parvovirus B19 infection in patients with rheumatoid arthritis. J Rheumatol. 2008;35:1265–70.
121. Chen Y-S, et al. Parvovirus B19 infection in patients with rheumatoid arthritis in Taiwan. J Rheumatol. 2006;33:887–91.
122. Meron MK, et al. Infectious aspects and the Etiopathogenesis of rheumatoid arthritis. Clin Rev Allergy Immunol. 2010;38:287–91.
123. Catrina AI, Deane KD, Scher JUG. Environment, microbiome and mucosal immune tolerance in rheumatoid arthritis. Rheumatology (Oxford). 2016;55:391–402.
124. Scher JU, et al. Periodontal disease and the oral microbiota in new-onset rheumatoid arthritis. Arthritis Rheum. 2012;64:3083–94.
125. Mikuls TR, et al. Periodontitis and Porphyromonas gingivalis in patients with rheumatoid arthritis. Arthritis Rheumatol. 2014;66:1090–100.
126. Wegner N, et al. Peptidylarginine deiminase from Porphyromonas gingivalis citrullinates human fibrinogen and α-enolase: implications for autoimmunity in rheumatoid arthritis. Arthritis Rheum. 2010;62:2662–72.
127. Valesini G, et al. Citrullination and autoimmunity. Autoimmun Rev. 2015;14:490–7.
128. Lerner A, Aminov R, Matthias T. Dysbiosis may trigger autoimmune diseases via inappropriate post-translational modification of host proteins. Front Microbiol. 2016;7:84.

Part III
Genome/Epigenome

Chapter 9
Introduction to Epigenetic Inheritance: Definition, Mechanisms, Implications and Relevance

Maximilian Lassi and Raffaele Teperino

Keywords Genetics · Epigenetics · Gene/environment interaction · Complex disease · DoHAD · DNA methylation · Histone post-translational modification · Non-coding RNAs · Disease risk · Phenotypic adaptation

9.1 Introduction to the Concept of Epigenetic Inheritance

According to Charles Darwin – the father of the most famous theory of evolution – complex organisms have evolved from less complex ancestors as a consequence of accumulating spontaneous genetic mutations, which, occurring under environmental pressure and transmitted across generations, result in different organisms with acquired survival advantages. This theory is best known as *natural selection*, else defined as the natural process allowing the preservation of a functional advantage (acquired from spontaneous genetic mutations across generations) that enables a species to survive by being more competitive in the wild (eg. mutant flies with wings pass on this advantageous trait to their offspring and outcompete those without wings and thus unable to fly and survive) [1]. Although Darwin's theory of evolution is the most famous, many pre-darwinian naturalists had also conceptualized evolution. The best known of those is Jean-Baptiste Lamarck, who anticipated Darwin in his theory of natural selection, while pointing at the environment – independently from occurring genetic mutations – as the major driving force of

M. Lassi
Helmholtz Zentum München, German Research Center for Environmental Health -, Institute of Experimental Genetics, Neuherberg, Germany

German Center for Diabetes Research (DZD) -, Neuherberg, Germany

R. Teperino (✉)
Head of the Environmental Epigenetics Group, Institute of Experimental Genetics & German Center for Diabetes Research (DZD), Helmholtz Zentrum München GmbH – German Research Center for Environmental Health, Neuherberg, Bayern, Germany
e-mail: raffaele.teperino@helmholtz-muenchen.de

R. Teperino (ed.), *Beyond Our Genes*,
https://doi.org/10.1007/978-3-030-35213-4_9

evolution. In other words, when the environment changes, living organisms have to change their behavior (and body) to survive (eg. the giraffe's neck, whose length increases to allow the animal to eat from the highest leaves). These acquired characteristics are passed on to the next generations to improve their adaptation and survival [2] Being the first to postulate the transmission of non-genetic (or acquired) information across generations, Lamarckian's theory has been strongly rejected by the community of mendelian geneticists and modern molecular biologists until slightly more than a decade ago, when it has been rekindled by the discovery of acquired epigenetic inheritance. In line with Lamarckian's theory, epigenetic inheritance is defined as the transmission of acquired environmental information across generations through the germline via environmentally modified epigenetic tags on the genome, which modifies offspring developmental and phenotypic trajectories.

The term "Epigenetics" was introduced by the British biologist Conrad Hal Waddington in the early 1940s. He defined epigenetics as *the branch of biology, which studies the causal interactions between genes and their products which bring the phenotype into being* [3]. Based on our current molecular understandings, Waddington's definition of epigenetics has been changed into *the group of molecular processes on DNA, which regulate genome activity independently from the DNA sequence*. Importantly, epigenetic processes are mitotically stable – and so heritable – and environmentally induced, thus constituting a way to incorporate Lamarckian's theory into Mendel's laws of inheritance.

In other words, according to the theory of epigenetic inheritance, certain environmental challenges (eg. food availability, stress, exposure to chemicals and toxicants) can affect developmental and phenotypic trajectories across several unexposed generations via heritable epigenetic modifications (and not genetic alterations) in the exposed parental generation. Interestingly, while genetic inheritance is deep determined in one's genes and therefore barely modifiable by lifestyle choices, acquired epigenetic traits are plastic and at least partially reversible.

A growing body of human epidemiological studies supports the relevance of acquired inheritance for the emergence of complex phenotypes (eg. increased/reduced basal energy expenditure, glucose tolerance, and behaviour) and susceptibility to complex diseases (such as diabetes, obesity and cancer). The most famous of those studies are the Dutch Hunger Winter and the Överkalix study.

During the winter of the 1944 – towards the end of the world war II – rations to the west part of The Netherlands including Amsterdam were limited by the German occupation and people from all social classes and including pregnant women, received as little as 400–800 calories/day. Known as the Dutch Hunger Winter, this famine period was followed by growing prosperity in the postwar. Later analysis of offspring and grand-offspring health trajectories has shown that children and grandchildren of women pregnant during the Dutch Hunger Winter presented a higher risk for metabolic and cardiovascular diseases [4] and pronounced age-associated decline of cognitive function [5] . Overall, the results of the Dutch Hunger Winter studies highlight the importance of the in-utero environment to the future development of phenotypic trajectories. While mother-to-offspring transmission due to environmental exposure during pregnancy is not strictly epigenetic as it doesn't involve epigenetic reprogramming of the gametes and germline to soma information transfer, evidence of transgenerational effects following the Dutch Hunger Winter suggest that alterations of the in-utero environment can reprogram offspring germline during development.

While the Dutch Hunger Winter study reports only maternal effects, the Överkalix study highlights the transgenerational effects of various parental environmental exposures. Överkalix is a small isolated municipality in the northeast of Sweden characterised by alternating period of feast and famine and rigorous tracking of individuals' health records, such as nutrition, BMIs, smoking and alcohol consumption among others. Retrospective analyses of the collected data have shown that grandparental nutrition during the slow growth phase (SGP - right before puberty) is associated to sex-specific transgenerational phenotypes. In particular, increased food consumption in paternal grandfathers leads to increased cardiovascular and diabetes risk in grandsons [6], while the same in paternal grandmothers leads to increased mortality in granddaughters [6, 7]. Few limitations in the study exist that prevent the Överkalix to be presented as an example of transgenerational epigenetic inheritance. To name two of those, since the region of Överkalix was characterised by alternating periods of feast and famine, the data was essentially collected from many cohorts during the years and - most importantly - the observed transgenerational phenotypes are not robustly replicated across the cohorts, and no independent replication cohort exists to date. Also, while interesting, the gender specificity of the phenotypes still hasn't found a definitive molecular explanation.

To conclude, we define epigenetic inheritance as **the transmission of acquired phenotypic informations across generations** as a consequence of **environmentally-induced and transmittable epigenetic modifications to the parental germline**.

Transmission of acquired phenotypes can be inter- or trans-generational depending on the number of generations presenting the phenotype with no direct exposure to the triggering environmental challenge. **In the case of paternal transmission, transgenerational epigenetic inheritance is defined as the persistence of the inherited phenotypes at least to the second unexposed generation (F2)**, or in other words, if the phenotype is still evident in grand-children. **This definition remains valid for maternal transmission, when the exposure to the environmental challenge is pre-conceptional** (i.e. outside of pregnancy) with no mother-child physical interaction. Alternatively, **one additional generation has to be counted in when exposure to the triggering environmental challenge happens during pregnancy.** In this case, in fact, the physical interaction between the mother and the child leads to indirect exposure of the primordial germ cells to the environmental challenge and makes the child a new F0 [8].

9.2 Developmental Windows of Susceptibility

Epigenetic or acquired inheritance is one of the non-genetic mechanisms to define individuals' phenotypic trajectories and susceptibility to complex – non mendelian – diseases [9].

Not every period of life is equally sensitive to environmental challenges for the establishment of environmentally induced heritable epigenetic alterations. The Dutch Hunger Winter, the Överkalix and the ever-growing number of studies in mammalian animal models have identified several critical windows of susceptibility. They include the **preconceptional**, the **early developmental**, the **gestational** and the **early life** periods. Among those – and in keeping with the definition of epigenetic inheritance given above – only the preconceptional and, to some extent, the early developmental periods can lead to parentally acquired epigenetically inherited phenotypes.

Acquired parental information embedded into oocyte and/or sperm epigenomes is to a certain extent transferred to the offspring at conception. Evidence indeed exist from studies that used in vitro fertilisation (IVF) that gametes are sufficient to intergenerationally transfer phenotypes acquired through exposures to different dietary [10–12], traumatic [13] and temperature [14] challenges.

For example, maternal and paternal exposure to a chronic high fat diet (HFD) challenge, which makes them overly obese and metabolically compromised, increases offspring susceptibility to diet-induced obesity and metabolic syndrome exclusively via oocyte and sperm-embedded information, in a gender and parent-of-origin specific fashion [10]. A similar approach has been used in a more recent study to demonstrate that alteration of DNA methylation in sperm induced by preconceptional cold exposure improves offspring basal energy expenditure and metabolic homeostasis and protects them from diet-induced obesity and metabolic syndrome [14].

Differently from preconceptional exposures, environmental insults during early development and gestational periods – exemplified in humans by the Dutch Hunger Winter study – affect phenotypic trajectories and disease susceptibility by inducing mechanisms of developmental programming aimed at preparing the fetus to an expected environment. This concept is central to the **DOHaD** (Developmental Origins of Health and Disease) hypothesis, by which *environmental factors acting during periods of developmental plasticity* (as eg. early development and gestation) *interact with genotypic information* (and variation) *to change the capacity of the organism to cope with its environment in later life* [15]. Studies in animals models recapitulating the DOHaD concept tend to mimic the Dutch Hunger Winter by restricting food access (or calories availability) to pregnant dams and/or inducing in-utero growth retardation (IUGR) [16–20]. In all cases, growth restriction leads to offspring characterised by low birth weight, catch-up growth during early life and eventually obesity and metabolic syndrome in adulthood. Although some of these phenotypes are linked to epigenetic reprogramming, they do not constitute canonical examples of epigenetic inheritance, as they do not involve parental germline epigenome reprogramming. While since its introduction by Barker et al. in 1993, the DOHaD hypothesis has been mostly used to explain the developmental origin of adult metabolic and cardiovascular diseases [21], different studies have shown that early development and gestation are sensitive periods for a dozen of environmental stimuli, including endocrine disrupting chemicals (EDCs) [22], and stress [23].

A third, important window of susceptibility is the early life period. During the first years of life, newborns are exquisitely sensitive to developmental insults, mostly related to physical and psychological stress. For example, maternal stress either during pregnancy or soon after birth significantly increases offspring risk of developing anxiety, depression and Attention Deficit Hyperactivity Disorder (ADHD) [24–26]. In experimental animals, pre-weaning stress as achieved by forced isolation, reduced maternal care or over/undernutrition as a consequence of imbalanced litter size, leads to adult development of neurobehavioral phenotypes and metabolic disorders [27]. Interestingly, ancestral early life stress may lead to intergenerational effects. Recent studies have indeed shown that different paternal stress reprogram offspring HPA axis [28, 29] and improve neuronal plasticity and behavioral flexibility [30], either via sperm micro RNAs [29] or via epigenetic reprogramming of the locus encoding the Mineralcorticoid receptor in the hippocampus of adult animals [30].

Thus, **environmental insults** on the parents, during early development or during early life **influence adult phenotypes and individual susceptibility to complex non-mendelian diseases.** While epigenetic mechanisms constitute common molecular underpinnings, **only preconceptional, parental insults result in germline epigenome reprogramming and epigenetic inheritance.**

9.3 Overview of the Mechanisms of Epigenetic Inheritance in Mammals

Epigenetic inheritance has been originally discovered and is well documented in plants [31] and lower organisms such as *C. elegans* [32]. The situation is somewhat different in mammals, where the concept of epigenetic inheritance – as defined before – is still cloudy to the scientific community. In particular, while experimental evidence suggest that parental health is important in defining offspring health and disease susceptibility, the underlying molecular mechanisms were – at least until less than a decade ago – grossly unknown. This gap in knowledge is most likely due to the fact that, while the definition of epigenetic inheritance entails the intergenerational transfer of epigenetic information from the gametes to the developing embryo, the inherited epigenome is in facts heavily reprogrammed upon fertilisation and at the early stages of embryonic development (eg. when the primordial germ cells are formed) [33]. Interestingly, these reprogramming events are not complete and escaping loci/molecules constitute potential signals for epigenetic inheritance.

To date, three main mediators of epigenetic information have been characterised as potential signals for epigenetic, intergenerational inheritance: **DNA methylation** (and in particular genomic imprinting), **histone post-translational modifications** and most and foremost, **small non-coding RNAs**.

Since each of these categories is separately reported in the following chapters, I will give here a general overview of each of them.

9.3.1 DNA Methylation and Genomic Imprinting

DNA methylation is defined as the enzymatic addition of one methyl group to the carbon in position 5 of a cytidine residue in a CpG dinucleotide. The addition of the methyl mark is catalysed by a family of enzymes called DNA methyl-transferases (DNMTs) and - while until recently thought to be a quite stable if not irreversible epigenetic modification - enzymatically removed by the family of TET (Ten Eleven Translocation) de-oxygenases [34, 35]. DNA methylation prevents gene activation by facilitating chromatin compaction and by masking the DNA negative charge thereby reducing DNA/Transcription Factor (and DNA/Protein in general) affinity. One of the most famous silencing events that involve DNA methylation is *genomic imprinting*, which indicates the characteristic of mammals to express a portion of their genes (~125 in humans and 100 in mice) in a parent-of-origin specific manner. Imprinted genes are expressed during early development and are important for the formation and the function of the placenta [36], as well as for proper offspring development and health. Problems during the establishment of imprinting indeed lead to severe human diseases, such as the *Prader-Willi* and the *Angelman* syndromes (to mention two) characterised by severe hyperphagia and obesity, primary and secondary derangement of metabolic homeostasis, mental retardation and altered behavior [37]. In isogenic mice, a specific set of paternally expressed genes defines an epigenetic on/off switch for obesity [38].

Imprinted genes are regulated through the imprinting control regions (ICR), cis-regulatory elements that define parent-specific allele expression. Importantly, ICRs are resistant to the post-fertilization DNA methylation reprogramming, therefore being potential carriers of epigenetic information from the germline to the developing embryo and forming the base for epigenetic inheritance [39]. Apart from ICRs, post-fertilization DNA methylation reprogramming is not 100% efficient and some genomic elements escape it. DNA methylation indeed persists on some retroelements, such as SVA (SINE-variable number of tandem repeats-Alu elements) [40] and IAPs (Intracisternal-A-Particle) [41], and on some loci associated with metabolic and neurological disorders [40].

Although the identification of loci that escape DNA methylation reprogramming suggests that DNA methylation might play a role in epigenetic inheritance, this remains largely unknown and rather controversial. Further details on this topic are reported in the following chap. 10.

9.3.2 Histone Post-Translational Modifications

Histones are DNA binding proteins and constitute – together with the associated DNA – the core components of the chromatin. Histone proteins are heavily modified in both somatic cells and in the germline and different patterns of post-translational modifications (PTM) exert different functions on DNA/protein interaction, chromatin stability, gene expression and cellular and organismal phenotypes [42]. Recently, the role of histone PTMs in epigenetic inheritance in lower organisms and in mammals has gained importance. In mammals, germline patterns of histone PTMs regulate early embryonic development and the assembly of embryonic heterochromatin [43–46], thus potentially influencing proper embryonic development and adult phenotypes.

In mature spermatozoa, the vast majority of histone-containing nucleosomes are replaced by protamines to allow sperm DNA to fit the small and hydrodynamic shape of the nucleus. Few loci (~1% in the mouse and 10% in humans) retain the nucleosomal structures and therefore can potentially represent the site of action of the environment to establish epimutations and induce intergenerational epigenetic inheritance. In mouse spermatozoa, retained nucleosomes are located at non-methylated CpG islands [47] and contain histones that are mostly modified on the lysine 4 and/or 27 of the histone H3 (H3K4me or H3K27me) [48]. Interestingly, genetic studies in mice have shown that perturbation of either H3K4me or H3K27me leads to intergenerational phenotypes. In particular, perturbation of H3K4me during spermatogenesis by overexpressing the K4-specific demethylase LSD1 (Lysine Specific Demethylase 1) alters development in wild-type offspring across several generations [49]; while alteration of H3K27me induces intergenerational effects secondary to loss of transposon silencing in the germ line and re-expression of retroelement pseudogenes in adult tissues [50]. More recently, paternal genetic deletion of the K27-specific demethylase KDM6 (or Utx) has shown that proper paternal H3K27me is important for establishment and maintenance of DNA methylation in the offspring, and that alterations of this process lead to increased susceptibility to several forms of cancer [51].

Interestingly, maternal H3K4me and H3K27me are mostly maintained during embryogenesis [52, 53] and are also important for proper embryonic development and offspring phenotypes [54]. Oocyte derived H3K27me, for example, plays an essential role in the establishment of DNA-methylation independent genomic imprints [55], and genetic perturbation of maternally-provided H3K27me results in offspring overgrowth [56]. Oocyte chromatin is characterised by broad H3K4me domains, which are essential for transcriptional silencing during oogenesis [57] and reactivation in late two-cell embryos [53]. As shown for paternally derived H3K4me, genetic alterations of maternally provided H3K4me leads to altered zygotic genome activation and early embryonic development [53, 54, 57, 58].

These pieces of evidence, together with others in lower organisms which we haven't reported here but that have been extensively and recently reported [8], indicate that germline chromatin constitutes an interesting potential platform for epigenetic inheritance, for which the direct proof is still missing. Further details on this topic are reported in the following chap. 12.

9.3.3 Small Non-coding RNAs

A substantial amount of data suggests a prominent role for small non-coding RNAs, maternally or paternally provided, in pre-implantation development and inheritance of acquired phenotypes [59]. Zygotic injection of "challenged" RNAs, indeed reproduces parentally acquired phenotypes, thus indicating small RNAs as important carriers of acquired epigenetic information [11, 12, 60]. Several environmental challenges, such as psychological stress [29, 61–64], diet and exercise [11, 12, 65–67] can alter the abundance and composition of RNAs carried by the germ cells. Both microRNAs and tRNA fragments (tRFs) have been shown to mediate epigenetic inheritance of acquired phenotypes [11–13, 62, 68–72] and, although the exact molecular mechanisms are still largely unknown, evidence indicate that inherited small RNAs control the expression of retroelements and their "target" genes [11] during early development. More recently, post-transcriptional RNA modifications, and in particular methylation of tRNA fragments, have been demonstrated to be necessary for paternal inheritance of diet-induced metabolic phenotypes [60].

While tRNA fragments seem to be predominant determinants of paternal intergenerational effects, mouse genetics have shown miRNAs as important mediators of both paternal and maternal transmission. Conditional deletion of Dicer1 in male and female germline indeed impairs early embryonic development leading to highly penetrant embryonic lethality [70, 72].

These and several other pieces of experimental evidence support an important (to date the most likely) role for small-RNAs in epigenetic inheritance of parentally acquired phenotypes. Further details on this topic are reported in the following chap. 11.

9.4 The Relevance of Epigenetic Inheritance to Phenotypic Adaptation and Disease Risk

Genome Wide Association Studies (GWAS) have identified more than 70.000 disease-associated Single Nucleotide Polymorphisms (SNPs) [73], that have defined a genetic framework for our understanding of complex disease pathogenesis. Nevertheless, of the identified SNPs, only few have a clear clinical meaning and can explain disease heritability. Indeed, the heritability explained by the identified SNP-trait associations does not match the observed heritability of complex human diseases (such as diabetes, obesity, cancer and immunological disorders) [74]. This phenomenon – known as the "*Missing Heritability*" problem – has motivated the scientific community to look for alternative or complementary pathogenetic components [75, 76]. Prominent candidates, supported by epidemiological data on familial phenotypic aggregation and studies on twins, are gene/environment interaction and epigenetic mechanisms. Indeed, one potential explanation for the missing heritability problem is that the observed heritability of disease risk reflects not only Mendelian inheritance, but also inheritance of epigenetic and/or environmental states [77–79]. Mathematical and statistical models have been developed to try to account for gene/environment interaction [80, 81], gene/gene interaction [82], epigenetic inheritance [83, 84] and inheritance of environmental states [81]. Of all the models, the one developed by Furrow R. et al. [81] incorporates epigenetic and environmental contributions, and tries to explain familial aggregation (by means of shared environmental conditions) and phenotypic inheritance uncoupled from genetic inheritance (phenotypes in non-carrier individuals within a family), which result from inherited epigenetic alterations (or epimutations). The most likely scenario is indeed three-ways. Individuals with or without an **inborn genetic risk** are exposed, during their life and in critical windows of susceptibility, to **environmental challenges** (eg. diet, toxicants, smoke to name few), which exacerbate phenotypic manifestations and - by reprogramming the germline epigenome – expose the offspring to **inherit an acquired disease risk**, which could be genotype independent and further exacerbated by a continuous environmental challenge (eg. bad familial dietary habits). This scenario is supported by experimental evidence in model organisms (reported here and in other chapters of this volume) indicating epigenetic inheritance as a prominent mechanism for inheritance of acquired phenotypes and susceptibility to complex diseases.

Apart from inheritance of disease risk and susceptibility, epigenetic inheritance might have a potential important role in adaptive evolution [85]. Experimental evidence exist that different parental environmental challenges, such as under and overnutrition, lead to similar phenotypic manifestation in the offspring [11, 12, 16, 86–88], suggesting that the environmental stimuli might trigger similar adaptive responses in the offspring via different epigenetic mechanisms, such as DNA methylation [87] or sperm-carried smallRNAs [11, 12]. Although in line with the Thrifty Phenotype hypothesis [89, 90], whether this contributes to a survival advantage (and thus proper adaptive evolution in keeping with Darwin and Lamarck) remains a point of question, which cannot be easily solved neither in humans (where experimental times will be too long to properly assess evolution), nor in mammalian experimental models (where the environmental conditions are so tightly controlled that the results lose physiological relevance).

An interesting aspect of environment-induced epigenetic alterations is their potential contribution to genetic variation [91]. W. Burggren, who hypothesised this scenario, called it *genetic takeover*, whereby a change that exists in a population as a pure epigenetic alteration might be fixed as a genetic change, with the epigenetic effect either being dispensable or stabilised by the genetic alteration [91]. Interestingly, experimental evidence exist supporting that epimutations have the potential to directly or indirectly promote mutations of the underlying DNA sequence. For example, removal of DNA methylation by cytosine deamination constitutes an important source of spontaneous DNA damage that, if improperly repaired, could lead to DNA mutation [92]. Also, mutation frequency greatly depends on the chromatin environment, with heterochromatin showing the highest and euchromatin the lowest [93]. While DNA methylation/demethylation and chromatin environments are directly associated to DNA mutations, perturbations of smallRNAs can indirectly affect genetic variation. For example, diet-induced changes in sperm tRNA fragments lead to re-expression (or mis-expression) of transposable elements in early embryos [11], thus exposing them to an extensive source of mutations and genetic polymorphisms [94].

Taken together, although still not extensive, evidence exist that epigenetic inheritance can contribute to individual disease susceptibility, can explain, at least part of the heritability missed by Genome-Wide Association Studies and can actively contribute to increasing genetic and phenotypic variation within the population. These three aspects are good premises for an important role of epigenetic inheritance in adaptive phenotypic evolution, which would need to be properly experimentally proven.

9.5 Conclusion

The most common questions I get asked at meetings and by students is: *why are we what we are? how much information we incorporate and manifest?*

There is no clear answer to these questions, but we are undoubtedly the results of many layers of informations, inherited or acquired. First, our **genotype**, the DNA blueprint we inherit from our parents that defines all our mendelian traits (eg. height, eye color, hair color); second, the **environment** we experience during the lifetime, which affects genome function and physiology; third, the **environment we experienced in the womb**, which influences our adult phenotypes by influencing our development; fourth, the **acquired information we epigenetically inherit from our parents**, which defines the trajectories of our complex - non mendelian - traits (eg. Body Mass Index, disease risk and susceptibility); and fifth, the **interaction** of all the aforementioned layers, which is likely the most relevant to complete the picture.

The take-home message of this introductory chapter is that we are complicated organisms and research is only starting to understand where we come from and why we are what we are. A deeper understanding of epigenetic inheritance and its implications for human development and adult phenotypic trajectories holds promises, but years of high quality research are still ahead of us.

References

1. Darwin C. The Origin of The Species; 1859. p. 1–247.
2. Lamarck. Recherches sur l'organisation des corps vivans et particulièrement sur son origine, sur la cause de ses développemens et des progrès de sa composition; 2003. p. 1–223.
3. Dupont C, Armant D, Brenner C. Epigenetics: definition, mechanisms and clinical perspective. Semin Reprod Med. 2009;27:351–7. https://doi.org/10.1055/s-0029-1237423.
4. Heijmans BT, Tobi EW, Stein AD, et al. Persistent epigenetic differences associated with prenatal exposure to famine in humans. Proc Natl Acad Sci U S A. 2008;105:17046–9. https://doi.org/10.1073/pnas.0806560105.
5. de Rooij SR, Wouters H, Yonker JE, et al. Prenatal undernutrition and cognitive function in late adulthood. Proc Natl Acad Sci U S A. 2010;107:16881–6. https://doi.org/10.1073/pnas.1009459107.

6. Kaati G, Bygren LO, Edvinsson S. Cardiovascular and diabetes mortality determined by nutrition during parents" and grandparents" slow growth period. Eur J Hum Genet. 2002;10:682–8. https://doi.org/10.1038/sj.ejhg.5200859.
7. Kaati G, Bygren LO, Pembrey M, Sjöström M. Transgenerational response to nutrition, early life circumstances and longevity. Eur J Hum Genet. 2007;15:784–90. https://doi.org/10.1038/sj.ejhg.5201832.
8. Skvortsova K, Iovino N, Bogdanović O. Functions and mechanisms of epigenetic inheritance in animals. Nat Rev Mol Cell Biol. 2018;19:1–17. https://doi.org/10.1038/s41580-018-0074-2.
9. Panzeri I, Pospisilik JA. Epigenetic control of variation and stochasticity in metabolic disease. Mol Metab. 2018;14:26–38. https://doi.org/10.1016/j.molmet.2018.05.010.
10. Sass S, Wu M, Dyckhoff D, et al. Epigenetic germline inheritance of diet-induced obesity and insulin resistance. Nat Genet. 2016;48:1–4. https://doi.org/10.1038/ng.3527.
11. Sharma U, Conine CC, Shea JM, et al. Biogenesis and function of tRNA fragments during sperm maturation and fertilization in mammals. Science. 2015;351:1–9. https://doi.org/10.1126/science.aad6780.
12. Chen Q, Yan M, Cao Z, et al. Sperm tsRNAs contribute to intergenerational inheritance of an acquired metabolic disorder. Science. 2015;351:1–8. https://doi.org/10.1126/science.aad7977.
13. Gapp K, Jawaid A, Sarkies P, et al. Implication of sperm RNAs in transgenerational inheritance of the effects of early trauma in mice. Nat Neurosci. 2014;17:667–9. https://doi.org/10.1038/nn.3695.
14. Sun W, Dong H, Becker AS, et al. Cold-induced epigenetic programming of the sperm enhances brown adipose tissue activity in the offspring. Nat Med. 2018;24:1–18. https://doi.org/10.1038/s41591-018-0102-y.
15. Heindel JJ, Balbus J, Birnbaum L, et al. Developmental origins of health and disease: integrating environmental influences. Endocrinology. 2015;156:3416–21. https://doi.org/10.1210/EN.2015-1394.
16. Jimenez-Chillaron JC, Jimenez-Chillaron JC, Isganaitis E, et al. Intergenerational transmission of glucose intolerance and obesity by in utero undernutrition in mice. Diabetes. 2009;58:460–8. https://doi.org/10.2337/db08-0490.
17. Simmons RA, Templeton LJ, Gertz SJ. Intrauterine growth retardation leads to the development of type 2 diabetes in the rat. Diabetes. 2001;50:2279–86. https://doi.org/10.2337/diabetes.50.10.2279.
18. Ozanne SE, Hales CN. The long-term consequences of intra-uterine protein malnutrition for glucose metabolism. Proc Nutr Soc. 2007;58:615–9. https://doi.org/10.1017/S0029665199000804.
19. Berends LM, Fernandez-Twinn DS, Martin-Gronert MS, et al. Catch-up growth following intra-uterine growth-restriction programmes an insulin-resistant phenotype in adipose tissue. Int J Obes. 2012;37:1051–7. https://doi.org/10.1038/ijo.2012.196.
20. Hardikar AA, Satoor SN, Karandikar MS, et al. Multigenerational Undernutrition increases susceptibility to obesity and diabetes that is not reversed after dietary recuperation. Cell Metab. 2015;22:1–9. https://doi.org/10.1016/j.cmet.2015.06.008.
21. Barker DJP. The origins of the developmental origins theory. J Intern Med. 2007;261:412–7. https://doi.org/10.1111/j.1365-2796.2007.01809.x.
22. Montrose L, Padmanabhan V, Goodrich JM, et al. Maternal levels of endocrine disrupting chemicals in the first trimester of pregnancy are associated with infant cord blood DNA methylation. Epigenetics. 2018;13:301–9. https://doi.org/10.1080/15592294.2018.1448680.
23. Berghänel A, Heistermann M, Schülke O, Ostner J. Prenatal stress accelerates offspring growth to compensate for reduced maternal investment across mammals. Proc Natl Acad Sci U S A. 2017;114:E10658–66. https://doi.org/10.1073/pnas.1707152114.
24. Ronald A, Pennell CE, Whitehouse AJO. Prenatal maternal stress associated with ADHD and autistic traits in early childhood. Front Psychology. 2011;1:1–8. https://doi.org/10.3389/fpsyg.2010.00223.

25. Rosenqvist MA, Sjölander A, Ystrom E, et al. Adverse family life events during pregnancy and ADHDsymptoms in five-year-old offspring. J Child Psychol Psychiatr. 2018;60:665–75. https://doi.org/10.1111/jcpp.12990.
26. Christine H, Nemeroff C. The role of childhood trauma in the neurobiology of mood and anxiety disorders: preclinical and clinical studies. BPS. 2001;49:1023–39. https://doi.org/10.1016/S0006-3223(01)01157-X.
27. Kappil M, Wright RO, Sanders AP. Developmental Origins of Common Disease: Epigenetic Contributions to Obesity. 2016; https://doi.org/10.1146/annurev-genom-090314-050057.
28. Rodgers AB, Morgan CP, Leu NA, Bale TL. Transgenerational epigenetic programming via sperm microRNA recapitulates effects of paternal stress. Proc Natl Acad Sci U S A. 2015;112:13699–704. https://doi.org/10.1073/pnas.1508347112.
29. Rodgers AB, Morgan CP, Bronson SL, et al. Paternal stress exposure alters sperm MicroRNA content and reprograms offspring HPA stress Axis regulation. J Neurosci. 2013;33:9003–12. https://doi.org/10.1523/JNEUROSCI.0914-13.2013.
30. Gapp K, Soldado-Magraner S, Alvarez-Sánchez M, et al. Early life stress in fathers improves behavioural flexibility in their offspring. Nat Commun. 2014;5:5466. https://doi.org/10.1038/ncomms6466.
31. Annacondia ML, Martinez G. Chapter 12 – plant models of transgenerational epigenetic inheritance. Transgenerational Epigenetics. 2019;263–82. https://doi.org/10.1016/B978-0-12-816363-4.00012-2.
32. Weiser NE, Kim JK.Multigenerational Regulation of the Caenorhabditis elegansChromatin Landscape by Germline Small RNAs. Annu Rev Genet 2019;53:annurev–genet–112618–043505–23. doi: https://doi.org/10.1146/annurev-genet-112618-043505
33. Eckersley-Maslin MA, Alda-Catalinas C, Reik W. Dynamics of the epigenetic landscape during the maternal-to-zygotic transition. Nat Rev Mol Cell Biol. 2018;19:436–50. https://doi.org/10.1038/s41580-018-0008-z.
34. Ito S, D'Alessio AC, Taranova OV, et al. Role of Tet proteins in 5mC to 5hmC conversion, ES-cell self-renewal and inner cell mass specification. Nature. 2010;466:1129–33. https://doi.org/10.1038/nature09303.
35. Kriaucionis S, Heintz N. The nuclear DNA base 5-hydroxymethylcytosine is present in Purkinje neurons and the brain. Science. 2009;324:929–30. https://doi.org/10.1126/science.1169786.
36. Tunster SJ, Jensen AB, John RM. Imprinted genes in mouse placental development and the regulation of fetal energy stores. Reproduction. 2013;145:R117–37. https://doi.org/10.1530/REP-12-0511.
37. Nicholls RD, Knepper JL. Genome Organization, Function, and Imprinting in Prader-Willi and Angelman Syndromes. 2003; https://doi.org/10.1146/annurev.genom.2.1.153.
38. Dalgaard K, Landgraf K, Heyne S, et al. Trim28 Haploinsufficiency triggers bi-stable epigenetic obesity. Cell. 2016;164:353–64. https://doi.org/10.1016/j.cell.2015.12.025.
39. Adalsteinsson B, Ferguson-Smith A. Epigenetic control of the genome—lessons from genomic imprinting. Genes (Basel). 2014;5:635–55. https://doi.org/10.3390/genes5030635.
40. Tang WWC, Dietmann S, Irie N, et al. A unique gene regulatory network resets the human Germline Epigenome for development. Cell. 2015;161:1453–67. https://doi.org/10.1016/j.cell.2015.04.053.
41. Kazachenka A, Bertozzi TM, Sjoberg-Herrera MK, et al. Identification, characterization, and heritability of murine metastable Epialleles: implications for non- genetic inheritance. Cell. 2018;175:1–27. https://doi.org/10.1016/j.cell.2018.09.043.
42. Peterson CL, Laniel M-A. Histones and histone modifications. Curr Biol. 2004;14:R546–51. https://doi.org/10.1016/j.cub.2004.07.007.
43. van de Werken C, van der Heijden GW, Eleveld C, et al (1AD) Paternal heterochromatin formation in human embryos is H3K9/HP1 directed and primed by sperm-derived histone modifications. Nature Communications 5:1–15. doi: https://doi.org/10.1038/ncomms6868
44. Puschendorf M, Terranova R, Boutsma E, et al. PRC1 and Suv39h specify parental asymmetry at constitutive heterochromatin in early mouse embryos. Nat Genet. 2008;40:411–20. https://doi.org/10.1038/ng.99.

45. Lesch BJ, Dokshin GA, Young RA, et al. A set of genes critical to development is epigenetically poised in mouse germ cells from fetal stages through completion of meiosis. Proc Natl Acad Sci U S A. 2013;110:16061–6. https://doi.org/10.1073/pnas.1315204110.
46. Lesch BJ, Silber SJ, McCarrey JR, Page DC. Parallel evolution of male germline epigenetic poising and somatic development in animals. Nat Genet. 2016;48:1–10. https://doi.org/10.1038/ng.3591.
47. Erkek S, Hisano M, Liang C-Y, et al. Molecular determinants of nucleosome retention at CpG-rich sequences in mouse spermatozoa. Nat Struct Mol Biol. 2013;20:868–75. https://doi.org/10.1038/nsmb.2599.
48. Brykczynska U, Hisano M, Erkek S, et al. Repressive and active histone methylation mark distinct promoters in human and mouse spermatozoa. Nat Publ Group. 2010;17:679–87. https://doi.org/10.1038/nsmb.1821.
49. Siklenka K, Erkek S, Godmann M, et al. Disruption of histone methylation in developing sperm impairs offspring health transgenerationally. Science. 2015;350:aab2006–aab2006. doi: https://doi.org/10.1126/science.aab2006
50. Stringer JM, Forster SC, Qu Z, et al. Reduced PRC2 function alters male germline epigenetic programming and paternal inheritance. BMC Biol. 2018;16:1–20. https://doi.org/10.1186/s12915-018-0569-5.
51. Lesch BJ, Tothova Z, Morgan EA, et al. Intergenerational epigenetic inheritance of cancer susceptibility in mammals. elife. 2019;1–29. https://doi.org/10.7554/eLife.39380.001.
52. Zheng H, Huang B, Zhang B, et al. Resetting epigenetic memory by reprogramming of histone modifications in mammals. Mol Cell. 2016;63:1066–79. https://doi.org/10.1016/j.molcel.2016.08.032.
53. Zhang B, Zheng H, Huang B, et al. Allelic reprogramming of the histone modification H3K4me3 in early mammalian development. Nature. 2016;537:1–18. https://doi.org/10.1038/nature19361.
54. Liu X, Wang C, Liu W, et al. Distinct features of H3K4me3 and H3K27me3 chromatin domains in pre-implantation embryos. Nat Publ Group. 2016;537:558–62. https://doi.org/10.1038/nature19362.
55. Inoue A, Jiang L, Lu F, et al. Maternal H3K27me3 controls DNA methylation-independent imprinting. Nat Publ Group. 2017;547:419–24. https://doi.org/10.1038/nature23262.
56. Prokopuk L, Stringer JM, White CR, et al. Loss of maternal EED results in postnatal overgrowth. Clin Epigenetics. 2018;10:1–14. https://doi.org/10.1186/s13148-018-0526-8.
57. Andreu-Vieyra CV, Chen R, Agno JE, et al. MLL2 is required in oocytes for bulk histone 3 lysine 4 Trimethylation and transcriptional silencing. PLoS Biol. 2010;8:e1000453. https://doi.org/10.1371/journal.pbio.1000453.
58. Dahl JA, Jung I, Aanes H, et al. Broad histone H3K4me3 domains in mouse oocytes modulate maternal-to-zygotic transition. Nature. 2016;537:548–52. https://doi.org/10.1038/nature19360.
59. Schoelz JM, Riddle NC. Chapter 9 - heritable generational epigenetic effects through small noncoding RNA. Transgenerational Epigenetics. 2019;185–212. https://doi.org/10.1016/B978-0-12-816363-4.00009-2.
60. Zhang Y, Zhang X, Shi J, et al. Dnmt2 mediates intergenerational transmission of paternally acquired metabolic disorders through sperm small non-coding RNAs. Nat Cell Biol. 2018;20:535–40. https://doi.org/10.1038/s41556-018-0087-2.
61. Wu L, Lu Y, Jiao Y, et al. Paternal psychological stress reprograms hepatic gluconeogenesis in offspring. Cell Metab. 2016;23:735–43. https://doi.org/10.1016/j.cmet.2016.01.014.
62. Gapp K, Steenwyk G, Germain PL, et al. Alterations in sperm long RNA contribute to the epigenetic inheritance of the effects of postnatal trauma. Mol Psychiatry. 2018:1–13. https://doi.org/10.1038/s41380-018-0271-6.
63. Short AK, Fennell KA, Perreau VM, et al. Elevated paternal glucocorticoid exposure alters the small noncoding RNA profile in sperm and modifies anxiety and depressive phenotypes in the offspring. 2016;6:e837–12. https://doi.org/10.1038/tp.2016.109.

64. Dickson DA, Paulus JK, Mensah V, et al. Reduced levels of miRNAs 449 and 34 in sperm of mice and men exposed to early life stress. Transl Psychiatry. 2018;8:1–10. https://doi.org/10.1038/s41398-018-0146-2.
65. Sarker G, Sun W, Rosenkranz D, et al. Maternal overnutrition programs hedonic and metabolic phenotypes across generations through sperm tsRNAs. Proc Natl Acad Sci U S A. 2019;116:10547–56. https://doi.org/10.1073/pnas.1820810116.
66. Fullston T, Ohlsson Teague EMC, Palmer NO, et al. Paternal obesity initiates metabolic disturbances in two generations of mice with incomplete penetrance to the F2 generation and alters the transcriptional profile of testis and sperm microRNA content. FASEB J. 2013;27:4226–43. https://doi.org/10.1096/fj.12-224048.
67. Stanford KI, Rasmussen M, Baer LA, et al. Paternal exercise improves glucose metabolism in adult offspring. Diabetes. 2018;67:2530–40. https://doi.org/10.2337/db18-0667.
68. Yang Q, Lin J, Liu M, et al. Highly sensitive sequencing reveals dynamic modifications and activities of small RNAs in mouse oocytes and early embryos. Sci Adv. 2016;2:e1501482. https://doi.org/10.1126/sciadv.1501482.
69. Rassoulzadegan M, Grandjean V, Gounon P, et al. RNA-mediated non-mendelian inheritance of an epigenetic change in the mouse. Nature. 2006;441:469–74. https://doi.org/10.1038/nature04674.
70. Tang F, Kaneda M, O'Carroll D, et al. Maternal microRNAs are essential for mouse zygotic development. Genes Dev. 2007;21:644–8. https://doi.org/10.1101/gad.418707.
71. Yuan S, Oliver D, Schuster A, et al. Breeding scheme and maternal small RNAs affect the efficiency of transgenerational inheritance of a paramutation in mice. Sci Rep. 2015;5:9266–15. https://doi.org/10.1038/srep09266.
72. Yuan S, Schuster A, Tang C, et al. Sperm-borne miRNAs and endo-siRNAs are important for fertilization and preimplantation embryonic development. Development. 2016;143:635–47. https://doi.org/10.1242/dev.131755.
73. Buniello A, MacArthur JAL, Cerezo M, et al. The NHGRI-EBI GWAS Catalog of published genome-wide association studies, targeted arrays and summary statistics 2019. Nucleic Acids Res. 2018;47:D1005–12. https://doi.org/10.1093/nar/gky1120.
74. Goldstein D. Common genetic variation and human traits. N Engl J Med. 2009;360:1696–8.
75. Trerotola M, Relli V, Simeone P, Alberti S. Epigenetic inheritance and the missing heritability. Hum Genomics. 2015;9:17. https://doi.org/10.1186/s40246-015-0041-3.
76. Manolio TA, Collins FS, Cox NJ, et al. Finding the missing heritability of complex diseases. Nature. 2009;461:747–53. https://doi.org/10.1038/nature08494.
77. Maher B. Personal genomes: the case of the missing heritability. Nature News. 2008;456:18–21. https://doi.org/10.1038/456018a.
78. Eichler EE, Flint J, Gibson G, et al. Missing heritability and strategies for finding the underlying causes of complex disease. Nat Rev Genet. 2010;11:446–50. https://doi.org/10.1038/nrg2809.
79. Petronis A. Epigenetics as a unifying principle in the aetiology of complex traits and diseases. Nature. 2010;465:721–7. https://doi.org/10.1038/nature09230.
80. Danchin É, Wagner RH. Inclusive heritability: combining genetic and non-genetic information to study animal behavior and culture. Oikos. 2010;119:210–8. https://doi.org/10.1111/j.1600-0706.2009.17640.x.
81. Furrow RE, Christiansen FB, Feldman MW. Environment-sensitive epigenetics and the heritability of complex diseases. Genetics. 2011;189:1377–87. https://doi.org/10.1534/genetics.111.131912.
82. Bubb KL, Queitsch C. A Two-State Epistasis Model Reduces Missing Heritability of Complex Traits. bioRxiv. 2016;p. 1–31. https://doi.org/10.1101/017491.
83. Slatkin M. Epigenetic inheritance and the missing heritability problem. Genetics. 2009;182:845–50. https://doi.org/10.1534/genetics.109.102798.
84. Tal O, Kisdi E, Jablonka E. Epigenetic contribution to covariance between relatives. Genetics. 2010;184:1037–50. https://doi.org/10.1534/genetics.109.112466.

85. Sarkies P. Molecular mechanisms of epigenetic inheritance_ possible evolutionary implications. Semin Cell Dev Biol. 2019;1–0. doi: https://doi.org/10.1016/j.semcdb.2019.06.005.
86. Öst A, Lempradl A, Casas E, et al. Paternal diet defines offspring chromatin state and intergenerational obesity. Cell. 2014;159:1352–64. https://doi.org/10.1016/j.cell.2014.11.005.
87. Radford EJ, Radford EJ, Ito M, et al. In utero effects. In utero undernourishment perturbs the adult sperm methylome and intergenerational metabolism. Science. 2014;345:1255903. https://doi.org/10.1126/science.1255903.
88. Carone BR, Fauquier L, Habib N, et al. Paternally induced Transgenerational environmental reprogramming of metabolic gene expression in mammals. Cell. 2010;143:1084–96. https://doi.org/10.1016/j.cell.2010.12.008.
89. Hales CN, bulletin DBBM, 2001 thrifty phenotype hypothesis | British Medical Bulletin | Oxford Academic. doi:https://doi.org/10.1093/bmb/60.1.5
90. Vaag AA, Grunnet LG, Arora GP, Brøns C. The thrifty phenotype hypothesis revisited. Diabetologia. 2012;55:2085–8. https://doi.org/10.1007/s00125-012-2589-y.
91. Burggren W. Epigenetic inheritance and its role in evolutionary biology: re-evaluation and new perspectives. Biology. 2016;5:24–2. https://doi.org/10.3390/biology5020024.
92. Alexandrov LB, Jones PH, Wedge DC, et al. Clock-like mutational processes in human somatic cells. Nat Publ Group. 2015;47:1402–7. https://doi.org/10.1038/ng.3441.
93. Schuster-Böckler B, Lehner B. Chromatin organization is a major influence on regional mutation rates in human cancer cells. Nature. 2012;488:504–7. https://doi.org/10.1038/nature11273.
94. Bourque G, Burns KH, Gehring M, et al. Ten things you should know about transposable elements. Genome Biol. 2018;19:1–12. https://doi.org/10.1186/s13059-018-1577-z.

Chapter 10
The (not so) Controversial Role of DNA Methylation in Epigenetic Inheritance Across Generations

Martin Irmler, Daniela Kaspar, Martin Hrabě de Angelis, and Johannes Beckers

Keywords Epigenetic inheritance · Intergenerational · Transgenerational · DNA methylation · 5`-methylcytosine · Reprogramming · Primordial germ cells · Epimutation · Epiallele · CpG island · Hypermethylation · Hypomethylation

10.1 Methylation of Cytosine Represses Gene Expression in Mammals

5`-methylcytosine (m^5C, more specifically 5-methyldesoxycytidin) is a chemical variant of the pyrimidine base cytosine and was to our knowledge first suggested as a natural constituent of mammalian genomic DNA by Hotchkiss in his chromatographic analysis of nucleic acids isolated from calf thymus in 1948 [1]. In fact, DNA methylation in mammals occurs mainly at cytosine residues of CpG dinucleotides of which on average 70% to 80% are methylated. About 60% of human gene promoters contain CpG-rich regions known as CpG islands (CGIs), with variable lengths of a few hundred basepairs to several kilobasepairs (kb) [2]. Methylation levels of CGIs are often inversely correlated with gene expression in mammals such that hypermethylation is associated with gene silencing and lower levels of

M. Irmler · D. Kaspar
Helmholtz Zentrum München GmbH, Institute of Experimental Genetics, Neuherberg, Germany

M. Hrabě de Angelis · J. Beckers (✉)
Helmholtz Zentrum München GmbH, Institute of Experimental Genetics, Neuherberg, Germany

Chair of Experimental Genetics, Technische Universität München, Freising, Germany

Deutsches Zentrum für Diabetesforschung, Neuherberg, Germany
e-mail: beckers@helmholtz-muenchen.de

R. Teperino (ed.), *Beyond Our Genes*,
https://doi.org/10.1007/978-3-030-35213-4_10

transcripts and proteins [3, 4]. In particular, methylation at transcriptional start sites is usually strongly correlated with long-term repression and has governed the view of functional consequences of DNA methylation, which is only one of several players involved in the regulation of gene expression (reviewed in [5]).

Genomic patterns of m^5C are tissue-specific and evolutionary conserved between mammals [6, 7]. Methyltransferases that write or erase m^5Cs, and proteins that recognize and bind DNA depending on its methylation status have been identified in mammals. As examples, genes like *Dnmt3a* and *Dnmt3b* are involved in *de novo* DNA methylation and *Tet* enzymes in their conversion back to normal cytosines. *DNA methyltransferase 1* (*Dnmt1*) recognizes hemi-methylated DNA and methylates the newly replicated DNA strand, thus maintaining the methylation status in daughter cells [8]. The functional deletion of the *Dnmt1* gene in mice reduced genome-wide methylation levels to approximately one-third to that of wildtype mice and resulted in embryonic lethality before mid-gestation [9]. In addition to its role in tissue-specific gene regulation and embryogenesis, numerous studies have demonstrated that proper regulation of DNA methylation patterns is required for inactivation of X chromosomes, genomic imprinting, chromosome stability, frequency of recombination and silencing of transposable elements [10–12]. Furthermore, there is a growing number of human diseases, including cancer, that are caused by the disturbance of proper DNA methylation maintenance and/or establishment [13]. Accordingly, DNA methylation has also become a prominent target, for example, for anti-cancer therapies [14].

DNA methylation in mammals can also occur as adenine N^6-methylation (m^6A). In mouse and man, m^6A levels are regulated by the methyltransferase *N6amt1* and the demethylase *Alkbh1* [15, 16]. High adenine methylation levels in promoter regions have been associated with increased expression levels of the corresponding genes [16], but in mammals they are much less studied than cytosine methylation. In contrast, m^6A is the most prominent methylation in prokaryotes and some other non-vertebrates like *C. elegans* [17], where adenine methylation is part of a transgenerational epigenetic signal for mitochondrial stress adaptation [18].

Thus, there is no doubt today that DNA methylation has an important functional role in the regulation of mammalian gene expression. Moreover, there is increasing evidence in the literature of the last decade, that acquired phenotypes can be inherited across generations in mammals and we will elucidate this issue in some more detail below [19]. However, clear answers to the question, whether DNA methylation may represent or contribute to the epigenetic code that is responsible for the inheritance of acquired traits over generations, are hard to find. Yet, it was previously suggested - based on the scientific knowledge of the time - that DNA methylation would be functionally important exclusively for the somatic lineage, but would have no role in early embryonic cells, including primordial germ cells and gametes [20]. Following more than twenty years of mammalian epigenetic research, we make an attempt to overview the relevant literature in order to re-address this question.

10.2 Epigenetic Inheritance – It All Began in Plants ...

Before addressing this questions, we briefly want to acknowledge that the initial discoveries of such important epigenetic phenomena – most of which involve DNA methylation - like transposable element inactivation, imprinting, paramutation, transgene silencing, and co-suppression were originally made in plants. Just a very few historical examples shall follow.

One fundamental implication of Mendel's laws is that genetic factors segregate unchanged from heterozygotes. This rule of *integrity of genes* in heterozygotes was ultimately confirmed in many experiments. However, in some genetic experiments in the 1950ies it appeared that there are rare exceptions from the rule of gene integrity. It was observed in maize that different pigmentation phenotypes in the offspring could be produced by R^r alleles depending on the allelic composition of the parental plants (R^rR^r or R^rR^{st}) [21]. The heterozygous allele affected the expression of the R^r allele *in trans*, and this change in expression was heritable and altered the offspring's phenotype. The inheritance of this so called *paramutation* was maintained over multiple generations, but was also reversible depending on the heterozygous allele. Although this represents a clear case of transgenerational epigenetic inheritance of a paramutation, the underlying mechanism rather involves RNA-mediated gene silencing instead of DNA methylation [22].

Another classical example for transgenerational epigenetic inheritance is provided by the natural variation of the toadflax's (*Linaria vulgaris*) floral symmetry: The morphological appearance of its flowers varies between radial or bilateral symmetry. The radial symmetric flower of *Linaria* is associated with extensive DNA methylation and reduced expression of the *Linaria cycloidea* (*Lcyc*) gene. The bilateral flower variant, in contrast, is caused by low methylation and strong expression of the *Lcyc* gene [23]. This epigenetic mutation (*epimutation*) involving DNA methylation can be inherited over many generations in *Linaria.*

Several arguments have been brought forward to explain why epigenetic inheritance may be more pervasive in plants than in animals. For example, it has been suggested that epigenetic inheritance in plants may provide a selective advantage as a means of adaptation to the environment that is faster than evolutionary genetic changes. Since parental plants and their offspring usually share the same environmental conditions, plants may be more dependent on changes in gene expression than animals [24, 25]. Also, epigenetic inheritance as adaptation to shared environments may be more common in organisms with a short life cycle as compared to, for example, long lived mammals [26]. With regards to the question how epigenetic information could be transferred from somatic cells to the germ line, it is important to consider that in plants there is no early separation between somatic cells and germ line in plants as is seen in mammals.

Despite such arguments in support of the idea that epigenetic inheritance may be particularly advantageous for plants, phenomena like paramutation and DNA methylation signatures that persist across generations were soon discovered also in rodents. The early and most prominent examples showed that either imprinting control regions or transgenic sequences could ectopically function as enhancers of DNA methylation and they also acted *in trans* – apparently against Mendel's rule of the integrity of genes [27, 28].

Before returning to epigenetic inheritance and the role of DNA methylation in mammals we would like to point out that the important finding that methylated DNA is more prone to mutation was most likely first made in fungi [29]. Therefore, many of the pathways and molecular mechanisms used by mammals to establish, maintain and regulate DNA methylation patterns were originally discovered because they are widely conserved in fungi and plants (reviewed in [30, 31]).

10.3 Detection of DNA Methylation: Current and Coming Technologies

The significant progress that has been made in our understanding of the functional role of DNA methylation is tightly linked to the technological progress that was made in the detection of m^5C and other chemical modifications of DNA. While immunofluorescent-based methods using antibodies against m^5C are instrumental for the large-scale screening of genome-wide methylation [32], sequencing-based methods allow the determination of methylated cytosines at nucleotide resolution. So far, whole-genome bisulfite sequencing (WGBS) is often considered as the gold standard for the genome-wide detection of m^5C [33]. Briefly, bisulfite-treated genomic DNA undergoes complete conversion of unmethylated cytosines to uracil. Subsequently, methylation of CpGs can be determined by comparing sequencing data of treated versus non-treated genomic DNA. A conceptual limitation of bisulfite sequencing, however, is that 5-hydroxymethylcytosine (hm^5C) cannot be distinguished from m^5C, since both, methylation as well as hydroxymethylation, prevent the bisulfite-driven chemical conversion of cytosine to uracil. *In vivo,* oxidation of m^5C to hm^5C is thought to initiate the conversion of the methylated cytosine back to the unmethylated cytosine. This conversion is catalyzed by the methylcytosine dioxygenase *Tet1* and is thought to mark the transition from the repressed to expressed gene [34]. Furthermore, WGBS usually requires high sequencing depth (20-fold or more) to achieve sufficient coverage of the genome making it still an expensive method. Modifications of the original WGBS protocol and new methods have been developed to overcome some of these limitations. Reduced representation bisulfite sequencing (RRBS), for example, uses restriction enzymes like *MspI* to reducc the required number of reads and still captures about 60 % of gene promoters [35]. Similarly, methylation-sensitive restriction enzyme bisulfite sequencing (MREBS) reduces costs for sequencing, but provides a higher level of CpG coverage by selecting for hypomethylated regions [36]. Tet-assisted bisulfite sequencing (TAB-Seq) allows to distinguish m^5C from hm^5C by specifically protecting hm^5C but not m^5C and cytosine from chemical conversion by bisulfite [37]. Also, the general limitations of next-generation shot gun sequencing are inherent to any WGBS approach. In particular, repeat regions are difficult to sequence and therefore the methylation status of repetitive sequences often remains unclear [38]. Besides, the generation of averaged signals from pooled cells is likely to mask mosaicism across cells and tissues in addition to allele-specific differences [39]. Single-cell bisulfite sequencing (scBS-Seq) is one of the recent methods that helps to address the heterogeneity in methylation patterns in populations of cells [40].

Third generation sequencing technologies like single molecule real-time (SMRT) sequencing from Pacific Biosciences (PacBio, [41]) and nanopore sequencing by Oxford Nanopore Technologies [42] provide methods for long read lengths and, in addition, the detection of epigenetic modifications [43, 44]. Although these methods have been used in combination with bisulfite treatment, there is no inherent requirement for base conversions in order to detect epigenetic modifications with third generation sequencing. In principle, these can be detected due to altered kinetics of base addition or processing during the normal course of sequencing. Both methods were used to detect methylated nucleotides and were applied separately or in combination to detect genome-wide methylation levels, although yet restricted to smaller genomes [45, 46].

10.4 Reprogramming the Early Mammalian Embryo

The genome-wide demethylation that has been observed in mammalian preimplantation embryos is often referred to as *reprogramming*. Since the question whether DNA methylation may account for epigenetic inheritance across generations is intrinsically linked to reprogramming, we briefly summarize this important embryonic process.

10.4.1 Hit Me Once: Reprogramming the Pre-implantation Embryo

Based on immunofluorescent data using an antibody against m^5C, demethylation of the male pronucleus in the mouse embryo is completed already 4h post fertilization. This results in transiently asymmetric methylation between the maternal and paternal pronucleus [47]. Subsequently, it was shown by reduced representation bisulfite sequencing (RRBS) that both maternal and paternal genomes undergo widespread active and passive demethylation in zygotes before the first mitotic division [48–50]. Active demethylation is initiated via m^5C oxidation by Tet dioxygenases and passive demethylation results from the first replication of the genome before the first cleavage. By the morula stage, little genomic methylation remains. It should also be pointed out that - at least to our knowledge - there is no data suggesting that the mammalian genome would be fully and completely de-methylated at any developmental stage. *De novo* methylation occurs by the blastocyst stage but is restricted to the inner cell mass (ICM) from which the embryo proper develops subsequently. The trophectoderm remains in the de-methylated state [47]. Analyses of the human DNA methylomes during early embryonic development rather support the idea that global methylation changes in human zygotes are very similar to those in mouse embryos [51, 52]. With regard to reprogramming and demethylation processes, mammals appear to be distinct among vertebrates. For example in the zebrafish embryo, it is the paternal methylation that is stably maintained and the relatively hypomethylated maternal genome undergoes drastic reprogramming to establish a sperm-like methylome [53, 54].

Unfortunately, we know very little about how genomic DNA re-methylation is regulated at specific loci. Is the DNA methylation information simply lost or erased during reprogramming? Or is the locus-specific DNA methylation information probably transferred to a different molecular entity, such as the chromatin, which may serve as a blueprint or mask for the subsequent re-methylation? Interestingly, recent data in zebrafish provided evidence that DNA methylation in gametes can be maintained or reprogrammed in the early embryo through the installation of so called placeholder nucleosomes, containing histone H2A variant H2A.Z(FV) and H3K4me1, during stages of transcriptional quiescence [55]. We would also like to point out that histones in mammalian sperm are widely replaced by protamines [56]. However, subsequent work revealed that a small fraction of nucleosomes in sperm and histone modifications are retained in specific genetic loci that function during embryogenesis [57, 58], suggesting that constituents of chromatin and their modifications may also contribute to epigenetic inheritance.

10.4.2 Hit Me Twice: Reprogramming from Primordial Germ Cells to Gonocytes

The proliferation and migration of primordial germ cells (PGCs) and the genome-wide changes of DNA methylation were studied in the mouse before they were examined also in humans. This second round of reprogramming is essential to generate totipotency of developing PGCs in mammals. Progenitors of the PGCs in mammals originate from the posterior mesoderm of the epiblast [59–61] where they transiently translocate from the epiblast to an extra-embryonic position. During their specification in the epiblast, PGCs originally inherit the methylation patterns of somatic cells. In mice, PGCs differentiate from the epiblast at around E6.5, and form a small cluster of cells in the extra-embryonic mesoderm at E7.25 [62]. This small group of cells then starts proliferation and migrates towards the dorsal aspect of the hindgut where they start to separate in left and right halves following E9.5. From there they migrate laterally along the dorsal body wall until they reach the area in which they contribute to the formation of the genital ridges (E10.5 to E12.5).

At this time, sex-specific differentiation begins. Proliferation stops in mice at around E13.5 when the group of PGCs has reached approximately 26.000 cells and male germ cells go into mitotic and female germ cells enter meiotic arrest. At the same time, genome-wide loss of m^5C in the PGCs is most prevalent [63].

Ultimately, CpG dinucleotide methylation levels in mice decreased from approximately 71% in the E6.5 epiblast to 14% in the male and 7% in the female PGCs at E13.5 [64, 65]. Erasure of m^5C in PGCs occurs via conversion to 5-hydroxymethylcytosine (hm^5C) and is driven by strong expression of *Tet1* and *Tet2* [66]. While these low methylation levels persist in female PGCs, de novo methylation increases methylation levels in male PGCs to about 50% by E16.5. This global de-methylation during PGC development [66] was not associated with promiscuous activation of gene transcription, but rather with a controlled and transient expression of the gene network that controls pluripotency [64]. In addition to intracisternal A particles (IAPs, endogenous retroviral elements that were recently acquired to the mouse genome), which are rather resistant to the global hypomethylation, it was also shown that a limited number of genes either resist global methylation reprogramming [66, 67] or include CpG islands (CGIs) with methylation levels that are variably erased between developmental stages, sexes and potentially between individuals [64, 68, 69]. It was suggested that such variably erased CGIs (VECs) may potentially act as carriers of transgenerational epigenetic inheritance in mammals [64, 66]. In particular, in human PGCs it was found that loci previously associated with essential functions in metabolic and neurological disorders were among those genes that were resistant to DNA demethylation [70]. Such loci that appear to escape epigenetic reprogramming in the germline reveal potential for transgenerational epigenetic inheritance that may have phenotypic consequences across generations [71]. Indeed, the number of such escapees was significant: Of more than 2,700 escapees in human PGCs and more than 3,500 hypermethylated genes at E13.5 in mouse PGCs, almost 800 were common in both species [64, 66, 70].

However, as pointed out more recently, the central function of reprogramming PGCs is to ensure the timely and efficient activation of those genes that are functionally required for the progression towards gametogenesis. Also, reprogramming of PGCs is a complex and integrated process that involves the interplay of DNA sequence characteristics, the polycomb (PRC1) complex, and dynamic changes in chromatin constituents in addition to DNA de-methylation and re-methylation [72].

Following this outline of embryonic and PGC reprogramming it should be apparent that the question of whether or to which extend DNA methylation in mammals is a means of epigenetic inheritance across generations is directly linked to the question how genetic regions may either escape from reprogramming or how epigenetic information is transiently transferred to another molecular entity - most likely to the chromatin [73]. This observation also leads us to a rational understanding of what makes epigenetic inheritance either inter- or transgenerational.

10.5 Inter- and Transgenerational Epigenetic Inheritance in Mammals

In order to better understand mechanisms of epigenetic inheritance, it is necessary to distinguish inter- from transgenerational inheritance. When we use the term *intergenerational epigenetic inheritance*, we refer to the inheritance of a parentally (F0) acquired phenotype that causes a distinct phenotypic change also in the next generation (F1). This type of inheritance occurs even when the environmental factor that triggered the phenotype in the parental generation is absent in the F1 generation. In order to classify as intergenerational epigenetic inheritance, the phenotype must be transmitted via any type of epimutation (DNA methylation, RNAs of any kind, histone modification or similar) in maternal and/or paternal gametes and the *epimutation* (or the resulting *epiallele*) is not encoded within the sequence of the genomic DNA. Intergenerational epigenetic inheritance is thus often hard to distinguish from other forms of epigenetic inheritance that may be caused, for example, by exposure of the embryo to the *in utero* environment or by maternal nutrition that the newborn receives during lactation. Such alternative "soft forms" of inheritance are often referred to as confounding factors of epigenetic inheritance. Intergenerational epigenetic inheritance implies that the causative epimutation(s) persist(s) through or are somewhat resistant against the first cycle of epigenetic reprogramming that strikes the pre-implantation mammalian embryo. Thus, we are not in favor of a definition for intergenerational inheritance that includes phenotypic effects that result, for example, from a direct exposure of the embryo *in utero* [74]. From the geneticist's point of view such a direct and immediate effect of an exposure of an individual (even if it is an embryo) to an environmental stimulus is not related to inheritance.

Transgenerational epigenetic inheritance includes the requirements for the definition of intergenerational epigenetic inheritance but, in addition, transgenerational epimutations also pass through the second round of reprogramming that strikes the primordial germ cells later during embryogenesis. Consequently, transgenerational epigenetic inheritance leads to the inheritance of acquired phenotypes from the parental generation to the generation of the grand-children (F2) and probably also subsequent generations (F3 and beyond).

In cases where the distinction between inter- and transgenerational epigenetic inheritance is either not relevant for the point we want to make, or when we discuss experiments where the distinction was not clearly considered, or when we mean both types of epigenetic inheritance, we sometimes take the freedom to generically refer to *epigenetic inheritance across generations*.

Research in the last two decades has provided compelling evidence that epigenetic inheritance of acquired traits across generations is probably more pervasive in animals than previously perceived. Although transgenerational epigenetic inheritance in mammals was previously reviewed [75, 76], we briefly update and summarize the current evidence for this mode of epigenetic inheritance in mammals below. Nutrients, stress, trauma, and enriched environments are among the factors that can affect metabolic health, anxiety and cognitive plasticity across generations. Whereas a functional role of small non-coding RNAs as causative epimutations has been substantiated by experimental data, a role for DNA methylation in epigenetic inheritance is still more controversial and will be discussed further below.

10.5.1 Metabolic Health and Epigenetic Inheritance

Nutrition can lead to acquired metabolic phenotypes that can be inherited across generations. For example, in Drosophila, a paternal high-sugar diet led to obesity in offspring through changes in the chromatin structure in sperm [77]. In rodents, it was shown that parental high-fat diet (HFD) could increase susceptibility to develop obesity and glucose intolerance in offspring that was generated via natural fecundation [78–81]. Streptozotocin-induced prediabetes in male mice affected DNA methylation patterns in their sperm as well as pancreatic islets of F1 and F2 offspring [82]. Additionally, intergenerational inheritance was evident when offspring was produced via *in vitro* fertilization (IVF) from gametes of parents fed from HFD and when healthy foster mothers were used to carry out the progeny. In this case, the use of IVF excluded that confounding factors during gestation and lactation, or transfer of parental microbiomes, or behavior acquired from parents could contribute to the observed epigenetic inheritance [83]. Similarly, low-protein diet [84] and over- [85] or undernutrition [86] were shown to have an impact on offspring's hedonistic behavior and metabolism. Phenotypic changes caused by the nutritional environment of the mother can be epigenetically transmitted on the one hand by the mother via the oocytes [83, 87] or during intrauterine exposure of the embryo and during lactation [81, 86, 88–90]. On the other hand, they may be transmitted by the father via sperm cells or factors such as hormones that are present in seminal fluids [91–94]. Intriguingly, several studies identified sperm-derived small RNAs as potential mediators of epigenetic inheritance. Either specific microRNAs or isolated fractions of sperm RNAs were experimentally injected into untreated zygotes and metabolic phenotypes of specific diets were reproduced in the offspring [92, 94, 95]. Although these experiments provided the proof that specific small RNAs are sufficient to reproduce the respective phenotypes, they did not demonstrate that the examined small RNAs represent the causative epimutation also *in vivo*. For the latter, it would be required to eliminate (or repress) these small RNAs in gametes and assess whether this leads to the loss of the phenotype in the resulting offspring.

At the same time, changes in the sperm methylome were not found accountable for this epigenetic inheritance. For example, paternal low-protein and HFD did impact on the metabolic phenotype in offspring, but sperm methylomes were not consistently affected [96]. Another study showed that a maternal HFD affected female body weights in the F3 generation via the paternal lineage. However, DNA methylation in F1 and F2 sperm was not significantly different depending on the parental or grand-parental diets [81]. Furthermore, maternal HFD during gestation had no effect on offspring's DNA methylation patterns in the liver [97].

In man, epidemiological studies from the Dutch hunger winter, the Danish National Birth Cohort and the Överkalix cohort also supported the concept of intergenerational inheritance of acquired metabolic disorders. Prenatal undernourishment during late gestation, was associated to decreased glucose tolerance and increased rates of cardiovascular disease in adults [98, 99]. Also, offspring of prenatally undernourished fathers, but not mothers, displayed higher rates of obesity than offspring of prenatally well-nourished parents [100]. Risk of overweight and obesity in 7 years old children was positively associated with maternal refined-grain intake during pregnancy [90]. Nutrition during the slow growth period of childhood had a profound effect on cardiovascular and diabetes-related deaths on offspring and grand-offspring through the male line [101]. Moreover, it was shown that obesity in man could affect epigenetic profiles like altered small RNA content and DNA methylation in human spermatozoa [102] and DNA methylation in offspring [103]. Unfortunately, it is very difficult if not impossible to distinguish confounding factors from veritable epigenetic inheritance across generations based on epidemiological data in humans. Yet, the correlations that can be made in man between phenotypes and epigenetic marks are at least in agreement with the hypothesis that causal relations found in other mammals (like mouse and rat) with regards to epigenetic inheritance across generations may also exist for man [104].

Taken together, there is abundant proof that parentally acquired diet-induced metabolic phenotypes can be intergenerationally inherited in mammals. There is also clear evidence, in particular from IVF-based approaches, that such inheritance can be epigenetically transmitted via maternal and paternal germ cells. And finally, whereas current observations provide some substantial evidence that small RNAs, including miRNAs and fragmented tRNAs, may represent the causative epimutations, there is only very limited evidence for DNA methylation as source for epimutations caused by malnutrition, so far. But this may also be due to the fact that methylation in these models simply has not been sufficiently examined until today.

10.5.2 Anxiety and Epigenetic Inheritance

Several other studies have shown that stress or trauma represent environmental factors that can affect behavioral traits over generations [105]. For example, data from rats demonstrated that an impairment in maternal care – measurable as reduced licking and grooming of pups and arched-back nursing – at early postnatal stages resulted in differential anxiety-like behaviors later in adulthood. Concomitantly, the *Nr3c1* promoter became hypermethylated at specific CpG sites resulting in reduced expression of the glucocorticoid receptor (GR) in the hippocampus and enhanced activation of the hypothalamic–pituitary–adrenal (HPA) axis under stress [106–108]. Intriguingly, these maternal care related phenotypes could be reversed by cross-fostering offspring, but also by treatment with an HDAC inhibitor [108]. Conversely, in control rats anxiety related phenotypes could be established by infusion of a methyl donor to the hippocampus [107]. These studies lend support to the hypothesis of a causal correlation between DNA methylation and an anxiety phenotype. Similarly, repeated random separation of mouse pups from their mothers resulted in behavioral traits that were reminiscent of depression and that affected the behavioral response to aversive environments later in adult life [109]. Interestingly, these traits were transgenerationally inherited in a complex and sex-specific fashion through the paternal lineage with some traits present in the third filial generation. At the same time, altered DNA methylation of several candidate genes in sperm of stressed males was observed and partly maintained in sperm of male offspring as well as in brains of female offspring [109]. Furthermore, it was shown that individual miRNAs were altered in sperm of mice that passed an acquired anxiety phenotype (paternal trauma by unpredictable maternal separation and maternal stress) to the next generation [110], indicating a role of small non-coding RNAs as potential mediators of epigenetic inheritance [111]. Behavioral changes were also associated with increased expression of GR and decreased methylation of the GR promoter. The transgenerational transmission of such behavioral symptoms was prevented by environmental enrichment (EE) [112].

Also, different kinds of stress during early gestation did lead to higher stress sensitivity in male mice and their male offspring, probably through changes in paternal miRNAs in sperm [113]. Similarly, paternal stress exposure during puberty or adulthood resulted in significant increase of nine miRNAs in sperm and impacted on the regulation of the HPA stress axis in offspring. However, no behavioral changes in the offspring of stressed sires regarding anxiety were detected [114].

In adult male mice, chronic social defeat stress resulted in anxiety- and depression-like behavior in both male and female offspring as well as increased levels of corticosterone in plasma and decreased levels of vascular endothelial growth factor (VEGF) in male offspring. Importantly, offspring that was generated through IVF did not display most of these behavioral changes pointing towards modes of intergenerational inheritance that were not associated with information transmitted via germ cells. Instead, mechanisms such as learning of behaviors from parents appeared to dominate the inheritance [115].

Odor fear conditioning in mice was used to demonstrate that olfactory experiences can be transgenerationally inherited and that this inheritance affects methylation of the responsible odorant receptor [116]. In this experiment, mice (F0) were pre-conceptionally conditioned using acetophenone, which activates the known odorant receptor *Olfr151*. The F1 generation and grandchildren (F2) showed increased behavioral sensitivity and an enhanced neuroanatomical representation of the *Olfr151* pathway. In addition, sperm of F0 and unexposed F1 males were characterized by hypomethylation of CpG sites in the *Olfr151* gene. Also using *in vitro* fertilization for some of the experiments, this work provides some of the most convincing data that DNA methylation in gametes can contribute to epigenetic inheritance across generations [116]. However, it might be worth mentioning that other modes of epigenetic gene regulation, such as histone methylation marks and inherited small or large transcripts, that could also affect olfactory receptor gene expression were not investigated in the latter study [117]. Interestingly, reversible intergenerational inheritance by odor fear conditioning was also observed with the odor Lyral (Hydroxymethylpentylcyclohexenecarboxaldehyde, a commercial fragrance) and its receptor *Olfr16* and was also associated with changes in the methylation status of the odor receptor gene [118].

Not only stress in early life or adulthood but also dietary composition impacted on anxiety- and depression-like behaviors in offspring of treated animals via both parental germlines as several studies in rats demonstrated. Grand-maternal HFD exposure during pregnancy increased anxiety- and depression-like behavior in granddaughters through the maternal line [119]. Dietary methyl donor depletion led to exacerbated anxiety- and depression-like behaviors in male rats [120] and their male offspring [121].

Bisphenol A (BPA) is a chemical used in many daily-use plastics and which acts as an endocrine disruptor in mammals by mimicking the action of some steroids. BPA is detectable in most humans that are exposed to this chemical and has been suspected to reduce fertility, increase diabetes risk and obesity and several other diseases in man. In mice, it has been shown that exposure to BPA also affects many behaviors including increased anxiety. In particular, it was shown that fetal exposure to BPA transgenerationally led to gene expression changes in specific brain regions and changes in gene methylation of an imprinted gene [122]. However, the epigenetic information in gametes responsible for the transgenerational effects of BPA still needs to be discovered. For other environmental endocrine disruptors, such as vinclozolin (used as fungicide in some countries) it has been demonstrated in rats that the transgenerational effect on male fertility correlated with altered DNA methylation patterns in the germ line [123, 124] and the developmental origin of differentially methylated regions (DMRs) throughout gametogenesis in the third generation was analyzed in great detail [125]. However, in some experiments the transgenerational effects of vinclozolin were not reproduced [126–128]. Potential causes for such discrepancies include specific experimental designs, such as exposure periods, routes of administration, or the use of different genetic animal strains [129, 130].

It was also suggested that intraperitoneal injection of glyphosate in pregnant rats (F0) induced the transgenerational inheritance of various pathologies. Surprisingly, altered phenotypes were only detected in the F2 and F3 generations but not in the F0 and F1 generations [131]. Although this study identified differentially methylated regions (DMRs) in F1, F2, and F3 sperm (albeit without any common DMRs), the data would also be in agreement with genetic (and not epigenetic) inheritance.

Taken together, regardless of the kind of stress or chemical compound that led to an altered behavioral phenotype, there is evidence for resulting epimutations for example in sperm as well as tissues in offspring. However, with rare exceptions most of these studies used natural fecundation instead of IVF to generate the offspring. Thus, confounding factors such as microbiomes, parental behavior or hormones in seminal fluids were not excluded as potential cause.

In humans, traumatic events can cause severe health problems including anxiety disorders and posttraumatic stress disorder (PTSD). Holocaust survivors who developed PTSD did transmit an enhanced PTSD risk and low cortisol levels to their offspring [132, 133] (also reviewed in [134, 135]). Later studies showed that maternal, but not paternal, PTSD was associated with increased glucocorticoid sensitivity in offspring [136, 137] concomitant with lower methylation at the glucocorticoid receptor gene (NR3C1) 1F promoter region [138]. Although these findings are in agreement with the hypothesis that DNA methylation may play a role in intergenerational epigenetic inheritance of anxiety related behavior in humans, they only represent correlations and no causal link. In particular, progeny of holocaust survivors was exposed to many confounding factors such as parental PTSD and altered behavior.

10.5.3 Cognitive Plasticity and Epigenetic Inheritance

Epigenetic transgenerational inheritance was identified as relevant biological process also for the development and function of the mammalian brain. Recent reviews highlighted the existing evidence on epigenetic processes that affect normal brain development [139], behavioral traits [140], and memory function [141]. Environmental enrichment (EE) refers to animal housing conditions, in which rodents are exposed to novel and stimulatory objects that support cognitive training, have enhanced social interactions, and have access to devices for physical activity, such as running wheels. An early study in 1987 already found that rat progeny showed enhanced learning ability when the dams had been exposed to EE before pregnancy [142]. Furthermore, a large number of studies in mice and rats demonstrated that such EE significantly improves brain functions and neuronal plasticity, and ameliorates mutant phenotypes in genetic models of disorders of the nervous system such as Huntington's and Alzheimer's disease (for review see [143]). For example, it was demonstrated that exposure to EE enhances long-term potentiation (LTP) in the hippocampus of mice. Interestingly, this strengthening of hippocampal synapses was also detected in offspring (F1) that were never exposed to EE if their mothers were exposed to EE at a juvenile age [144]. Since offspring was procreated via natural fecundation, this study did not answer the question whether the environmentally induced cognitive phenotype was inherited via oocytes or whether interactions *in utero* between mother and fetus were responsible for the inherited phenotype. In contrast, when male mice were exposed to EE for an extended period during adulthood, they also transmitted enhanced hippocampal LTP and memory function to their male and female offspring, but not to their grand-offspring (F2) [145]. In addition, injection of isolated sperm RNA from either EE exposed or non-exposed fathers into zygotes from naïve parents provided evidence that the over-expression of miR212/132 in sperm is sufficient to confer enhanced hippocampal LTP in the offspring, but did not suffice to provide a cognitive benefit. In addition, alterations in dietary methyl donors induced effects in offspring, such that a paternal diet enriched in methyl donors negatively influenced cognitive and neural functions in the offspring generation [146].

In summary, although nutrients, stress and other environmental exposures can directly alter DNA methylation levels and lead to epigenetic inheritance across generations, it is still unclear if DNA methylation is the causal epigenetic information for the inheritance of metabolism- and behavior-related disorders. The existing data in favor of DNA methylation as the epigenetically transmitted information is mostly based on correlations of phenotypes and epimutations. Instead, there is more functional data that support a role of small non-coding RNAs in epigenetic inheritance.

10.5.4 Hepatic Wound Healing and Epigenetic Inheritance

A study in rats showed that ancestral liver damage induced by the hepatotoxin CCl_4 promoted adaptation of hepatic wound healing by suppressing fibrogenesis in subsequent generations [147]. For example, grandsons of injured grandparents showed increased expression of the master repressor of hepatic stellate cell transdifferentiation, *Pparg*, and reduced expression of the fibrogenic growth factor *Tgfb1* following liver injury. Although expression of these two genes correlated well with the DNA methylation changes in the respective gene promoters in liver samples, it was not shown – as in most other studies above - that such DNA methylation patterns were inherited through germ cells. Instead, it was suggested that modulation of chromatin remodeling in sperm or germ cells could contribute to epigenetic inheritance across generations. In particular, the *Pparg* genomic region in sperm from males with liver fibrosis was enriched for the histone variant H2A.Z and H3K27me3 [147]. These data support the hypothesis that, in addition to small non-coding RNAs, also modulations of the chromatin may represent epigenetically inherited information in germ cells.

10.5.5 Parental Imprinting: Making Male and Female Gametes Compatible for Embryogenesis

Genomic imprinting is a rather well-studied epigenetic mechanism that restricts the expression of a gene to one of the parental chromosomes (reviewed in [148–151]). Microsurgical transplantation experiments in which maternal and paternal pronuclei were exchanged in mouse zygotes demonstrated that imprinting distinctly programs both parental genomes for embryogenesis and that normal embryogenesis requires the interplay of a maternally and a paternally imprinted genome [152]. The majority of our roughly 24.000 protein coding genes are expressed from both the maternal and paternal alleles. However, there is an estimated number between a hundred and two hundred genes in the mammalian genome that display imprinting characterized by parental origin-specific mono-allelic expression [153]. Most imprinted genes are organized in distinct clusters across the genome [154] and one major mechanism that regulates expression of these genes is DNA methylation of *cis*-acting imprinting control elements or regions (ICEs or ICRs) [155]. Usually, low methylation levels inversely correlate to higher expression of the associated genes and vice versa. As an exception from this general rule it was shown that high mRNA transcript levels in mouse oocytes positively correlated with high methylation levels in bodies of imprinted genes, and there was no correlation between gene expression and methylation levels in promoter regions (near transcription start sites) of oocytes. In contrast, in mouse sperm there was an inverse correlation between methylation in promoter regions and gene expression levels [68]. Several deletion experiments of known ICR sequences in mice were accompanied by the loss of parent-of-origin specific expression of most genes within the corresponding gene clusters [14, 156]. Methylation of ICRs by DNA methyltransferases (*Dnmt3a* and *Dnmt3l*) is first established during maturation of gametes in adult mammals [157, 158]. Following fertilization, methylation of imprinted loci is then retained in somatic and extra-embryonic cells (but not in PGCs) throughout embryogenesis and is to some extend persistent throughout the entire life span. This is in contrast to most other differentially methylated regions (DMRs) that are not associated with imprinted gene clusters (see above), and which are usually erased during pre-implantation embryogenesis [68, 159, 160]. The ICRs that are differentially methylated in mature gametes are considered as primary ICRs [150]. Secondary ICRs may be established after fertilization at imprinted gene clusters via mechanisms that require, for example, the expression of non-coding RNAs [161, 162]. A study in patients with Prader-Willi or Angelman syndrome without an apparent genetic contribution suggested that a failure to erase imprinting during gametogenesis may contribute to the transgenerational inheritance of these neurogenetic disorders [163].

Thus ICRs and imprinted genes provide a clear case for a mechanism of epigenetic inheritance in which genetic loci in mammals can somehow escape the reprogramming through global demethylation that occurs in the pre-implantation embryo. However, the mechanism by which ICRs escape global demethylation is still not fully understood. It has been suggested that sequence-specific binding factors such as ZFP57 and ZFP445 might recruit epigenetic modifiers that retain methylation patterns [164, 165]. In particular, mouse mutants carrying loss-of-function alleles for both of these genes lose methylation at almost all ICRs [166, 167]. Similarly, it was shown that oocyte-expressed Uhrf1 has a critical role in maintaining methylation at ICRs of pre-implantation mouse embryos [168]. It binds to hemi-methylated CpG sites and recruits the maintenance methyltransferase Dnmt1 [169, 170].

Yet, there is still much to learn in order to fully understand why and how ICRs can apparently escape global demethylation in pre-implantation embryos and why other differentially methylated regions (usually) cannot escape reprogramming. Nevertheless, the persistent methylation of imprinted genes provides a clear case that the so-called global demethylation (or reprogramming) in pre-implantation embryos is not as genome-wide and extensive as occasionally anticipated. Instead, it appears that reprogramming in the early embryo may be a highly regulated process that affects most germline DMRs but specifically leaves out others.

Maintenance of methylation at imprinted genes and the parent-of-origin specific mono-allelic expression in general are highly regulated processes, as also supported by the observation that genomic patterns of parentally imprinted methylation are not identical throughout all somatic cells. Instead, they show distinct variation in the comparison of differentiated somatic cells and at different developmental stages [171]. Also, the exact definition of genomic imprinting as exclusive mono-allelic expression has come under debate, since genes with only biased expression of the paternal and maternal alleles are known and the preferential expression of maternal and paternal alleles can also vary between tissues and developmental stages [172]. These findings open up a grey zone between classical genomic imprinting of up to 200 genes [148] and the broader phenomenon of allele-specific expression (ASE). The latter has been described to affect the expression levels of several hundreds or thousands of genes across tissues and species [173–175]. It is tempting to speculate that the thousands of DMRs between maternal and paternal genes provide a reservoir of genomic marks, of which more than currently known may survive embryonic development. Once we understand the mechanisms that regulate gene specific methylation in gametes, we may be able to manipulate them more specifically and test their functional requirement for intergenerational epigenetic inheritance.

Methylation marks at imprinted loci are finally erased during embryogenesis between E10.5 and E13.5 in mice when a second round of global demethylation strikes the PGCs. This occurs first in an active process that involves Tet1/Tet2-dependet conversion of m^5C to hm^5C, the repression of de novo and maintenance methylation machineries including *Dnmt3a/b* and *Uhrf1*, followed by passive depletion of m^5C. Most interestingly, it was shown that at least some genes (*Vmn2r29*, *Sfi1*, and *Srm2*) maintain their methylation throughout reprogramming of PGCs [66]. These again present an exception from the general rule of reprogramming and may hint towards an involvement of DNA methylation also in transgenerational inheritance.

10.5.6 The Showcase Examples for DNA Methylation Associated Epigenetic Inheritance: The Agouti and Axin Epialleles

The murine *Agouti viable yellow* (A^{vy}) and the *Axin Fused* (*Axin*^{Fu}) alleles represent some of the best studied gene-focused cases of non-genetic inheritance across generations in mammals that have been associated with DNA methylation [176]. In these alleles, the insertion of an intracisternal A particle (IAP) retrotransposon modulates the expression of the adjacent genes *Agouti* (IAP insertion up-stream) and *Axin* (IAP insertion intergenic) [177]. Hypomethylation of the long terminal repeat (LTR) promoter of the IAP correlates with higher expression of the *Agouti* gene and a yellow coat color. Hypermethylation is associated with agouti mice and intermediate methylation levels with mottled mice [178]. In case of the *Axin*^{Fu} allele, methylation levels of the IAP's LTR promoter correlate with the occurrence of a kinked tail phenotype. The methylation status of these promoters has been termed metastable due to its stable methylation within the cells of an individual and its high variability between individual mice, and this status can be passed to the next generation [179]. For both epialleles, the methylation state in sperm has been reported to reflect the methylation state in somatic tissues of the male individual that produced the sperm (i.e. the father) [180, 181], although the A^{vy} epimutation is only inherited maternally. In contrast, the $Axin^{Fu}$ epimutation can be inherited transgenerationally through the maternal and paternal germ lines. Some experimental data support the idea that the parent-of-origin specific epigenetic inheritance of the A^{vy} epimutation may be due to differences in the reprogramming of maternal and paternal *Agouti* alleles in the pre-implantation embryo [181]. Whereas the paternally inherited A^{vy} allele is rapidly demethylated during the first cell divisions after fertilization, the maternal allele is demethylated more slowly. Intriguingly, it was observed that even the maternal epi-allele was completely erased at the blastocyst stage. This may suggest that either DNA methylation does not represent the inherited epigenetic mark of the A^{vy} epimutation or that the epigenetic information encoded as DNA methylation may be transiently transferred to a different molecule such as (a) histone modification(s) during the pre-implantation reprogramming [181, 182]. However, numerous studies have provided additional evidence that *in utero* or pre-gestational exposure to environmental factors, such as methyl donors or ethanol can alter the methylation status of the A^{vy} epiallele and affect the associated coat color phenotype [183–187]. Although, more recent data suggests that methyl donor supplementation acts through an indirect mechanism to silence the A^{vy} epiallele [188].

Since endogenous retroviruses (ERVs) make up 12% of the mouse genome and approximately 12,000 ERVs are of the IAP subclass [189], the question whether the IAP associated mechanisms of the *Agouti* and *Axin* epialleles may represent a more general mechanism of epigenetic inheritance across generations is evident. Also, the observation that the methylation of most IAPs was resistant to reprogramming during embryogenesis lend further support to this idea [190]. However, a genome-wide study found that the majority of variably methylated IAPs does not function as heterologous promoters and does not initiate transcription of an adjacent gene [191]. Furthermore, newly identified variably methylated IAPs were all fully methylated in sperm and, in general, neither maternal nor paternal methylation levels significantly affected methylation levels in the offspring. Thus, the impact of variably methylated IAPs on gene expression in the A^{vy} and $Axin^{Fu}$ loci and the inheritance across generations of their methylation levels rather appear to be exceptions than the general rule.

Despite the fact that numerous inherited and spontaneous diseases in man are caused by mutations that originate from retro-transposition, we are not aware of epimutations inherited in man across generations that would be caused by mechanisms similar to the *Agouti* and *Axin* epialleles described above.

10.5.7 Epimutations That Are Not Primary Epimutations: Methylation as Secondary Mode of Action

Secondary epimutations are caused by (an) adjacent or remote primary genetic mutation(s) that may initially remain undetected. Thus, in order to identify DNA methylation as original cause for epigenetic inheritance across generations, care must be taken to distinguish primary epimutations from secondary epimutations. This is particularly true for studies in man where genetic heterogeneity between individuals and variations in the genomic sequences may complicate or obscure the identification of primary causative mutations.

A specific example of a secondary epimutation was described in cblC patients that are unable to process vitamin B_{12} (cobalamin, cbl) and which causes multiple severe health problems. In most cases, cblC is caused either by homozygous or compound heterozygous mutations in the paternal and maternal *MMACHC* genes. However, in some cases the causative mutation in the *MMACHC* gene is found only in one parental gene and the second *MMACHC* gene is silenced by extensive methylation. This mechanism was termed epi-cblC. It was then discovered that the epimutation was indeed secondary to a primary and causative genetic mutation in the neighboring *PRDX1* gene, resulting in a transcriptional run-through to the *MMACHC* gene [192]. Since *PRDX1* is also expressed in gametes, in this particular case the secondary epimutation of the *MMACHC* gene was also detected in the germline.

Similarly, mutations in human DNA mismatch repair genes such as *MLH1* and *MSH2*, can cause dominant familial forms of Lynch syndrome, which is characterized by colorectal, and other types of cancers in affected families [193, 194]. However, for about one third of these patients no underlying genetic mutation was initially identified in the mismatch repair genes. Instead, some of them were described as carriers of epimutations, in which the *MLH1* or *MSH2* genes were transcriptionally silenced due to hypermethylation within their promoters. But also here in several cases it was suggested that epimutations were secondary to linked primary mutations in *cis*. For example, the deletion of the last two exons of the *TACSTD1* gene, which is located immediately up-stream of *MSH2*, extends transcription into the DNA mismatch repair gene [195]. This read-through in the sense direction leads to silencing and hypermethylation of the MSH2 promoter.

A primary adjacent mutation may also cause secondary epimutations by read-through via antisense transcription. For example, in some familial cases of alpha-thalassemia it was shown that a genomic deletion leads to the juxtaposition of a widely expressed gene (*LUC7L*) to a normal and unmutated alpha-globin (*HBA2*) gene. The transcriptional orientation of both genes are directed towards each other, such that transcription of the *LUC7L* gene runs through the *HBA2* gene in reverse orientation. This causes hypermethylation of CpG islands, silencing of the *HBA2* gene and the anemia phenotype [196]. Such antisense RNA-mediated CpG hypermethylation and silencing are reminiscent of the mechanisms of genomic imprinting and X-chromosome inactivation.

In cases where secondary epimutations are linked to a causative primary mutation in *cis* such epimutations are rather stably transmitted across generations. Even in these cases it is possible that the methylation signature of the secondary epimutation could be erased in gametes and subsequently re-established later in adult somatic cells due to the inherited primary genetic mutation. In contrast, the inheritance across generations of primary epimutations can be more variable [197]. Alternatively, epimutations may be caused by undiscovered primary mutations in *cis* or *trans* somewhere in the genome.

10.6 Conclusion

Today there is plenty of evidence that acquired phenotypes can be inherited in rodents across generations via gametes – inter- as well as transgenerationally [198]. We have summarized above examples for epigenetic inheritance across generations of phenotypes resulting from diets, enriched environments, traumatic events, and endocrine disruptors. Confounding factors, for example, from social interaction, exposure *in utero* or during lactation, shared microbiomes and others were excluded by the use of *in vitro* fertilization to generate offspring from healthy foster mothers at least in some of the cited experiments. The study of epigenetic inheritance across generations in humans is usually aggravated by environmental and cultural inheritance [199], as well as genomic heterogeneity and uncontrolled envirotypes [200]. However, analyses of epidemiological and family studies following famines or traumatic events are at least in agreement with epigenetic inheritance also in humans.

We have also described above that imprinting of a limited number of genes in mammals is a natural process of epigenetic inheritance that involves the escape from global reprogramming and demethylation. Malfunctions of this process lead to severe diseases in man, such as the Prader-Willi and Angelman syndromes. Furthermore, based on today's knowledge it appears that the epigenetic inheritance associated with DNA methylation of phenotypes caused by *Agouti* or *Axin* alleles due to the rather recent nearby insertion of retrotransposons represent special cases (even in rodents). Current data suggests that they do not point towards a more general mechanism of epigenetic inheritance involving DNA methylation and retrotransposons.

That small non-coding RNAs functionally contribute to epigenetic inheritance, also has been demonstrated in rodents [201]. In several cases, the injection of either over-expressed miRNAs or fragmented tRNAs into fertilized mouse eggs was sufficient to reproduce epigenetically inherited phenotypes [202] – additional examples were given above.

Currently, we are lacking similar functional experiments that would demonstrate that the hyper- or hypomethylation of (a) specific gene(s) in either gametes or the zygote would be sufficient to reproduce an epigenetically inherited phenotype. So far, we know several examples in which specific changes in envirotypes can alter DNA methylation and gene expression in somatic cells. The altered methylation can sometimes be found in gametes (sperm in particular) and also in somatic tissues in (the) offspring generation(s) [203]. These findings may correlate with epigenetic inheritance of altered phenotypes, but, so far, do not proof a causal relationship.

Recently developed methods for epigenome editing exploit the CRISPR-Cas9 system and may allow a more direct study of functional requirements of DNA methylation in epigenetic inheritance across generations (reviewed in [204]). For example, a *deactivated Cas9* (*dCas9*) nuclease coupled to the catalytic domain of *Dnmt3a* was used to specifically target the promoter regions of IL6ST and BACH2. This resulted in their methylation and decreased gene expression [205]. In a similar approach, *Dnmt3a* was targeted to methylate and silence the *Hoxa5* gene. Here, an antibody fragment fused to the DNA methylase was used to bind to the epitope-tagged dCas9 [206]. Specific promoter demethylation could be achieved by dCas9-mediated targeting of the catalytic domain of *Tet1* [207]. Such delivery systems were successfully used *in vivo* in mouse fetuses and adult animals [207, 208]. DNA methylation was also edited in single mammalian oocytes [209], which should establish new avenues to study maternally inherited epimutations. We are confident that these or related emerging technologies will soon help us to uncover causal mechanisms of epigenetic inheritance involving DNA methylation in mammals.

References

1. Hotchkiss RD. The quantitative separation of purines, pyrimidines, and nucleosides by paper chromatography. J Biol Chem. 1948;175(1):315-332. Epub 1948/08/01. PubMed PMID: 18873306.
2. Bird AP. CpG-rich islands and the function of DNA methylation. Nature. 1986;321(6067):209-213. Epub 1986/05/15. doi: https://doi.org/10.1038/321209a0. PubMed PMID: 2423876.
3. Esteller M. CpG island hypermethylation and tumor suppressor genes: a booming present, a brighter future. Oncogene. 2002;21(35):5427-5440. Epub 2002/08/03. doi: https://doi.org/10.1038/sj.onc.1205600. PubMed PMID: 12154405.
4. Feng W, Zhou L, Wang H, Hu Z, Wang X, Fu J, et al. Functional analysis of DNA methylation of the PACSIN1 promoter in pig peripheral blood mononuclear cells. J Cell Biochem. 2019;120(6):10118-10127. Epub 2018/12/12. doi: https://doi.org/10.1002/jcb.28295. PubMed PMID: 30537176.
5. Jones PA. Functions of DNA methylation: islands, start sites, gene bodies and beyond. Nat Rev Genet. 2012;13(7):484-492. Epub 2012/05/30. doi: https://doi.org/10.1038/nrg3230. PubMed PMID: 22641018.
6. Pai AA, Bell JT, Marioni JC, Pritchard JK, Gilad Y. A genome-wide study of DNA methylation patterns and gene expression levels in multiple human and chimpanzee tissues. PLoS Genet. 2011;7(2):e1001316. Epub 2011/03/09. doi: https://doi.org/10.1371/journal.pgen.1001316. PubMed PMID: 21383968; PubMed Central PMCID: PMCPMC3044686.
7. Zhou J, Sears RL, Xing X, Zhang B, Li D, Rockweiler NB, et al. Tissue-specific DNA methylation is conserved across human, mouse, and rat, and driven by primary sequence conservation. BMC Genomics. 2017;18(1):724. Epub 2017/09/14. doi: https://doi.org/10.1186/s12864-017-4115-6. PubMed PMID: 28899353; PubMed Central PMCID: PMCPMC5596466.
8. Goll MG, Bestor TH. Eukaryotic cytosine methyltransferases. Annu Rev Biochem. 2005;74:481-514. Epub 2005/06/15. doi: https://doi.org/10.1146/annurev.biochem.74.010904.153721. PubMed PMID: 15952895.

9. Li E, Bestor TH, Jaenisch R. Targeted mutation of the DNA methyltransferase gene results in embryonic lethality. Cell. 1992;69(6):915-926. Epub 1992/06/12. PubMed PMID: 1606615.
10. Jones PA, Takai D. The role of DNA methylation in mammalian epigenetics. Science. 2001;293(5532):1068-1070. Epub 2001/08/11. doi: https://doi.org/10.1126/science.1063852. PubMed PMID: 11498573.
11. Smith ZD, Meissner A. DNA methylation: roles in mammalian development. Nat Rev Genet. 2013;14(3):204-220. Epub 2013/02/13. doi: https://doi.org/10.1038/nrg3354. PubMed PMID: 23400093.
12. Hsieh CL, Lieber MR. CpG methylated minichromosomes become inaccessible for V(D)J recombination after undergoing replication. EMBO J. 1992;11(1):315-325. Epub 1992/01/01. PubMed PMID: 1371250; PubMed Central PMCID: PMCPMC556452.
13. Robertson KD. DNA methylation and human disease. Nat Rev Genet. 2005;6(8):597-610. Epub 2005/09/02. doi: https://doi.org/10.1038/nrg1655. PubMed PMID: 16136652.
14. Yang X, Lay F, Han H, Jones PA. Targeting DNA methylation for epigenetic therapy. Trends Pharmacol Sci. 2010;31(11):536-546. Epub 2010/09/18. doi: https://doi.org/10.1016/j.tips.2010.08.001. PubMed PMID: 20846732; PubMed Central PMCID: PMCPMC2967479.
15. Wu TP, Wang T, Seetin MG, Lai Y, Zhu S, Lin K, et al. DNA methylation on N(6)-adenine in mammalian embryonic stem cells. Nature. 2016;532(7599):329-333. Epub 2016/03/31. doi: https://doi.org/10.1038/nature17640. PubMed PMID: 27027282; PubMed Central PMCID: PMCPMC4977844.
16. Xiao CL, Zhu S, He M, Chen, Zhang Q, Chen Y, et al. N(6)-Methyladenine DNA Modification in the Human Genome. Mol Cell. 2018;71(2):306-318 e7. Epub 2018/07/19. doi: https://doi.org/10.1016/j.molcel.2018.06.015. PubMed PMID: 30017583.
17. Greer EL, Blanco MA, Gu L, Sendinc E, Liu J, Aristizabal-Corrales D, et al. DNA Methylation on N6-Adenine in C. elegans. Cell. 2015;161(4):868-878. Epub 2015/05/06. doi: https://doi.org/10.1016/j.cell.2015.04.005. PubMed PMID: 25936839; PubMed Central PMCID: PMCPMC4427530.
18. Ma C, Niu R, Huang T, Shao LW, Peng Y, Ding W, et al. N6-methyldeoxyadenine is a transgenerational epigenetic signal for mitochondrial stress adaptation. Nat Cell Biol. 2019;21(3):319-327. Epub 2018/12/05. doi: https://doi.org/10.1038/s41556-018-0238-5. PubMed PMID: 30510156.
19. Miska EA, Ferguson-Smith AC. Transgenerational inheritance: Models and mechanisms of non-DNA sequence-based inheritance. Science. 2016;354(6308):59-63. Epub 2016/11/16. doi: https://doi.org/10.1126/science.aaf4945. PubMed PMID: 27846492.
20. Jaenisch R. DNA methylation and imprinting: why bother? Trends Genet. 1997;13(8):323-329. Epub 1997/08/01. PubMed PMID: 9260519.
21. Brink RA. A Genetic Change Associated with the R Locus in Maize Which Is Directed and Potentially Reversible. Genetics. 1956;41(6):872-889. Epub 1956/11/01. PubMed PMID: 17247669; PubMed Central PMCID: PMCPMC1224369.
22. Arteaga-Vazquez MA, Chandler VL. Paramutation in maize: RNA mediated trans-generational gene silencing. Curr Opin Genet Dev. 2010;20(2):156-163. Epub 2010/02/16. doi: https://doi.org/10.1016/j.gde.2010.01.008. PubMed PMID: 20153628; PubMed Central PMCID: PMCPMC2859986.
23. Cubas P, Vincent C, Coen E. An epigenetic mutation responsible for natural variation in floral symmetry. Nature. 1999;401(6749):157-161. Epub 1999/09/18. doi: https://doi.org/10.1038/43657. PubMed PMID: 10490023.
24. Jablonka E, Raz G. Transgenerational epigenetic inheritance: prevalence, mechanisms, and implications for the study of heredity and evolution. Q Rev Biol. 2009;84(2):131-176. Epub 2009/07/18. PubMed PMID: 19606595.
25. Uller T, English S, Pen I. When is incomplete epigenetic resetting in germ cells favoured by natural selection? Proc Biol Sci. 2015;282(1811). Epub 2015/07/03. doi: https://doi.org/10.1098/rspb.2015.0682. PubMed PMID: 26136447; PubMed Central PMCID: PMCPMC4528548.

26. Houri-Zeevi L, Rechavi O. A Matter of Time: Small RNAs Regulate the Duration of Epigenetic Inheritance. Trends Genet. 2017;33(1):46-57. Epub 2016/12/13. doi: https://doi.org/10.1016/j.tig.2016.11.001. PubMed PMID: 27939252.
27. Herman H, Lu M, Anggraini M, Sikora A, Chang Y, Yoon BJ, et al. Trans allele methylation and paramutation-like effects in mice. Nat Genet. 2003;34(2):199-202. Epub 2003/05/13. doi: https://doi.org/10.1038/ng1162. PubMed PMID: 12740578; PubMed Central PMCID: PMCPMC2744043.
28. Rassoulzadegan M, Magliano M, Cuzin F. Transvection effects involving DNA methylation during meiosis in the mouse. EMBO J. 2002;21(3):440-450. Epub 2002/02/02. doi: https://doi.org/10.1093/emboj/21.3.440. PubMed PMID: 11823436; PubMed Central PMCID: PMCPMC125843.
29. Cambareri EB, Jensen BC, Schabtach E, Selker EU. Repeat-induced G-C to A-T mutations in Neurospora. Science. 1989;244(4912):1571-1575. Epub 1989/06/30. doi: https://doi.org/10.1126/science.2544994. PubMed PMID: 2544994.
30. Feng S, Jacobsen SE, Reik W. Epigenetic reprogramming in plant and animal development. Science. 2010;330(6004):622-7. Epub 2010/10/30. doi: 10.1126/science.1190614. PubMed PMID: 21030646; PubMed Central PMCID: PMCPMC2989926.
31. Law JA, Jacobsen SE. Establishing, maintaining and modifying DNA methylation patterns in plants and animals. Nat Rev Genet. 2010;11(3):204-20. Epub 2010/02/10. doi: 10.1038/nrg2719. PubMed PMID: 20142834; PubMed Central PMCID: PMCPMC3034103.
32. Santos F, Dean W. Using immunofluorescence to observe methylation changes in mammalian preimplantation embryos. Methods Mol Biol. 2006;325:129-137. Epub 2006/06/10. doi: https://doi.org/10.1385/1-59745-005-7:129. PubMed PMID: 16761724.
33. Cokus SJ, Feng S, Zhang X, Chen Z, Merriman B, Haudenschild CD, et al. Shotgun bisulphite sequencing of the Arabidopsis genome reveals DNA methylation patterning. Nature. 2008;452(7184):215-9. Epub 2008/02/19. doi: 10.1038/nature06745. PubMed PMID: 18278030; PubMed Central PMCID: PMCPMC2377394.
34. Bocker MT, Tuorto F, Raddatz G, Musch T, Yang FC, Xu M, et al. Hydroxylation of 5-methylcytosine by TET2 maintains the active state of the mammalian HOXA cluster. Nat Commun. 2012;3:818. Epub 2012/05/10. doi: 10.1038/ncomms1826. PubMed PMID: 22569366; PubMed Central PMCID: PMCPMC3576573.
35. Gu H, Smith ZD, Bock C, Boyle P, Gnirke A, Meissner A. Preparation of reduced representation bisulfite sequencing libraries for genome-scale DNA methylation profiling. Nat Protoc. 2011;6(4):468-481. Epub 2011/03/18. doi: https://doi.org/10.1038/nprot.2010.190. PubMed PMID: 21412275.
36. Bonora G, Rubbi L, Morselli M, Ma F, Chronis C, Plath K, et al. DNA methylation estimation using methylation-sensitive restriction enzyme bisulfite sequencing (MREBS). PLoS One. 2019;14(4):e0214368. Epub 2019/04/05. doi: 10.1371/journal.pone.0214368. PubMed PMID: 30946758; PubMed Central PMCID: PMCPMC6448829.
37. Yu M, Hon GC, Szulwach KE, Song CX, Jin P, Ren B, et al. Tet-assisted bisulfite sequencing of 5-hydroxymethylcytosine. Nat Protoc. 2012;7(12):2159-70. Epub 2012/12/01. doi: 10.1038/nprot.2012.137. PubMed PMID: 23196972; PubMed Central PMCID: PMCPMC3641661.
38. Treangen TJ, Salzberg SL. Repetitive DNA and next-generation sequencing: computational challenges and solutions. Nat Rev Genet. 2011;13(1):36-46. Epub 2011/11/30. doi: 10.1038/nrg3117. PubMed PMID: 22124482; PubMed Central PMCID: PMCPMC3324860.
39. van der Maarel SM, Deidda G, Lemmers RJ, van Overveld PG, van der Wielen M, Hewitt JE, et al. De novo facioscapulohumeral muscular dystrophy: frequent somatic mosaicism, sex-dependent phenotype, and the role of mitotic transchromosomal repeat interaction between chromosomes 4 and 10. Am J Hum Genet. 2000;66(1):26-35. Epub 2000/01/13. doi: 10.1086/302730. PubMed PMID: 10631134; PubMed Central PMCID: PMCPMC1288331.
40. Smallwood SA, Lee HJ, Angermueller C, Krueger F, Saadeh H, Peat J, et al. Single-cell genome-wide bisulfite sequencing for assessing epigenetic heterogeneity. Nat Methods. 2014;11(8):817-20. Epub 2014/07/22. doi: 10.1038/nmeth.3035. PubMed PMID: 25042786; PubMed Central PMCID: PMCPMC4117646.

41. Eid J, Fehr A, Gray J, Luong K, Lyle J, Otto G, et al. Real-time DNA sequencing from single polymerase molecules. Science. 2009;323(5910):133-138. Epub 2008/11/22. doi: https://doi.org/10.1126/science.1162986. PubMed PMID: 19023044.
42. Clarke J, Wu HC, Jayasinghe L, Patel A, Reid S, Bayley H. Continuous base identification for single-molecule nanopore DNA sequencing. Nat Nanotechnol. 2009;4(4):265-270. Epub 2009/04/08. doi: https://doi.org/10.1038/nnano.2009.12. PubMed PMID: 19350039.
43. Flusberg BA, Webster DR, Lee JH, Travers KJ, Olivares EC, Clark TA, et al. Direct detection of DNA methylation during single-molecule, real-time sequencing. Nat Methods. 2010;7(6):461-5. Epub 2010/05/11. doi: 10.1038/nmeth.1459. PubMed PMID: 20453866; PubMed Central PMCID: PMCPMC2879396.
44. Jain M, Olsen HE, Paten B, Akeson M. The Oxford Nanopore MinION: delivery of nanopore sequencing to the genomics community. Genome Biol. 2016;17(1):239. Epub 2016/11/27. doi: 10.1186/s13059-016-1103-0. PubMed PMID: 27887629; PubMed Central PMCID: PMCPMC5124260.
45. Chen Y, Wu Y, Liu L, Feng J, Zhang T, Qin S, et al. Study of the whole genome, methylome and transcriptome of Cordyceps militaris. Sci Rep. 2019;9(1):898. Epub 2019/01/31. doi: 10.1038/s41598-018-38021-4. PubMed PMID: 30696919; PubMed Central PMCID: PMCPMC6351555.
46. McIntyre ABR, Alexander N, Grigorev K, Bezdan D, Sichtig H, Chiu CY, et al. Single-molecule sequencing detection of N6-methyladenine in microbial reference materials. Nat Commun. 2019;10(1):579. Epub 2019/02/06. doi: 10.1038/s41467-019-08289-9. PubMed PMID: 30718479; PubMed Central PMCID: PMCPMC6362088.
47. Santos F, Hendrich B, Reik W, Dean W. Dynamic reprogramming of DNA methylation in the early mouse embryo. Dev Biol. 2002;241(1):172-182. Epub 2002/01/11. doi: https://doi.org/10.1006/dbio.2001.0501. PubMed PMID: 11784103.
48. Guo F, Li X, Liang D, Li T, Zhu P, Guo H, et al. Active and passive demethylation of male and female pronuclear DNA in the mammalian zygote. Cell Stem Cell. 2014;15(4):447-459. Epub 2014/09/16. doi: https://doi.org/10.1016/j.stem.2014.08.003. PubMed PMID: 25220291.
49. Shen L, Inoue A, He J, Liu Y, Lu F, Zhang Y. Tet3 and DNA replication mediate demethylation of both the maternal and paternal genomes in mouse zygotes. Cell Stem Cell. 2014;15(4):459-71. Epub 2014/10/04. doi: 10.1016/j.stem.2014.09.002. PubMed PMID: 25280220; PubMed Central PMCID: PMCPMC4201500.
50. Wang L, Zhang J, Duan J, Gao X, Zhu W, Lu X, et al. Programming and inheritance of parental DNA methylomes in mammals. Cell. 2014;157(4):979-91. Epub 2014/05/13. doi: 10.1016/j.cell.2014.04.017. PubMed PMID: 24813617; PubMed Central PMCID: PMCPMC4096154.
51. Guo H, Zhu P, Yan L, Li R, Hu B, Lian Y, et al. The DNA methylation landscape of human early embryos. Nature. 2014;511(7511):606-610. Epub 2014/08/01. doi: https://doi.org/10.1038/nature13544. PubMed PMID: 25079557.
52. Smith ZD, Chan MM, Humm KC, Karnik R, Mekhoubad S, Regev A, et al. DNA methylation dynamics of the human preimplantation embryo. Nature. 2014;511(7511):611-5. Epub 2014/08/01. doi: 10.1038/nature13581. PubMed PMID: 25079558; PubMed Central PMCID: PMCPMC4178976.
53. Jiang L, Zhang J, Wang JJ, Wang L, Zhang L, Li G, et al. Sperm, but not oocyte, DNA methylome is inherited by zebrafish early embryos. Cell. 2013;153(4):773-84. Epub 2013/05/15. doi: 10.1016/j.cell.2013.04.041. PubMed PMID: 23663777; PubMed Central PMCID: PMCPMC4081501.
54. Potok ME, Nix DA, Parnell TJ, Cairns BR. Reprogramming the maternal zebrafish genome after fertilization to match the paternal methylation pattern. Cell. 2013;153(4):759-72. Epub 2013/05/15. doi: 10.1016/j.cell.2013.04.030. PubMed PMID: 23663776; PubMed Central PMCID: PMCPMC4030421.
55. Murphy PJ, Wu SF, James CR, Wike CL, Cairns BR. Placeholder Nucleosomes Underlie Germline-to-Embryo DNA Methylation Reprogramming. Cell. 2018;172(5):993-1006 e13. Epub 2018/02/20. doi: https://doi.org/10.1016/j.cell.2018.01.022. PubMed PMID: 29456083.

56. Ward WS, Coffey DS. DNA packaging and organization in mammalian spermatozoa: comparison with somatic cells. Biol Reprod. 1991;44(4):569-574. Epub 1991/04/01. doi: https://doi.org/10.1095/biolreprod44.4.569. PubMed PMID: 2043729.
57. Hammoud SS, Nix DA, Zhang H, Purwar J, Carrell DT, Cairns BR. Distinctive chromatin in human sperm packages genes for embryo development. Nature. 2009;460(7254):473-8. Epub 2009/06/16. doi: 10.1038/nature08162. PubMed PMID: 19525931; PubMed Central PMCID: PMCPMC2858064.
58. Wykes SM, Krawetz SA. The structural organization of sperm chromatin. J Biol Chem. 2003;278(32):29471-29477. Epub 2003/05/31. doi: https://doi.org/10.1074/jbc.M304545200. PubMed PMID: 12775710.
59. Falconer DS, Avery PJ. Variability of chimaeras and mosaics. J Embryol Exp Morphol. 1978;43:195-219. Epub 1978/02/01. PubMed PMID: 632737.
60. Gardner RL. The relationship between cell lineage and differentiation in the early mouse embryo. Results Probl Cell Differ. 1978;9:205-241. Epub 1978/01/01. PubMed PMID: 373038.
61. McMahon A, Fosten M, Monk M. Random X-chromosome inactivation in female primordial germ cells in the mouse. J Embryol Exp Morphol. 1981;64:251-258. Epub 1981/08/01. PubMed PMID: 7310304.
62. Ginsburg M, Snow MH, McLaren A. Primordial germ cells in the mouse embryo during gastrulation. Development. 1990;110(2):521-528. Epub 1990/10/01. PubMed PMID: 2133553.
63. Hajkova P, Erhardt S, Lane N, Haaf T, El-Maarri O, Reik W, et al. Epigenetic reprogramming in mouse primordial germ cells. Mech Dev. 2002;117(1-2):15-23. Epub 2002/09/03. PubMed PMID: 12204247.
64. Seisenberger S, Andrews S, Krueger F, Arand J, Walter J, Santos F, et al. The dynamics of genome-wide DNA methylation reprogramming in mouse primordial germ cells. Mol Cell. 2012;48(6):849-62. Epub 2012/12/12. doi: 10.1016/j.molcel.2012.11.001. PubMed PMID: 23219530; PubMed Central PMCID: PMCPMC3533687.
65. Cowley M, Oakey RJ. Resetting for the next generation. Mol Cell. 2012;48(6):819-821. Epub 2013/01/01. doi: https://doi.org/10.1016/j.molcel.2012.12.007. PubMed PMID: 23273741.
66. Hackett JA, Sengupta R, Zylicz JJ, Murakami K, Lee C, Down TA, et al. Germline DNA demethylation dynamics and imprint erasure through 5-hydroxymethylcytosine. Science. 2013;339(6118):448-52. Epub 2012/12/12. doi: 10.1126/science.1229277. PubMed PMID: 23223451; PubMed Central PMCID: PMCPMC3847602.
67. Guibert S, Forne T, Weber M. Global profiling of DNA methylation erasure in mouse primordial germ cells. Genome research. 2012;22(4):633-41. Epub 2012/02/24. doi: 10.1101/gr.130997.111. PubMed PMID: 22357612; PubMed Central PMCID: PMCPMC3317146.
68. Kobayashi H, Sakurai T, Imai M, Takahashi N, Fukuda A, Yayoi O, et al. Contribution of intragenic DNA methylation in mouse gametic DNA methylomes to establish oocyte-specific heritable marks. PLoS Genet. 2012;8(1):e1002440. Epub 2012/01/14. doi: 10.1371/journal.pgen.1002440. PubMed PMID: 22242016; PubMed Central PMCID: PMCPMC3252278.
69. Kobayashi H, Sakurai T, Miura F, Imai M, Mochiduki K, Yanagisawa E, et al. High-resolution DNA methylome analysis of primordial germ cells identifies gender-specific reprogramming in mice. Genome research. 2013;23(4):616-27. Epub 2013/02/16. doi: 10.1101/gr.148023.112. PubMed PMID: 23410886; PubMed Central PMCID: PMCPMC3613579.
70. Tang WW, Dietmann S, Irie N, Leitch HG, Floros VI, Bradshaw CR, et al. A Unique Gene Regulatory Network Resets the Human Germline Epigenome for Development. Cell. 2015;161(6):1453-67. Epub 2015/06/06. doi: 10.1016/j.cell.2015.04.053. PubMed PMID: 26046444; PubMed Central PMCID: PMCPMC4459712.
71. Illum LRH, Bak ST, Lund S, Nielsen AL. DNA methylation in epigenetic inheritance of metabolic diseases through the male germ line. J Mol Endocrinol. 2018;60(2):R39-R56. Epub 2017/12/06. doi: https://doi.org/10.1530/JME-17-0189. PubMed PMID: 29203518.
72. Hill PWS, Leitch HG, Requena CE, Sun Z, Amouroux R, Roman-Trufero M, et al. Epigenetic reprogramming enables the transition from primordial germ cell to gonocyte. Nature.

2018;555(7696):392-6. Epub 2018/03/08. doi: 10.1038/nature25964. PubMed PMID: 29513657; PubMed Central PMCID: PMCPMC5856367.
73. Baxter FA, Drake AJ. Non-genetic inheritance via the male germline in mammals. Philos Trans R Soc Lond B Biol Sci. 2019;374(1770):20180118. Epub 2019/04/11. doi: 10.1098/rstb.2018.0118. PubMed PMID: 30966887; PubMed Central PMCID: PMCPMC6460076.
74. Heard E, Martienssen RA. Transgenerational epigenetic inheritance: myths and mechanisms. Cell. 2014;157(1):95-109. Epub 2014/04/01. doi: 10.1016/j.cell.2014.02.045. PubMed PMID: 24679529; PubMed Central PMCID: PMCPMC4020004.
75. Daxinger L, Whitelaw E. Transgenerational epigenetic inheritance: more questions than answers. Genome research. 2010;20(12):1623-8. Epub 2010/11/03. doi: 10.1101/gr.106138.110. PubMed PMID: 21041414; PubMed Central PMCID: PMCPMC2989988.
76. Daxinger L, Whitelaw E. Understanding transgenerational epigenetic inheritance via the gametes in mammals. Nat Rev Genet. 2012;13(3):153-162. Epub 2012/02/01. doi: https://doi.org/10.1038/nrg3188. PubMed PMID: 22290458.
77. Ost A, Lempradl A, Casas E, Weigert M, Tiko T, Deniz M, et al. Paternal diet defines offspring chromatin state and intergenerational obesity. Cell. 2014;159(6):1352-1364. Epub 2014/12/07. doi: https://doi.org/10.1016/j.cell.2014.11.005. PubMed PMID: 25480298.
78. de Castro Barbosa T, Alm PS, Krook A, Barres R, Zierath JR. Paternal high-fat diet transgenerationally impacts hepatic immunometabolism. FASEB J. 2019;33(5):6269-6280. Epub 2019/02/16. doi: https://doi.org/10.1096/fj.201801879RR. PubMed PMID: 30768368.
79. Fullston T, Ohlsson Teague EM, Palmer NO, DeBlasio MJ, Mitchell M, Corbett M, et al. Paternal obesity initiates metabolic disturbances in two generations of mice with incomplete penetrance to the F2 generation and alters the transcriptional profile of testis and sperm microRNA content. FASEB J. 2013;27(10):4226-4243. Epub 2013/07/13. doi: https://doi.org/10.1096/fj.12-224048. PubMed PMID: 23845863.
80. Ng SF, Lin RC, Laybutt DR, Barres R, Owens JA, Morris MJ. Chronic high-fat diet in fathers programs beta-cell dysfunction in female rat offspring. Nature. 2010;467(7318):963-966. Epub 2010/10/22. doi: https://doi.org/10.1038/nature09491. PubMed PMID: 20962845.
81. Sarker G, Berrens R, von Arx J, Pelczar P, Reik W, Wolfrum C, et al. Transgenerational transmission of hedonic behaviors and metabolic phenotypes induced by maternal overnutrition. Translational psychiatry. 2018;8(1):195. Epub 2018/10/14. doi: 10.1038/s41398-018-0243-2. PubMed PMID: 30315171; PubMed Central PMCID: PMCPMC6185972.
82. Wei Y, Yang CR, Wei YP, Zhao ZA, Hou Y, Schatten H, et al. Paternally induced transgenerational inheritance of susceptibility to diabetes in mammals. Proc Natl Acad Sci U S A. 2014;111(5):1873-8. Epub 2014/01/23. doi: 10.1073/pnas.1321195111. PubMed PMID: 24449870; PubMed Central PMCID: PMCPMC3918818.
83. Huypens P, Sass S, Wu M, Dyckhoff D, Tschop M, Theis F, et al. Epigenetic germline inheritance of diet-induced obesity and insulin resistance. Nat Genet. 2016;48(5):497-499. Epub 2016/03/15. doi: https://doi.org/10.1038/ng.3527. PubMed PMID: 26974008.
84. Carone BR, Fauquier L, Habib N, Shea JM, Hart CE, Li R, et al. Paternally induced transgenerational environmental reprogramming of metabolic gene expression in mammals. Cell. 2010;143(7):1084-96. Epub 2010/12/25. doi: 10.1016/j.cell.2010.12.008. PubMed PMID: 21183072; PubMed Central PMCID: PMCPMC3039484.
85. Peleg-Raibstein D, Sarker G, Litwan K, Kramer SD, Ametamey SM, Schibli R, et al. Enhanced sensitivity to drugs of abuse and palatable foods following maternal overnutrition. Translational psychiatry. 2016;6(10):e911. Epub 2016/10/05. doi: 10.1038/tp.2016.176. PubMed PMID: 27701408; PubMed Central PMCID: PMCPMC5315546.
86. Radford EJ, Ito M, Shi H, Corish JA, Yamazawa K, Isganaitis E, et al. In utero effects. In utero undernourishment perturbs the adult sperm methylome and intergenerational metabolism. Science. 2014;345(6198):1255903. Epub 2014/07/12. doi: 10.1126/science.1255903. PubMed PMID: 25011554; PubMed Central PMCID: PMCPMC4404520.
87. Ferey JLA, Boudoures AL, Reid M, Drury A, Scheaffer S, Modi Z, et al. A maternal high-fat, high-sucrose diet induces transgenerational cardiac mitochondrial dysfunction independently

of maternal mitochondrial inheritance. American journal of physiology Heart and circulatory physiology. 2019;316(5):H1202-H10. Epub 2019/03/23. doi: 10.1152/ajpheart.00013.2019. PubMed PMID: 30901280; PubMed Central PMCID: PMCPMC6580388.
88. Pankey CL, Walton MW, Odhiambo JF, Smith AM, Ghnenis AB, Nathanielsz PW, et al. Intergenerational impact of maternal overnutrition and obesity throughout pregnancy in sheep on metabolic syndrome in grandsons and granddaughters. Domestic animal endocrinology. 2017;60:67-74. Epub 2017/05/22. doi: https://doi.org/10.1016/j.domaniend.2017.04.002. PubMed PMID: 28527530.
89. Sharma SS, Jangale NM, Harsulkar AM, Gokhale MK, Joshi BN. Chronic maternal calcium and 25-hydroxyvitamin D deficiency in Wistar rats programs abnormal hepatic gene expression leading to hepatic steatosis in female offspring. The Journal of nutritional biochemistry. 2017;43:36-46. Epub 2017/02/22. doi: https://doi.org/10.1016/j.jnutbio.2017.01.008. PubMed PMID: 28219837.
90. Zhu Y, Olsen SF, Mendola P, Halldorsson TI, Yeung EH, Granstrom C, et al. Maternal dietary intakes of refined grains during pregnancy and growth through the first 7 y of life among children born to women with gestational diabetes. The American journal of clinical nutrition. 2017;106(1):96-104. Epub 2017/06/09. doi: 10.3945/ajcn.116.136291. PubMed PMID: 28592607; PubMed Central PMCID: PMCPMC5486192.
91. An T, Zhang T, Teng F, Zuo JC, Pan YY, Liu YF, et al. Long non-coding RNAs could act as vectors for paternal heredity of high fat diet-induced obesity. Oncotarget. 2017;8(29):47876-89. Epub 2017/06/10. doi: 10.18632/oncotarget.18138. PubMed PMID: 28599310; PubMed Central PMCID: PMCPMC5564612.
92. Chen Q, Yan M, Cao Z, Li X, Zhang Y, Shi J, et al. Sperm tsRNAs contribute to intergenerational inheritance of an acquired metabolic disorder. Science. 2016;351(6271):397-400. Epub 2016/01/02. doi: https://doi.org/10.1126/science.aad7977. PubMed PMID: 26721680.
93. Fullston T, McPherson NO, Owens JA, Kang WX, Sandeman LY, Lane M. Paternal obesity induces metabolic and sperm disturbances in male offspring that are exacerbated by their exposure to an "obesogenic" diet. Physiological reports. 2015;3(3). Epub 2015/03/26. doi: 10.14814/phy2.12336. PubMed PMID: 25804263; PubMed Central PMCID: PMCPMC4393169.
94. Grandjean V, Fourre S, De Abreu DA, Derieppe MA, Remy JJ, Rassoulzadegan M. RNA-mediated paternal heredity of diet-induced obesity and metabolic disorders. Sci Rep. 2015;5:18193. Epub 2015/12/15. doi: 10.1038/srep18193. PubMed PMID: 26658372; PubMed Central PMCID: PMCPMC4677355.
95. Sharma U, Conine CC, Shea JM, Boskovic A, Derr AG, Bing XY, et al. Biogenesis and function of tRNA fragments during sperm maturation and fertilization in mammals. Science. 2016;351(6271):391-6. Epub 2016/01/02. doi: 10.1126/science.aad6780. PubMed PMID: 26721685; PubMed Central PMCID: PMCPMC4888079.
96. Shea JM, Serra RW, Carone BR, Shulha HP, Kucukural A, Ziller MJ, et al. Genetic and Epigenetic Variation, but Not Diet, Shape the Sperm Methylome. Dev Cell. 2015;35(6):750-8. Epub 2015/12/26. doi: 10.1016/j.devcel.2015.11.024. PubMed PMID: 26702833; PubMed Central PMCID: PMCPMC4691283.
97. Cannon MV, Buchner DA, Hester J, Miller H, Sehayek E, Nadeau JH, et al. Maternal nutrition induces pervasive gene expression changes but no detectable DNA methylation differences in the liver of adult offspring. PLoS One. 2014;9(3):e90335. Epub 2014/03/07. doi: 10.1371/journal.pone.0090335. PubMed PMID: 24594983; PubMed Central PMCID: PMCPMC3940881.
98. Painter RC, Roseboom TJ, Bleker OP. Prenatal exposure to the Dutch famine and disease in later life: an overview. Reprod Toxicol. 2005;20(3):345-352. Epub 2005/05/17. doi: https://doi.org/10.1016/j.reprotox.2005.04.005. PubMed PMID: 15893910.
99. Ravelli AC, van der Meulen JH, Michels RP, Osmond C, Barker DJ, Hales CN, et al. Glucose tolerance in adults after prenatal exposure to famine. Lancet (London, England). 1998;351(9097):173-7. Epub 1998/02/05. doi: 10.1016/s0140-6736(97)07244-9. PubMed PMID: 9449872.

100. Veenendaal MV, Painter RC, de Rooij SR, Bossuyt PM, van der Post JA, Gluckman PD, et al. Transgenerational effects of prenatal exposure to the 1944-45 Dutch famine. BJOG: an international journal of obstetrics and gynaecology. 2013;120(5):548-553. Epub 2013/01/26. doi: https://doi.org/10.1111/1471-0528.12136. PubMed PMID: 23346894.
101. Kaati G, Bygren LO, Edvinsson S. Cardiovascular and diabetes mortality determined by nutrition during parents' and grandparents' slow growth period. European journal of human genetics: EJHG. 2002;10(11):682-688. Epub 2002/10/31. doi: https://doi.org/10.1038/sj.ejhg.5200859. PubMed PMID: 12404098.
102. Donkin I, Versteyhe S, Ingerslev LR, Qian K, Mechta M, Nordkap L, et al. Obesity and Bariatric Surgery Drive Epigenetic Variation of Spermatozoa in Humans. Cell Metab. 2016;23(2):369-378. Epub 2015/12/17. doi: https://doi.org/10.1016/j.cmet.2015.11.004. PubMed PMID: 26669700.
103. Soubry A, Murphy SK, Wang F, Huang Z, Vidal AC, Fuemmeler BF, et al. Newborns of obese parents have altered DNA methylation patterns at imprinted genes. Int J Obes (Lond). 2015;39(4):650-7. Epub 2013/10/26. doi: 10.1038/ijo.2013.193. PubMed PMID: 24158121; PubMed Central PMCID: PMCPMC4048324.
104. Ling C, Ronn T. Epigenetics in Human Obesity and Type 2 Diabetes. Cell Metab. 2019;29(5):1028-44. Epub 2019/04/16. doi: 10.1016/j.cmet.2019.03.009. PubMed PMID: 30982733; PubMed Central PMCID: PMCPMC6509280.
105. Stenz L, Schechter DS, Serpa SR, Paoloni-Giacobino A. Intergenerational Transmission of DNA Methylation Signatures Associated with Early Life Stress. Current genomics. 2018;19(8):665-75. Epub 2018/12/12. doi: 10.2174/1389202919666171229145656. PubMed PMID: 30532646; PubMed Central PMCID: PMCPMC6225454.
106. Weaver IC, Cervoni N, Champagne FA, D'Alessio AC, Sharma S, Seckl JR, et al. Epigenetic programming by maternal behavior. Nat Neurosci. 2004;7(8):847-854. Epub 2004/06/29. doi: https://doi.org/10.1038/nn1276. PubMed PMID: 15220929.
107. Weaver IC, Champagne FA, Brown SE, Dymov S, Sharma S, Meaney MJ, et al. Reversal of maternal programming of stress responses in adult offspring through methyl supplementation: altering epigenetic marking later in life. J Neurosci. 2005;25(47):11045-11054. Epub 2005/11/25. doi: https://doi.org/10.1523/JNEUROSCI.3652-05.2005. PubMed PMID: 16306417.
108. Weaver IC, Meaney MJ, Szyf M. Maternal care effects on the hippocampal transcriptome and anxiety-mediated behaviors in the offspring that are reversible in adulthood. Proc Natl Acad Sci U S A. 2006;103(9):3480-5. Epub 2006/02/18. doi: 10.1073/pnas.0507526103. PubMed PMID: 16484373; PubMed Central PMCID: PMCPMC1413873.
109. Franklin TB, Russig H, Weiss IC, Graff J, Linder N, Michalon A, et al. Epigenetic transmission of the impact of early stress across generations. Biol Psychiatry. 2010;68(5):408-415. Epub 2010/08/03. doi: https://doi.org/10.1016/j.biopsych.2010.05.036. PubMed PMID: 20673872.
110. Gapp K, Soldado-Magraner S, Alvarez-Sanchez M, Bohacek J, Vernaz G, Shu H, et al. Early life stress in fathers improves behavioural flexibility in their offspring. Nat Commun. 2014;5:5466. Epub 2014/11/19. doi: https://doi.org/10.1038/ncomms6466. PubMed PMID: 25405779.
111. Gapp K, Jawaid A, Sarkies P, Bohacek J, Pelczar P, Prados J, et al. Implication of sperm RNAs in transgenerational inheritance of the effects of early trauma in mice. Nat Neurosci. 2014;17(5):667-9. Epub 2014/04/15. doi: 10.1038/nn.3695. PubMed PMID: 24728267; PubMed Central PMCID: PMCPMC4333222.
112. Gapp K, Bohacek J, Grossmann J, Brunner AM, Manuella F, Nanni P, et al. Potential of Environmental Enrichment to Prevent Transgenerational Effects of Paternal Trauma. Neuropsychopharmacology. 2016;41(11):2749-58. Epub 2016/06/10. doi: 10.1038/npp.2016.87. PubMed PMID: 27277118; PubMed Central PMCID: PMCPMC5026744.
113. Morgan CP, Bale TL. Early prenatal stress epigenetically programs dysmasculinization in second-generation offspring via the paternal lineage. J Neurosci. 2011;31(33):11748-55.

Epub 2011/08/19. doi: 10.1523/JNEUROSCI.1887-11.2011. PubMed PMID: 21849535; PubMed Central PMCID: PMCPMC3164823.

114. Rodgers AB, Morgan CP, Bronson SL, Revello S, Bale TL. Paternal stress exposure alters sperm microRNA content and reprograms offspring HPA stress axis regulation. J Neurosci. 2013;33(21):9003-12. Epub 2013/05/24. doi: 10.1523/JNEUROSCI.0914-13.2013. PubMed PMID: 23699511; PubMed Central PMCID: PMCPMC3712504.
115. Dietz DM, Laplant Q, Watts EL, Hodes GE, Russo SJ, Feng J, et al. Paternal transmission of stress-induced pathologies. Biol Psychiatry. 2011;70(5):408-14. Epub 2011/06/18. doi: 10.1016/j.biopsych.2011.05.005. PubMed PMID: 21679926; PubMed Central PMCID: PMCPMC3217197.
116. Dias BG, Ressler KJ. Parental olfactory experience influences behavior and neural structure in subsequent generations. Nat Neurosci. 2014;17(1):89-96. Epub 2013/12/03. doi: 10.1038/nn.3594. PubMed PMID: 24292232; PubMed Central PMCID: PMCPMC3923835.
117. Magklara A, Yen A, Colquitt BM, Clowney EJ, Allen W, Markenscoff-Papadimitriou E, et al. An epigenetic signature for monoallelic olfactory receptor expression. Cell. 2011;145(4):555-70. Epub 2011/05/03. doi: 10.1016/j.cell.2011.03.040. PubMed PMID: 21529909; PubMed Central PMCID: PMCPMC3094500.
118. Aoued HS, Sannigrahi S, Doshi N, Morrison FG, Linsenbaum H, Hunter SC, et al. Reversing Behavioral, Neuroanatomical, and Germline Influences of Intergenerational Stress. Biol Psychiatry. 2019;85(3):248-56. Epub 2018/10/08. doi: 10.1016/j.biopsych.2018.07.028. PubMed PMID: 30292395; PubMed Central PMCID: PMCPMC6326876.
119. Winther G, Eskelund A, Bay-Richter C, Elfving B, Muller HK, Lund S, et al. Grandmaternal high-fat diet primed anxiety-like behaviour in the second-generation female offspring. Behavioural brain research. 2019;359:47-55. Epub 2018/10/20. doi: https://doi.org/10.1016/j.bbr.2018.10.017. PubMed PMID: 30336180.
120. McCoy CR, Jackson NL, Day J, Clinton SM. Genetic predisposition to high anxiety- and depression-like behavior coincides with diminished DNA methylation in the adult rat amygdala. Behavioural brain research. 2017;320:165-78. Epub 2016/12/15. doi: 10.1016/j.bbr.2016.12.008. PubMed PMID: 27965039; PubMed Central PMCID: PMCPMC5239722.
121. McCoy CR, Jackson NL, Brewer RL, Moughnyeh MM, Smith DL, Jr., Clinton SM. A paternal methyl donor depleted diet leads to increased anxiety- and depression-like behavior in adult rat offspring. Bioscience reports. 2018;38(4). Epub 2018/06/28. doi: 10.1042/BSR20180730. PubMed PMID: 29945927; PubMed Central PMCID: PMCPMC6153370.
122. Drobna Z, Henriksen AD, Wolstenholme JT, Montiel C, Lambeth PS, Shang S, et al. Transgenerational Effects of Bisphenol A on Gene Expression and DNA Methylation of Imprinted Genes in Brain. Endocrinology. 2018;159(1):132-44. Epub 2017/11/23. doi: 10.1210/en.2017-00730. PubMed PMID: 29165653; PubMed Central PMCID: PMCPMC5761590.
123. Anway MD, Cupp AS, Uzumcu M, Skinner MK. Epigenetic transgenerational actions of endocrine disruptors and male fertility. Science. 2005;308(5727):1466-1469. Epub 2005/06/04. doi: https://doi.org/10.1126/science.1108190. PubMed PMID: 15933200.
124. Guerrero-Bosagna C, Covert TR, Haque MM, Settles M, Nilsson EE, Anway MD, et al. Epigenetic transgenerational inheritance of vinclozolin induced mouse adult onset disease and associated sperm epigenome biomarkers. Reprod Toxicol. 2012;34(4):694-707. Epub 2012/10/09. doi: 10.1016/j.reprotox.2012.09.005. PubMed PMID: 23041264; PubMed Central PMCID: PMCPMC3513496.
125. Skinner MK, Nilsson E, Sadler-Riggleman I, Beck D, Ben Maamar M, McCarrey JR. Transgenerational sperm DNA methylation epimutation developmental origins following ancestral vinclozolin exposure. Epigenetics. 2019;14(7):721-39. Epub 2019/05/14. doi: 10.1080/15592294.2019.1614417. PubMed PMID: 31079544; PubMed Central PMCID: PMCPMC6557599.
126. Iqbal K, Tran DA, Li AX, Warden C, Bai AY, Singh P, et al. Deleterious effects of endocrine disruptors are corrected in the mammalian germline by epigenome reprogramming. Genome

Biol. 2015;16:59. Epub 2015/04/09. doi: 10.1186/s13059-015-0619-z. PubMed PMID: 25853433; PubMed Central PMCID: PMCPMC4376074.
127. Schneider S, Kaufmann W, Buesen R, van Ravenzwaay B. Vinclozolin--the lack of a transgenerational effect after oral maternal exposure during organogenesis. Reprod Toxicol. 2008;25(3):352-360. Epub 2008/05/20. doi: https://doi.org/10.1016/j.reprotox.2008.04.001. PubMed PMID: 18485663.
128. Schneider S, Marxfeld H, Groters S, Buesen R, van Ravenzwaay B. Vinclozolin--no transgenerational inheritance of anti-androgenic effects after maternal exposure during organogenesis via the intraperitoneal route. Reprod Toxicol. 2013;37:6-14. Epub 2013/01/15. doi: https://doi.org/10.1016/j.reprotox.2012.12.003. PubMed PMID: 23313085.
129. Latchney SE, Fields AM, Susiarjo M. Linking inter-individual variability to endocrine disruptors: insights for epigenetic inheritance. Mamm Genome. 2018;29(1-2):141-52. Epub 2017/12/09. doi: 10.1007/s00335-017-9729-0. PubMed PMID: 29218402; PubMed Central PMCID: PMCPMC5849504.
130. Nadeau JH. The nature of evidence for and against epigenetic inheritance. Genome Biol. 2015;16:137. Epub 2015/07/15. doi: 10.1186/s13059-015-0709-y. PubMed PMID: 26162370; PubMed Central PMCID: PMCPMC4499204.
131. Kubsad D, Nilsson EE, King SE, Sadler-Riggleman I, Beck D, Skinner MK. Assessment of Glyphosate Induced Epigenetic Transgenerational Inheritance of Pathologies and Sperm Epimutations: Generational Toxicology. Sci Rep. 2019;9(1):6372. Epub 2019/04/24. doi: 10.1038/s41598-019-42860-0. PubMed PMID: 31011160; PubMed Central PMCID: PMCPMC6476885.
132. Yehuda R, Schmeidler J, Giller EL, Jr., Siever LJ, Binder-Brynes K. Relationship between posttraumatic stress disorder characteristics of Holocaust survivors and their adult offspring. The American journal of psychiatry. 1998;155(6):841-843. Epub 1998/06/10. doi: https://doi.org/10.1176/ajp.155.6.841. PubMed PMID: 9619162.
133. Yehuda R, Teicher MH, Seckl JR, Grossman RA, Morris A, Bierer LM. Parental posttraumatic stress disorder as a vulnerability factor for low cortisol trait in offspring of holocaust survivors. Archives of general psychiatry. 2007;64(9):1040-1048. Epub 2007/09/05. doi: https://doi.org/10.1001/archpsyc.64.9.1040. PubMed PMID: 17768269.
134. Yehuda R, Bierer LM. Transgenerational transmission of cortisol and PTSD risk. Progress in brain research. 2008;167:121-135. Epub 2007/11/27. doi: https://doi.org/10.1016/S0079-6123(07)67009-5. PubMed PMID: 18037011.
135. Yehuda R, Lehrner A, Bierer LM. The public reception of putative epigenetic mechanisms in the transgenerational effects of trauma. Environmental epigenetics. 2018;4(2):dvy018. Epub 2018/07/25. doi: 10.1093/eep/dvy018. PubMed PMID: 30038801; PubMed Central PMCID: PMCPMC6051458.
136. Lehrner A, Bierer LM, Passarelli V, Pratchett LC, Flory JD, Bader HN, et al. Maternal PTSD associates with greater glucocorticoid sensitivity in offspring of Holocaust survivors. Psychoneuroendocrinology. 2014;40:213-20. Epub 2014/02/04. doi: 10.1016/j.psyneuen.2013.11.019. PubMed PMID: 24485493; PubMed Central PMCID: PMCPMC3967845.
137. Yehuda R, Bell A, Bierer LM, Schmeidler J. Maternal, not paternal, PTSD is related to increased risk for PTSD in offspring of Holocaust survivors. Journal of psychiatric research. 2008;42(13):1104-11. Epub 2008/02/19. doi: 10.1016/j.jpsychires.2008.01.002. PubMed PMID: 18281061; PubMed Central PMCID: PMCPMC2612639.
138. Yehuda R, Daskalakis NP, Lehrner A, Desarnaud F, Bader HN, Makotkine I, et al. Influences of maternal and paternal PTSD on epigenetic regulation of the glucocorticoid receptor gene in Holocaust survivor offspring. The American journal of psychiatry. 2014;171(8):872-80. Epub 2014/05/17. doi: 10.1176/appi.ajp.2014.13121571. PubMed PMID: 24832930; PubMed Central PMCID: PMCPMC4127390.
139. Bale TL. Epigenetic and transgenerational reprogramming of brain development. Nat Rev Neurosci. 2015;16(6):332-344. Epub 2015/04/30. doi: https://doi.org/10.1038/nrn3818. PubMed PMID: 25921815.

140. Bohacek J, Mansuy IM. Molecular insights into transgenerational non-genetic inheritance of acquired behaviours. Nat Rev Genet. 2015;16(11):641-652. Epub 2015/09/30. doi: https://doi.org/10.1038/nrg3964. PubMed PMID: 26416311.
141. Fischer A. Epigenetic memory: the Lamarckian brain. EMBO J. 2014;33(9):945-67. Epub 2014/04/11. doi: 10.1002/embj.201387637. PubMed PMID: 24719207; PubMed Central PMCID: PMCPMC4193930.
142. Dell PA, Rose FD. Transfer of effects from environmentally enriched and impoverished female rats to future offspring. Physiology & behavior. 1987;39(2):187-190. Epub 1987/01/01. PubMed PMID: 3575452.
143. Nithianantharajah J, Hannan AJ. Enriched environments, experience-dependent plasticity and disorders of the nervous system. Nat Rev Neurosci. 2006;7(9):697-709. Epub 2006/08/23. doi: https://doi.org/10.1038/nrn1970. PubMed PMID: 16924259.
144. Arai JA, Li S, Hartley DM, Feig LA. Transgenerational rescue of a genetic defect in long-term potentiation and memory formation by juvenile enrichment. J Neurosci. 2009;29(5):1496-502. Epub 2009/02/06. doi: 10.1523/JNEUROSCI.5057-08.2009. PubMed PMID: 19193896; PubMed Central PMCID: PMCPMC3408235.
145. Benito E, Kerimoglu C, Ramachandran B, Pena-Centeno T, Jain G, Stilling RM, et al. RNA-Dependent Intergenerational Inheritance of Enhanced Synaptic Plasticity after Environmental Enrichment. Cell Rep. 2018;23(2):546-54. Epub 2018/04/12. doi: 10.1016/j.celrep.2018.03.059. PubMed PMID: 29642011; PubMed Central PMCID: PMCPMC5912949.
146. Ryan DP, Henzel KS, Pearson BL, Siwek ME, Papazoglou A, Guo L, et al. A paternal methyl donor-rich diet altered cognitive and neural functions in offspring mice. Molecular psychiatry. 2018;23(5):1345-55. Epub 2017/04/05. doi: 10.1038/mp.2017.53. PubMed PMID: 28373690; PubMed Central PMCID: PMCPMC5984088.
147. Zeybel M, Hardy T, Wong YK, Mathers JC, Fox CR, Gackowska A, et al. Multigenerational epigenetic adaptation of the hepatic wound-healing response. Nat Med. 2012;18(9):1369-77. Epub 2012/09/04. doi: 10.1038/nm.2893. PubMed PMID: 22941276; PubMed Central PMCID: PMCPMC3489975.
148. Tucci V, Isles AR, Kelsey G, Ferguson-Smith AC, Erice Imprinting G. Genomic Imprinting and Physiological Processes in Mammals. Cell. 2019;176(5):952-965. Epub 2019/02/23. doi: https://doi.org/10.1016/j.cell.2019.01.043. PubMed PMID: 30794780.
149. Peters J. The role of genomic imprinting in biology and disease: an expanding view. Nat Rev Genet. 2014;15(8):517-530. Epub 2014/06/25. doi: https://doi.org/10.1038/nrg3766. PubMed PMID: 24958438.
150. Bartolomei MS, Ferguson-Smith AC. Mammalian genomic imprinting. Cold Spring Harb Perspect Biol. 2011;3(7). Epub 2011/05/18. doi: 10.1101/cshperspect.a002592. PubMed PMID: 21576252; PubMed Central PMCID: PMCPMC3119911.
151. Barlow DP, Bartolomei MS. Genomic imprinting in mammals. Cold Spring Harb Perspect Biol. 2014;6(2). Epub 2014/02/05. doi: 10.1101/cshperspect.a018382. PubMed PMID: 24492710; PubMed Central PMCID: PMCPMC3941233.
152. McGrath J, Solter D. Nuclear transplantation in the mouse embryo by microsurgery and cell fusion. Science. 1983;220(4603):1300-1302. Epub 1983/06/17. doi: https://doi.org/10.1126/science.6857250. PubMed PMID: 6857250.
153. Lee JT, Bartolomei MS. X-inactivation, imprinting, and long noncoding RNAs in health and disease. Cell. 2013;152(6):1308-1323. Epub 2013/03/19. doi: https://doi.org/10.1016/j.cell.2013.02.016. PubMed PMID: 23498939.
154. Cattanach BM, Jones J. Genetic imprinting in the mouse: implications for gene regulation. J Inherit Metab Dis. 1994;17(4):403-420. Epub 1994/01/01. PubMed PMID: 7967491.
155. Li E, Beard C, Jaenisch R. Role for DNA methylation in genomic imprinting. Nature. 1993;366(6453):362-365. Epub 1993/11/25. doi: https://doi.org/10.1038/366362a0. PubMed PMID: 8247133.
156. Fitzpatrick GV, Soloway PD, Higgins MJ. Regional loss of imprinting and growth deficiency in mice with a targeted deletion of KvDMR1. Nat Genet. 2002;32(3):426-431. Epub 2002/11/01. doi: https://doi.org/10.1038/ng988. PubMed PMID: 12410230.

157. Bourc'his D, Xu GL, Lin CS, Bollman B, Bestor TH. Dnmt3L and the establishment of maternal genomic imprints. Science. 2001;294(5551):2536-2539. Epub 2001/11/24. doi: https://doi.org/10.1126/science.1065848. PubMed PMID: 11719692.
158. Kaneda M, Okano M, Hata K, Sado T, Tsujimoto N, Li E, et al. Essential role for de novo DNA methyltransferase Dnmt3a in paternal and maternal imprinting. Nature. 2004;429(6994):900-903. Epub 2004/06/25. doi: https://doi.org/10.1038/nature02633. PubMed PMID: 15215868.
159. Smallwood SA, Tomizawa S, Krueger F, Ruf N, Carli N, Segonds-Pichon A, et al. Dynamic CpG island methylation landscape in oocytes and preimplantation embryos. Nat Genet. 2011;43(8):811-4. Epub 2011/06/28. doi: 10.1038/ng.864. PubMed PMID: 21706000; PubMed Central PMCID: PMCPMC3146050.
160. Sanchez-Delgado M, Court F, Vidal E, Medrano J, Monteagudo-Sanchez A, Martin-Trujillo A, et al. Human Oocyte-Derived Methylation Differences Persist in the Placenta Revealing Widespread Transient Imprinting. PLoS Genet. 2016;12(11):e1006427. Epub 2016/11/12. doi: 10.1371/journal.pgen.1006427. PubMed PMID: 27835649; PubMed Central PMCID: PMCPMC5106035.
161. Nowak K, Stein G, Powell E, He LM, Naik S, Morris J, et al. Establishment of paternal allele-specific DNA methylation at the imprinted mouse Gtl2 locus. Epigenetics. 2011;6(8):1012-20. Epub 2011/07/05. doi: 10.4161/epi.6.8.16075. PubMed PMID: 21725202; PubMed Central PMCID: PMCPMC3359488.
162. Sasaki H, Ferguson-Smith AC, Shum AS, Barton SC, Surani MA. Temporal and spatial regulation of H19 imprinting in normal and uniparental mouse embryos. Development. 1995;121(12):4195-4202. Epub 1995/12/01. PubMed PMID: 8575319.
163. Buiting K, Gross S, Lich C, Gillessen-Kaesbach G, el-Maarri O, Horsthemke B. Epimutations in Prader-Willi and Angelman syndromes: a molecular study of 136 patients with an imprinting defect. Am J Hum Genet. 2003;72(3):571-7. Epub 2003/01/25. doi: 10.1086/367926. PubMed PMID: 12545427; PubMed Central PMCID: PMCPMC1180233.
164. Quenneville S, Verde G, Corsinotti A, Kapopoulou A, Jakobsson J, Offner S, et al. In embryonic stem cells, ZFP57/KAP1 recognize a methylated hexanucleotide to affect chromatin and DNA methylation of imprinting control regions. Mol Cell. 2011;44(3):361-72. Epub 2011/11/08. doi: 10.1016/j.molcel.2011.08.032. PubMed PMID: 22055183; PubMed Central PMCID: PMCPMC3210328.
165. Mackay DJ, Callaway JL, Marks SM, White HE, Acerini CL, Boonen SE, et al. Hypomethylation of multiple imprinted loci in individuals with transient neonatal diabetes is associated with mutations in ZFP57. Nat Genet. 2008;40(8):949-951. Epub 2008/07/16. doi: https://doi.org/10.1038/ng.187. PubMed PMID: 18622393.
166. Takahashi N, Coluccio A, Thorball CW, Planet E, Shi H, Offner S, et al. ZNF445 is a primary regulator of genomic imprinting. Genes Dev. 2019;33(1-2):49-54. Epub 2019/01/04. doi: 10.1101/gad.320069.118. PubMed PMID: 30602440; PubMed Central PMCID: PMCPMC6317318.
167. Li X, Ito M, Zhou F, Youngson N, Zuo X, Leder P, et al. A maternal-zygotic effect gene, Zfp57, maintains both maternal and paternal imprints. Dev Cell. 2008;15(4):547-57. Epub 2008/10/16. doi: 10.1016/j.devcel.2008.08.014. PubMed PMID: 18854139; PubMed Central PMCID: PMCPMC2593089.
168. Maenohara S, Unoki M, Toh H, Ohishi H, Sharif J, Koseki H, et al. Role of UHRF1 in de novo DNA methylation in oocytes and maintenance methylation in preimplantation embryos. PLoS Genet. 2017;13(10):e1007042. Epub 2017/10/05. doi: 10.1371/journal.pgen.1007042. PubMed PMID: 28976982; PubMed Central PMCID: PMCPMC5643148.
169. Bostick M, Kim JK, Esteve PO, Clark A, Pradhan S, Jacobsen SE. UHRF1 plays a role in maintaining DNA methylation in mammalian cells. Science. 2007;317(5845):1760-1764. Epub 2007/08/04. doi: https://doi.org/10.1126/science.1147939. PubMed PMID: 17673620.
170. Sharif J, Muto M, Takebayashi S, Suetake I, Iwamatsu A, Endo TA, et al. The SRA protein Np95 mediates epigenetic inheritance by recruiting Dnmt1 to methylated DNA. Nature. 2007;450(7171):908-912. Epub 2007/11/13. doi: https://doi.org/10.1038/nature06397. PubMed PMID: 17994007.

171. Andergassen D, Dotter CP, Wenzel D, Sigl V, Bammer PC, Muckenhuber M, et al. Mapping the mouse Allelome reveals tissue-specific regulation of allelic expression. Elife. 2017;6. Epub 2017/08/15. doi: 10.7554/eLife.25125. PubMed PMID: 28806168; PubMed Central PMCID: PMCPMC5555720.
172. Perez JD, Rubinstein ND, Fernandez DE, Santoro SW, Needleman LA, Ho-Shing O, et al. Quantitative and functional interrogation of parent-of-origin allelic expression biases in the brain. Elife. 2015;4:e07860. Epub 2015/07/04. doi: 10.7554/eLife.07860. PubMed PMID: 26140685; PubMed Central PMCID: PMCPMC4512258.
173. Ge B, Pokholok DK, Kwan T, Grundberg E, Morcos L, Verlaan DJ, et al. Global patterns of cis variation in human cells revealed by high-density allelic expression analysis. Nat Genet. 2009;41(11):1216-1222. Epub 2009/10/20. doi: https://doi.org/10.1038/ng.473. PubMed PMID: 19838192.
174. Hasin-Brumshtein Y, Hormozdiari F, Martin L, van Nas A, Eskin E, Lusis AJ, et al. Allele-specific expression and eQTL analysis in mouse adipose tissue. BMC Genomics. 2014;15:471. Epub 2014/06/15. doi: 10.1186/1471-2164-15-471. PubMed PMID: 24927774; PubMed Central PMCID: PMCPMC4089026.
175. Onuchic V, Lurie E, Carrero I, Pawliczek P, Patel RY, Rozowsky J, et al. Allele-specific epigenome maps reveal sequence-dependent stochastic switching at regulatory loci. Science. 2018;361(6409). Epub 2018/08/25. doi: 10.1126/science.aar3146. PubMed PMID: 30139913; PubMed Central PMCID: PMCPMC6198826.
176. Dickies MM. A new viable yellow mutation in the house mouse. J Hered. 1962;53:84-86. Epub 1962/03/01. doi: https://doi.org/10.1093/oxfordjournals.jhered.a107129. PubMed PMID: 13886198.
177. Vasicek TJ, Zeng L, Guan XJ, Zhang T, Costantini F, Tilghman SM. Two dominant mutations in the mouse fused gene are the result of transposon insertions. Genetics. 1997;147(2):777-86. Epub 1997/10/23. PubMed PMID: 9335612; PubMed Central PMCID: PMCPMC1208197.
178. Michaud EJ, van Vugt MJ, Bultman SJ, Sweet HO, Davisson MT, Woychik RP. Differential expression of a new dominant agouti allele (Aiapy) is correlated with methylation state and is influenced by parental lineage. Genes Dev. 1994;8(12):1463-1472. Epub 1994/06/15. doi: https://doi.org/10.1101/gad.8.12.1463. PubMed PMID: 7926745.
179. Morgan HD, Sutherland HG, Martin DI, Whitelaw E. Epigenetic inheritance at the agouti locus in the mouse. Nat Genet. 1999;23(3):314-318. Epub 1999/11/05. doi: https://doi.org/10.1038/15490. PubMed PMID: 10545949.
180. Rakyan VK, Chong S, Champ ME, Cuthbert PC, Morgan HD, Luu KV, et al. Transgenerational inheritance of epigenetic states at the murine Axin(Fu) allele occurs after maternal and paternal transmission. Proc Natl Acad Sci U S A. 2003;100(5):2538-43. Epub 2003/02/26. doi: 10.1073/pnas.0436776100. PubMed PMID: 12601169; PubMed Central PMCID: PMCPMC151376.
181. Blewitt ME, Vickaryous NK, Paldi A, Koseki H, Whitelaw E. Dynamic reprogramming of DNA methylation at an epigenetically sensitive allele in mice. PLoS Genet. 2006;2(4):e49. Epub 2006/04/11. doi: 10.1371/journal.pgen.0020049. PubMed PMID: 16604157; PubMed Central PMCID: PMCPMC1428789.
182. Dolinoy DC, Weinhouse C, Jones TR, Rozek LS, Jirtle RL. Variable histone modifications at the A(vy) metastable epiallele. Epigenetics. 2010;5(7):637-44. Epub 2010/07/31. doi: 10.4161/epi.5.7.12892. PubMed PMID: 20671424; PubMed Central PMCID: PMCPMC3052847.
183. Cropley JE, Suter CM, Beckman KB, Martin DI. Germ-line epigenetic modification of the murine A vy allele by nutritional supplementation. Proc Natl Acad Sci U S A. 2006;103(46):17308-12. Epub 2006/11/15. doi: 10.1073/pnas.0607090103. PubMed PMID: 17101998; PubMed Central PMCID: PMCPMC1838538.
184. Kaminen-Ahola N, Ahola A, Maga M, Mallitt KA, Fahey P, Cox TC, et al. Maternal ethanol consumption alters the epigenotype and the phenotype of offspring in a mouse model. PLoS Genet. 2010;6(1):e1000811. Epub 2010/01/20. doi: 10.1371/journal.pgen.1000811. PubMed PMID: 20084100; PubMed Central PMCID: PMCPMC2797299.

185. Wolff GL, Kodell RL, Moore SR, Cooney CA. Maternal epigenetics and methyl supplements affect agouti gene expression in Avy/a mice. FASEB J. 1998;12(11):949-957. Epub 1998/08/26. PubMed PMID: 9707167.
186. Dolinoy DC, Weidman JR, Waterland RA, Jirtle RL. Maternal genistein alters coat color and protects Avy mouse offspring from obesity by modifying the fetal epigenome. Environ Health Perspect. 2006;114(4):567-72. Epub 2006/04/04. doi: 10.1289/ehp.8700. PubMed PMID: 16581547; PubMed Central PMCID: PMCPMC1440782.
187. Dolinoy DC, Huang D, Jirtle RL. Maternal nutrient supplementation counteracts bisphenol A-induced DNA hypomethylation in early development. Proc Natl Acad Sci U S A. 2007;104(32):13056-61. Epub 2007/08/03. doi: 10.1073/pnas.0703739104. PubMed PMID: 17670942; PubMed Central PMCID: PMCPMC1941790.
188. Cropley JE, Suter CM, Beckman KB, Martin DI. CpG methylation of a silent controlling element in the murine Avy allele is incomplete and unresponsive to methyl donor supplementation. PLoS One. 2010;5(2):e9055. Epub 2010/02/09. doi: 10.1371/journal.pone.0009055. PubMed PMID: 20140227; PubMed Central PMCID: PMCPMC2816220.
189. Smit AFA, Hubley R, Green P. RepeatMasker Open-4.0. http://wwwrepeatmaskerorg. 2013-2015.
190. Lane N, Dean W, Erhardt S, Hajkova P, Surani A, Walter J, et al. Resistance of IAPs to methylation reprogramming may provide a mechanism for epigenetic inheritance in the mouse. Genesis. 2003;35(2):88-93. Epub 2003/01/21. doi: https://doi.org/10.1002/gene.10168. PubMed PMID: 12533790.
191. Kazachenka A, Bertozzi TM, Sjoberg-Herrera MK, Walker N, Gardner J, Gunning R, et al. Identification, Characterization, and Heritability of Murine Metastable Epialleles: Implications for Non-genetic Inheritance. Cell. 2018;175(5):1259-71 e13. Epub 2018/11/21. doi: 10.1016/j.cell.2018.09.043. PubMed PMID: 30454646; PubMed Central PMCID: PMCPMC6242299.
192. Gueant JL, Chery C, Oussalah A, Nadaf J, Coelho D, Josse T, et al. APRDX1 mutant allele causes a MMACHC secondary epimutation in cblC patients. Nat Commun. 2018;9(1):67. Epub 2018/01/06. doi: 10.1038/s41467-017-02306-5. PubMed PMID: 29302025; PubMed Central PMCID: PMCPMC5754367.
193. Lynch HT, Lynch PM, Lanspa SJ, Snyder CL, Lynch JF, Boland CR. Review of the Lynch syndrome: history, molecular genetics, screening, differential diagnosis, and medicolegal ramifications. Clin Genet. 2009;76(1):1-18. Epub 2009/08/08. doi: 10.1111/j.1399-0004.2009.01230.x. PubMed PMID: 19659756; PubMed Central PMCID: PMCPMC2846640.
194. Hitchins MP. The role of epigenetics in Lynch syndrome. Fam Cancer. 2013;12(2):189-205. Epub 2013/03/07. doi: https://doi.org/10.1007/s10689-013-9613-3. PubMed PMID: 23462881.
195. Ligtenberg MJ, Kuiper RP, Chan TL, Goossens M, Hebeda KM, Voorendt M, et al. Heritable somatic methylation and inactivation of MSH2 in families with Lynch syndrome due to deletion of the 3' exons of TACSTD1. Nat Genet. 2009;41(1):112-117. Epub 2008/12/23. doi: https://doi.org/10.1038/ng.283. PubMed PMID: 19098912.
196. Tufarelli C, Stanley JA, Garrick D, Sharpe JA, Ayyub H, Wood WG, et al. Transcription of antisense RNA leading to gene silencing and methylation as a novel cause of human genetic disease. Nat Genet. 2003;34(2):157-165. Epub 2003/05/06. doi: https://doi.org/10.1038/ng1157. PubMed PMID: 12730694.
197. Hitchins MP, Wong JJ, Suthers G, Suter CM, Martin DI, Hawkins NJ, et al. Inheritance of a cancer-associated MLH1 germ-line epimutation. N Engl J Med. 2007;356(7):697-705. Epub 2007/02/16. doi: https://doi.org/10.1056/NEJMoa064522. PubMed PMID: 17301300.
198. Schaefer S, Nadeau JH. The Genetics of Epigenetic Inheritance: Modes, Molecules, and Mechanisms. Q Rev Biol. 2015;90(4):381-415. Epub 2015/12/31. PubMed PMID: 26714351.
199. Horsthemke B. A critical view on transgenerational epigenetic inheritance in humans. Nat Commun. 2018;9(1):2973. Epub 2018/08/01. doi: https://doi.org/10.1038/s41467-018-05445-5. PubMed PMID: 30061690; PubMed Central PMCID: PMCPMC6065375.

200. van Otterdijk SD, Michels KB. Transgenerational epigenetic inheritance in mammals: how good is the evidence? FASEB J. 2016;30(7):2457-2465. Epub 2016/04/03. doi: https://doi.org/10.1096/fj.201500083. PubMed PMID: 27037350.
201. Chen Q, Yan W, Duan E. Epigenetic inheritance of acquired traits through sperm RNAs and sperm RNA modifications. Nat Rev Genet. 2016;17(12):733-43. Epub 2016/11/01. doi: 10.1038/nrg.2016.106. PubMed PMID: 27694809; PubMed Central PMCID: PMCPMC5441558.
202. Wagner KD, Wagner N, Ghanbarian H, Grandjean V, Gounon P, Cuzin F, et al. RNA induction and inheritance of epigenetic cardiac hypertrophy in the mouse. Dev Cell. 2008;14(6):962-969. Epub 2008/06/10. doi: https://doi.org/10.1016/j.devcel.2008.03.009. PubMed PMID: 18539123.
203. Beckers J, Wurst W, de Angelis MH. Towards better mouse models: enhanced genotypes, systemic phenotyping and envirotype modelling. Nat Rev Genet. 2009;10(6):371-380. Epub 2009/05/13. doi: https://doi.org/10.1038/nrg2578. PubMed PMID: 19434078.
204. Xie N, Zhou Y, Sun Q, Tang B. Novel Epigenetic Techniques Provided by the CRISPR/Cas9 System. Stem Cells Int. 2018;2018:7834175. Epub 2018/08/21. doi: https://doi.org/10.1155/2018/7834175. PubMed PMID: 30123293; PubMed Central PMCID: PMCPMC6079388.
205. Vojta A, Dobrinic P, Tadic V, Bockor L, Korac P, Julg B, et al. Repurposing the CRISPR-Cas9 system for targeted DNA methylation. Nucleic Acids Res. 2016;44(12):5615-5628. Epub 2016/03/13. doi: https://doi.org/10.1093/nar/gkw159. PubMed PMID: 26969735; PubMed Central PMCID: PMCPMC4937303.
206. Huang YH, Su J, Lei Y, Brunetti L, Gundry MC, Zhang X, et al. DNA epigenome editing using CRISPR-Cas SunTag-directed DNMT3A. Genome Biol. 2017;18(1):176. Epub 2017/09/20. doi: https://doi.org/10.1186/s13059-017-1306-z. PubMed PMID: 28923089; PubMed Central PMCID: PMCPMC5604343.
207. Morita S, Noguchi H, Horii T, Nakabayashi K, Kimura M, Okamura K, et al. Targeted DNA demethylation in vivo using dCas9-peptide repeat and scFv-TET1 catalytic domain fusions. Nat Biotechnol. 2016;34(10):1060-1065. Epub 2016/08/30. doi: https://doi.org/10.1038/nbt.3658. PubMed PMID: 27571369.
208. Liao HK, Hatanaka F, Araoka T, Reddy P, Wu MZ, Sui Y, et al. In Vivo Target Gene Activation via CRISPR/Cas9-Mediated Trans-epigenetic Modulation. Cell. 2017;171(7):1495-1507 e15. Epub 2017/12/12. doi: https://doi.org/10.1016/j.cell.2017.10.025. PubMed PMID: 29224783; PubMed Central PMCID: PMCPMC5732045.
209. Wei Y, Lang J, Zhang Q, Yang CR, Zhao ZA, Zhang Y, et al. DNA methylation analysis and editing in single mammalian oocytes. Proc Natl Acad Sci U S A. 2019;116(20):9883-9892. Epub 2019/04/24. doi: https://doi.org/10.1073/pnas.1817703116. PubMed PMID: 31010926; PubMed Central PMCID: PMCPMC6525536.

Chapter 11
Small Non-Coding RNAs and Epigenetic Inheritance

Tiina Lehtiniemi, Margareeta Mäkelä, and Noora Kotaja

Keywords Non-coding RNAs · PiRNA · MiRNA · TRNA · TRFs · Spermatogenesis · Spermatozoa · Sperm RNAs · Testis · Epididymis · Epigenetic inheritance.

11.1 Introduction

In recent years, it has become evident that exposure to various environmental factors can induce epigenetic changes in gametes [15, 17, 28, 33, 84, 85]. This allows the parents to transmit traits they have acquired during their lifetime to their offspring. Both maternal and paternal transgenerational transmission of acquired disorders has been documented. While it is well-known that both parents contribute quite equally to the genetic content of the zygote, the contribution in the form of epigenetic information is less well understood. Due to technical limitations, much more progress has been made studying paternal than maternal epigenetic inheritance and this chapter will mostly focus on the former. Epigenetic studies on oocytes are challenged by the limited number of oocytes in the ovary, and to date only a few studies have addressed the effects of environmental factors on the oocyte epigenome [36, 37, 49]. The studies on paternal epigenetic inheritance require less experimental resources due to the massive daily production of spermatozoa compared to oocytes. In addition, confounding factors are easier to exclude as paternal contribution is limited to the content of one spermatozoon. As for the maternal contribution, a mother's physiological state affects the development of an oocyte, but also directly impacts the *in utero* environment for the growing embryo. Furthermore, the F2 generation in addition to the F1 may be influenced *in utero* through changes in the developing germ cells of the F1 embryo. Therefore, transgenerational studies on maternal inheritance have to be extended to the F3 generation further complicating the experimental set up.

T. Lehtiniemi · M. Mäkelä · N. Kotaja (✉)
Institute of Biomedicine, Research Centre for Integrative Physiology and Pharmacology, University of Turku, Turku, Finland
e-mail: noora.kotaja@utu.fi

R. Teperino (ed.), *Beyond Our Genes*,
https://doi.org/10.1007/978-3-030-35213-4_11

Suggested mechanisms for epigenetic inheritance include DNA methylation and histone modifications (discussed in more detail in other chapters), as well as non-coding RNAs (ncRNAs). DNA methylation and histone modifications are known to be passed to daughter cells during cell division and are therefore able to transmit specific gene expression patterns from one cell to another. However, during mammalian development, there is a near to complete erasure of DNA methylation in early germline cells [70], which is thought to allow the erasure of cellular memory and establishment of developmental totipotency. Subsequently, new DNA methylation marks are generated during germ cell development. Furthermore, in males, the majority of histone modifications are lost during the late steps of spermatogenesis when most of the histones are replaced by the sperm-specific chromatin-packing proteins, protamines [35]. The renewal of these epigenetic marks during gametogenesis and early embryogenesis confounds the mechanism of how these modifications could be transmitted to offspring and how they could contribute to epigenetic inheritance. However, as some histones are retained in the sperm epigenome and similarly, some genomic regions escape DNA demethylation, these regions and their modification may allow the transfer of environmentally-induced epigenetic changes from one generation to the next.

As the existence and the complexity of spermatozoal RNAs were recently revealed, RNA soon became a new likely carrier of epigenetic information from the father to offspring [40]. Sperm transcripts were first suggested to have a role in embryonic development in 2004 by Krawetz and colleagues, who showed that spermatozoa contain specific RNAs that are delivered to the oocyte at fertilization [78]. Another important study demonstrated that zygotic transfer of RNA molecules was responsible for the inheritance of a phenotype in mice [86]. After these studies, particularly the role of the spermatozoal small ncRNAs (sncRNAs) in epigenetic inheritance has been actively explored. Since sncRNAs have the ability to regulate gene expression both at the posttranscriptional and the chromatin level, they are functionally competent to act as intergenerational carriers of epigenetic information. Although the exact mechanism of how sperm sncRNAs transmit information to offspring is still unclear, according to the current hypothesis they could bring about changes in the zygotic gene expression. As discussed in detail later, the fast emerging data on small RNAs in paternal epigenetic inheritance has now been produced by several groups. In particular their role in inheritance of metabolic diseases and psychiatric conditions will be covered.

11.2 Gametogenesis

In order to transmit the environmentally-induced condition to offspring, information has to be programmed to the germline epigenome. First of all, the information must travel from the peripheral tissues to the reproductive tissues by still unknown mechanisms. It was recently proposed that RNA-containing nanovesicles released from somatic tissues could reach the reproductive tissues via the bloodstream [101].

Secondly, the information has to be transferred to the gamete, though it is not yet known which specific developmental phases during gamete differentiation and maturation are sensitive for receiving such epigenetic information. Female and male gametogenesis have very different dynamics, and therefore, the critical time window for environmental preconception exposures, as well as the mechanisms of germline-mediated epigenetic inheritance may vary between the sexes.

Both spermatogenesis and oogenesis initiate during embryonic development with early germline allocation [92], followed by migration and proliferation of the primordial germ cells (PGCs) in the gonadal ridge. A key event that occurs during this proliferative phase in male and female PGCs is epigenetic reprogramming, including a genome-wide DNA demethylation that is essential for restoring totipotency to the germ cell lineage. Demethylation is followed by sex-specific establishment of novel epigenetic marks, including genomic imprinting [30]. Interestingly, the second wave of DNA demethylation takes place shortly after fertilization in the early preimplantation embryo. This includes a combination of active hydroxymethylation and global DNA demethylation in the male chromatin, and passive global demethylation in the female chromatin. This process proceeds according to sex-specific programs, which spare the removal of DNA methylation on genomic imprints. Disturbances in the epigenetic reprogramming can contribute to epigenetic diseases and predispose the offspring, or even following generations, to cancer and other diseases related to epigenetic instability [44].

After arriving to the gonadal ridge, PGCs begin to follow either female- or male-specific programs of gametogenesis to produce oocytes or spermatozoa, respectively. The female gamete, oocyte, develops in the ovarian follicle during the process of oogenesis. Human fetal ovaries contain nests of primary oocytes that initiate the prophase of meiosis I but subsequently stop meiotic progression, which is resumed much later in the sexually adult female [1]. The second meiosis of oocytes takes place at fertilization. The current dogma is that oocytes form during the fetal period only, and as the adult ovaries lack stem cells, they are fully dependent on this prenatally produced pool of oocytes.

In contrast to female, male gametogenesis is an active, ongoing process with a massive daily production of spermatozoa during the entire length of a male's reproductive age. Male PGCs enter into mitotic arrest upon their arrival to the genital ridge and stay quiescent in the G0/G1 phase of the cell cycle for the remaining embryonic period [46, 107]. The production of male gametes begins postnatally when PGCs resume proliferation and initiate spermatogenesis. Some cells remain as spermatogonial stem cells (SSCs) that are maintained in the basal compartment of the seminiferous epithelium and provide source material for adult spermatogenesis. Upon a specific signal, SSCs enter the differentiation pathway. In mouse, the differentiation starts with the mitotic proliferation of type A, intermediate and type B spermatogonia [24]. The final mitotic division of type B spermatogonia produces primary spermatocytes that subsequently undergo a long and complicated process of meiosis I that lasts for several days. The following process, meiosis II is fast and results in haploid spermatids in a matter of hours. The final phase of spermatogenesis is known as haploid differentiation (spermiogenesis). During spermiogenesis, spermatids polarize and undergo dramatic morphological changes including tail and acrosome biogenesis, cytoplasmic exclusion and chromatin condensation, which results in compaction of the sperm head and transcriptional silencing [45]. Once spermiogenesis is complete and the sperm has obtained its sleek shape, it is released from the seminiferous epithelium to the lumen of the tubule and transported through the epididymis for post-testicular sperm maturation [98].

The full duration of spermatogenesis is 74 days in human and 35 days in mouse, followed by additional 2 to 6 or 10 to 13 days, respectively, as the sperm transits through the epididymis [19]. Taking into consideration the reasonably fast, cyclic turnover of the maturating germ cell population, environmentally-induced epigenetic changes in the male germline can be relatively transient. On the other hand, if changes become more permanently programmed in the spermatogonial stem cells, or in somatic cells along the reproductive tract the information may have an impact for a longer period of time. Early oocytes in primordial follicles are already formed during mid-gestation and remain in the ovaries until ovulation, thus remaining exposed to environmental information for a long time period.

11.3 Spermatozoal RNAs

It is becoming clear that spermatozoal RNA molecules have the ability to mediate intergenerational epigenetic information. As previously mentioned, due to their highly compacted genome, spermatozoa are largely transcriptionally inactive [57]. This is reflected by the lack of ribosomal RNAs and exceptionally diminished amounts of RNAs in general, making their identification and analysis much harder compared to other cells containing massive amounts of RNA. However, through new analyzing techniques, several different types of sperm RNAs have been discovered. In 1999, the first sequence-level characterization revealed a group of translationally quiescent RNAs in human sperm [71]. Progress in the development of microarray technology at the beginning of the 2000s allowed the identification of spermatozoal mRNAs [77], as well as the detection of sncRNAs such as microRNAs (miRNAs) [79]. Next Generation Sequencing (NGS) further boosted sperm RNA studies and identified the presence of a complex population of RNAs, including both coding and non-coding transcripts and several species of sncRNAs, such as miRNAs, piRNAs, repeat-derived small RNAs and tRNA-derived fragments [16, 53, 54, 61, 82, 95, 96]. These RNAs are known to have different functional roles in the regulation of spermatogenesis, fertilization and embryo development, and may also have an important consequence on the offspring's phenotype.

The spermatozoon is composed of a highly compacted nucleus, a flagellum and a very limited amount of cytoplasm (Fig. 11.1). The distribution of RNAs inside the sperm has been addressed by cellular fractionation of mouse spermatozoa combined with RNA sequencing. The analysis of longer transcripts (> 100 nt) in sonicated mouse spermatozoa (including separated tails and heads with intact membrane structures) and demembranated sperm heads revealed the localization of RNA

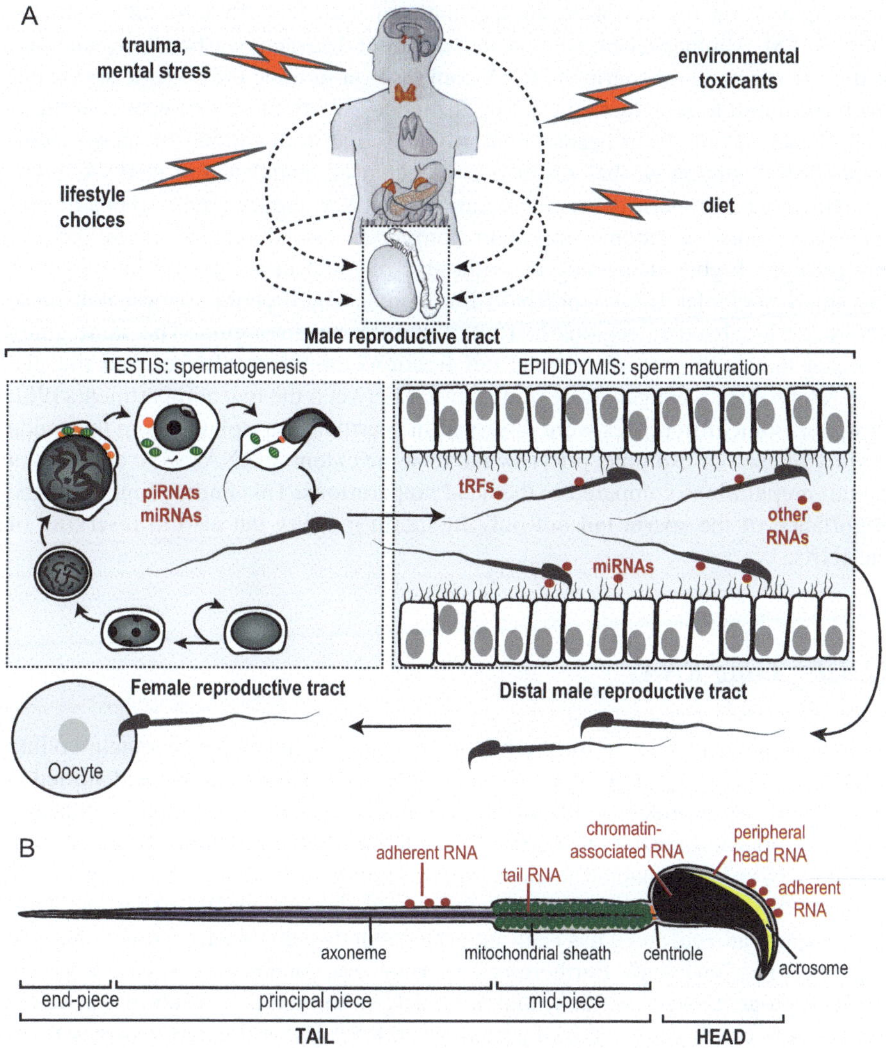

Fig. 11.1 Paternal small RNA-mediated epigenetic inheritance. (**A**) Environmental factors, such as diet, lifestyle choices, traumatic experiences and environmental toxicants, can modify the phenotype of an individual. Information about these acquired conditions travels to the reproductive tract, where it induces alterations in the sperm RNA profile. Some sperm RNAs originate in the testis during spermatogenesis as differentiating germ cells express cell type-specific RNAs, such as piRNAs and miRNAs that can be retained in transcriptionally inactive spermatozoa. During the travel through the epididymis, the RNA profile of spermatozoa changes when they receive sncRNAs, in particular tRFs as well as miRNAs from the epididymal fluid and small vesicular structures known as epididymosomes (red circles) that can fuse with spermatozoa to deliver their RNA content. The potential contribution of the distal male reproductive tract to the sperm RNA profile is currently unknown. In fertilization, sperm RNAs are released into a zygote where they contribute to the inheritance of acquired phenotypes. (**B**) RNAs have been shown to localize both in the head and the tail of spermatozoa. In the head, they can localize inside the nucleus (chromatin-associated RNA) or in the space between the nuclear membrane and the plasma membrane (peripheral head RNA). In addition, some RNAs may be found on the surface of spermatozoa as adherent epididymosomes or ribonucleoprotein complexes (adherent RNA)

predominately in the periphery of spermatozoa with less RNA found within the nucleus [55]. Another recent study tested different detergent washes (low, medium, or high stringency) on sperm sncRNA composition in mice [96]. While spermatozoa were found to be intact after a low stringency wash, more stringent conditions often detached tails from heads, most probably due to disruption of sperm membranes. All detergent washes resulted in a decreased abundance of miRNAs compared to unwashed sperm. This may be due to detergent-induced removal of adherent epididymosomes or ribonucleoprotein complexes bound on the sperm surface. Interestingly, higher-stringency washing also resulted in a decrease in the rRNA and small nucleolar RNA (snoRNA) fragments, which could suggest that these RNAs are localized in regions lost upon membrane disruption. The same study revealed that both sperm head and tail fractions contain sncRNAs, and that the sncRNA profiles were generally quite similar between the two compartments [96]. However, some differences were observed in spermatozoa isolated from the cauda epididymides. For example, piRNAs and to some extent miRNAs were enriched in the tail preparations compared to the head preparations. This finding highlights the importance of the sperm tail not only in sperm motility but also in reserving of sncRNAs.

11.3.1 Long RNAs

A heterogenous profile of long RNA transcripts, including both protein-coding mRNAs as well as different types of long ncRNAs (lncRNAs) has been identified in human [94] and in mouse [113] sperm. Some of these RNAs are intact, but a substantial proportion appears to be fragmented, and the functionality of these fragmented RNAs has remained unclear. Sperm lncRNAs include for example small nuclear RNAs (snRNAs) that are components of spliceosomes [94]. They are abundant in spermatozoa, and it has been suggested that they could have a functional role in postfertilization events. Furthermore, spermatozoa contain exonuclease resistant, stable circular RNAs (circRNA) that are ncRNAs formed as a result of a covalent linkage between 3′ and 5′ ends of a transcribed RNA [20, 94]. circRNAs may function in the cytoplasm, for example as miRNA "sponges" that repress the miRNA-mediated mRNA downregulation, or in the nucleus by controlling their parental genes through direct interactions [20]. Intronic retained elements represent yet another type of sperm lncRNAs. Interestingly, many of them appear to be full-length introns that have somehow escaped degradation and are retained in sperm [53, 54]. The functional importance of these sperm transcripts is largely undescribed. However lncRNAs have been identified as important regulators of gene expression in other tissues. They can act either by recruiting chromatin remodeling complexes on the gene regulatory elements or by affecting RNA processing or stability at the post-transcriptional level. Therefore, they have been suggested to contribute to epigenetic inheritance by affecting embryonic gene expression after being released into the oocyte in fertilization.

11.3.2 Small Non-Coding RNAs

The pool of sperm RNAs is composed of several classes of sncRNAs (< 40 nt) with a broad range of functions in the control of gene expression and genome integrity [61, 82]. miRNAs are the best-characterized subtype of sncRNAs. They are around 22 nt long and in fertile men cover approximately 7% of the total sperm sncRNAs [61]. miRNAs function as posttranscriptional regulators of gene expression. They form an RNA-induced silencing complex (RISC) together with Argonaute (AGO) proteins and guide the complex to its target mRNAs to repress mRNA expression [72]. One miRNA can have several target mRNAs and on the other hand, one mRNA can be targeted by several distinct miRNAs [32]. Therefore, the miRNA-mRNA regulatory networks may be extremely complex and provide powerful means to change cellular gene expression. miRNAs are transcribed from endogenous miRNA genes as hairpin-structured precursors that are first processed by a microprosessor complex containing DROSHA and DiGeorge syndrome critical region gene (DGCR8), and subsequently by DICER that makes the final cleavage to produce mature miRNAs. The importance of germ cell-derived miRNAs for spermatogenesis, epididymal epithelium-derived miRNAs for sperm maturation and sperm-derived miRNAs for postfertilization events in early embryo has been demonstrated by several *in vivo* knockout studies targeting miRNA biogenesis proteins [8, 9, 59, 60, 111, 115].

piRNAs form a unique class of sncRNAs (26–31 nucleotides) that are predominantly expressed in the germline [47]. piRNAs are synthesized in germ cells during the fetal life (fetal piRNAs) or during spermatogenesis (prepachytene and pachytene piRNAs). Some piRNA are retained in mature spermatozoa, and they have been identified from both human and mouse sperm [29, 40, 61, 96]. In differentiating germ cells, piRNAs function by forming complexes with PIWI proteins, which constitute a germline-expressed subfamily of the AGO family [21, 106]. A functional piRNA pathway is required for normal spermatogenesis and mutation of piRNA pathway genes in mice, including the genes for PIWI proteins MIWI, MILI and MIWI2, leads to male sterility [11, 25, 62, 106]. Two distinct piRNA biogenesis pathways exist. One produces primary piRNAs from genomic clusters. These piRNAs operate as silencing triggers with multiple functions including keeping resident mobile elements repressed and switching off genic transcripts during postnatal spermatogenesis [41, 42, 88, 105, 112]. The second is the adaptive piRNA pathway that is used to amplify piRNAs by a so-called ping-pong cycle. This secondary piRNA biogenesis pathway takes place particularly in fetal male germ cells during the global resetting of the male germline epigenome, which causes an induction of the transposon expression. The produced secondary piRNAs function by repressing active transposons and therefore protect the genomic integrity by preventing harmful transposon insertions [21]. Just like miRNAs, piRNAs are effective regulators of gene expression, and can target a broad range of RNAs, including mRNAs and lncRNAs in addition to transposons. Interestingly, they can also function at the chromatin level by inducing epigenetic gene silencing [4, 23, 63].

tRFs constitute the most abundant class of sncRNAs in spermatozoa. They are especially numerous in the mouse, where tRFs derived from the 5′ end of tRNAs cover almost 80% of the whole RNA profile in sperm [82]. However, in the testis tRFs are generally scarce, and their abundance in sperm increases when maturing spermatozoa pass through the long epididymal duct [95]. tRFs are quite heterogeneous in size, nucleotide composition and biogenesis, some of them being derived from precursor tRNA molecules and others derived from mature cytoplasmic tRNAs. Some fragments are constitutive components of all cells, whereas others are only produced in cells exposed to adverse conditions. We are still lacking many details on the function of tRFs, but they seem to have the capability to regulate gene expression in a similar manner to miRNAs and piRNAs [2, 93].

11.4 Origin of Sperm RNAs

During the maturation of sperm, RNA payload changes with a drastic drop of piRNAs and abundant growth of tRFs [95, 96]. Therefore, the current consensus is that while the spermatozoal piRNAs could have a testicular origin, the majority of tRFs are collected during the epididymal transit. Interestingly, the composition of the sperm RNA profile appears to differ between species. While tRNA fragments are the most abundant class of RNAs in mouse spermatozoa [82], in humans, piRNAs and miRNAs constitute abundant sncRNA groups alongside tRNA- and rRNA-derived fragments [40, 54]. These differences between mouse and human may have an evolutionary explanation; however, they may also originate from the differences in the sperm collection and purification methods. Mouse spermatozoa are harvested from cauda epididymides while human spermatozoa are purified from ejaculated semen. Because sperm RNA profiles have been shown to change significantly during the travel through the male genital tract [95], different collection points may indeed contribute to the observed interspecies differences.

11.4.1 Testis-Derived RNAs

Some spermatozoal RNAs originate from the testis and arise during different steps of spermatogenesis. It is currently unknown whether these RNAs are actively selected to be retained in spermatozoa, and whether the selection mechanism could be affected by environmental signals. However, it is clear that major epigenetic and transcriptomic remodeling events during male germ cell differentiation create a favorable setting for the translation of environmental signals to an epigenetic code. The progress of spermatogenesis is accompanied by orchestrated waves of gene expression. This results in a specific transcriptome profile for each differentiating cell type that consist of a diverse set of protein-coding mRNAs and their isoforms, but also a wide variety of ncRNAs, including lncRNAs and sncRNAs and poorly conserved intergenic transcripts [5, 13, 14, 64, 100, 103]. In fact, male germ cells, particularly meiotic spermatocytes and post-meiotic round spermatids, have exceptionally diverse transcriptomes compared to any other somatic cell type, which creates a high demand for accurate posttranscriptional regulation [100]. A specific feature in the RNA profile of differentiating male germ cells is the huge induction of piRNA production in pachytene spermatocytes [39]. At the time of the highest transcriptional activity and transcriptome diversity, germ cell-specific ribonucleoprotein (RNP) granules (germ granules), such as the chromatoid body (CB), appear in the cytoplasm of germ cells [66]. Germ granules provide important platforms for the biosynthesis and function of piRNAs and for the posttranscriptional coordination of the male germ cell's transcriptome.

A notable posttranscriptional challenge is faced in late steps of spermiogenesis when the chromatin of elongating spermatids condenses. In these cells, the replacement of the majority of histones by protamines enables genomic material to become particularly tightly packed. This compaction of chromatin effectively halts transcription, and therefore, the mRNAs encoding for the proteins required during the very late phases of spermatogenesis have to be transcribed earlier and repressed translationally until needed [51, 58]. Furthermore, haploid cells are challenged by the important task of eliminating unnecessary transcripts. A bulk of RNA is removed when the majority of the cytoplasm is discarded with the residual body during the late steps of spermatogenesis [31]. However, degradation mechanisms have also been revealed, including the pachytene piRNA-mediated elimination of meiotic and postmeiotic transcripts [41, 42, 88, 105, 112]. It is currently unknown by which mechanism some RNAs escape the degradation and are retained in mature spermatozoa. Interestingly, although the most prominent CB structure is present in round spermatids, remnants of the CB stay in the cytoplasm of germ cells during late steps of spermiogenesis at the time of chromatin compaction and transcriptional silencing. The function of this so-called late CB in RNA regulation has not been determined, but since it derives from the central piRNA-accumulating CB, it is tempting to speculate that it could have a role in preserving RNAs, including piRNAs, during spermiogenesis and passing them on to spermatozoa.

11.4.2 Epididymal Contribution to Sperm RNA Profile

The mammalian epididymis contains an exceptionally long, convoluted ductal system, which receives and concentrates immature spermatozoa for maturation and stores the functional sperm prior to release at the time of ejaculation. The epididymis can be divided into three anatomical regions: the caput, the corpus and the cauda [19, 56]. Each region of the epididymis has its own pattern of gene expression related to the physiological function it is meant to perform for sperm maturation [98]. Recent studies have provided compelling evidence for the existence of numerous sncRNAs in the epididymal fluid. Such entities appear predominantly, but perhaps not exclusively, to be associated with small vesicular structures, epididymosomes [6].

The miRNA content of the vesicles appears to change along the length of the epididymis [87]. There is also evidence that epididymosomes can convey their macromolecular payload to spermatozoa as well as to downstream epididymal epithelial cells [108]. Epididymosomes thus represent a likely player for the selective modification of the sperm proteome and epigenome during their post-testicular maturation [6].

RNA sequencing of sperm isolated either from the testis or the epididymis revealed that while tRFs are the primary sncRNA population in both caput and cauda epididymal sperm, testicular sperm only contains very low amounts of them [96]. Therefore the RNA cargo of sperm drastically changes during its journey through the epididymal ducts. Interestingly, the RNA cargo of isolated epididymosomes has been shown to include a very similar population of sncRNAs to that gained by spermatozoa during its transit [95]. Furthermore, isolated epididymosomes have been shown to deliver sncRNAs, such as miRNAs and tRFs, to testicular spermatozoa [96]. Using a mouse model with an epididymis-specific expression of uracil phosphoribosyltransferase (UPRT) which labels small RNAs with 4-thiouracil in the caput epididymis, cauda sperm was shown to contain small RNAs synthesized in the epididymal epithelium. Interestingly, intracytoplasmic sperm injection using sperm from caput versus cauda epididymis showed that embryos generated using caput sperm had significant gene expression changes, implanted inefficiently, and failed soon after implantation [18]. Remarkably, these defects were rescued by microinjection of small RNAs purified from cauda epididymis into caput sperm-derived embryos [18]. These findings support the important role of epididymal somatic cells in trafficking small RNAs to the germline in mammals.

11.5 Epigenetic Inheritance of Acquired Traits and Disorders Via Sperm RNAs

Epidemiological studies have demonstrated associations between parental exposures to toxicants, lifestyle, nutrition or even traumatic stress-induced conditions and the specific features in offspring. These environmental factors have been shown to induce alterations in the epigenome of the parental germline both in human and rodents. In animal studies the phenomenon of transgenerational epigenetic inheritance of acquired traits has been confirmed, with the sncRNAs discussed in previous chapters as likely carriers of transgenerational epigenetic information [15, 17, 28, 33, 84, 85].

The most thoroughly examined paternal exposures involve traumatic stress and metabolic stress that will be discussed in detail below. An increasing number of examples about other exposures also exists in the literature [74]. For example, environmental toxicants, such as endocrine disrupting chemicals (EDCs), have been shown to cause transgenerationally inherited phenotypes. In mammals, one of the first studies that addressed molecular epigenetic changes associated with transgenerational inheritance of disease investigated the effects of the agricultural fungicide vinclozolin treatment during pregnancy in rats [3]. Vinclozolin treatment carried a far reaching impact on fertility as the F3 generation, never exposed to vinclozolin, had reproductive abnormalities, and these defects were correlated with changes in DNA methylation in the F3 sperm. Several studies have revealed EDC-induced epigenetic changes in DNA methylation and chromatin modification in germ cells [74], and there is also accumulating evidence for the involvement of ncRNAs. For example, *in utero* exposure to vinclozolin caused a reduction in the number of PGCs accompanied by an increased expression of two specific miRNAs (miR-23b and miR-21). The altered miRNA expression in turn induced disequilibrium in the Lin28/let-7/Blimp1 pathway, a crucial regulator of PGC differentiation, in three successive generations of males not exposed to the compound themselves [10]. Another study demonstrated that transient vinclozolin exposure of gestating female rats (F0 generation) during fetal gonadal development induced alterations in both germline DNA methylation and ncRNAs, and these changes were still observed in F3 rats spermatozoa [7]. The fact that many toxicants affect not only the health of an exposed individual but also future generations reinforces the growing concern about the impact of environmental chemicals and toxicants on human health.

11.5.1 Metabolic Disorders

Obesity has become a major health problem due to its high prevalence and the fact that it increases the risk of other diseases such as diabetes, cancer, cardiovascular and neurodegenerative diseases. Obesity is a highly inherited condition, though the genetic variation alone cannot explain the whole heritability [81, 83]. It has been shown that paternal overnutrition that leads to obesity triggers the development of obesity in offspring in diverse organisms from fly to mammals [28, 50, 76]. In humans, the availability of food early in life has been shown to influence the risk of developing cardiovascular diseases and obesity in offspring [65, 75]. Although the mechanistic evidence is still missing, it is likely that at least a part of the unsolved heritability of obesity and type 2 diabetes is ascribed to the epigenetic inheritance of environmentally-induced acquired conditions.

Both nutritional status and the level of physical activity induce changes in the sperm epigenome, which suggests that these changes are responsible for transmitting the condition to offspring [26, 29, 52]. In addition to the DNA methylation changes discussed elsewhere, the RNA profile of sperm can respond to altered metabolic conditions as well. Not only is the sperm piRNA profile from obese but otherwise healthy men significantly different from those of lean men, but the profile has been shown to change in response to exercise [29, 52]. The expression of six different sperm piRNAs had altered levels in lean and healthy young men after 6 weeks of endurance training compared to the baseline before the training period. Interestingly, the changes in piRNA levels were almost completely reversed after a

3-month detraining period, which demonstrates that the sperm piRNA profile is subject to dynamic changes [52].

Well-controlled experimental setups using different model organisms have provided strong evidence that lifestyle and nutrition changes can affect the sperm epigenome and modulate the developmental programming of offspring [16, 17, 22, 28, 95]. The most convincing evidence about the involvement of sperm RNAs as intergenerational carriers of epigenetic information comes from studies on high-fat diet (HFD) and low-protein diet mouse models. Both diets have been shown to induce changes in the sperm RNA profile, the tRNA-derived tRFs in particular [16, 95]. These studies strongly suggest that tRFs can function as sensitive markers of environmental exposures. It was shown that zygotic injection of tRF-enriched fractions of sperm RNAs isolated from males with a diet-induced metabolic disorder triggered the disorder in offspring, therefore demonstrating the direct role of sperm sncRNAs in epigenetic inheritance [16] (Fig. 11.2). Sperm tRFs were also shown to contain numerous RNA modifications, and the levels of 5-methylcytidine (m5C) and N2-methylguanosine (m2G) modifications were significantly increased after paternal HFD consumption [16], thus providing yet another possible mechanism for the transmission of epigenetic information. The role of RNA modifications in

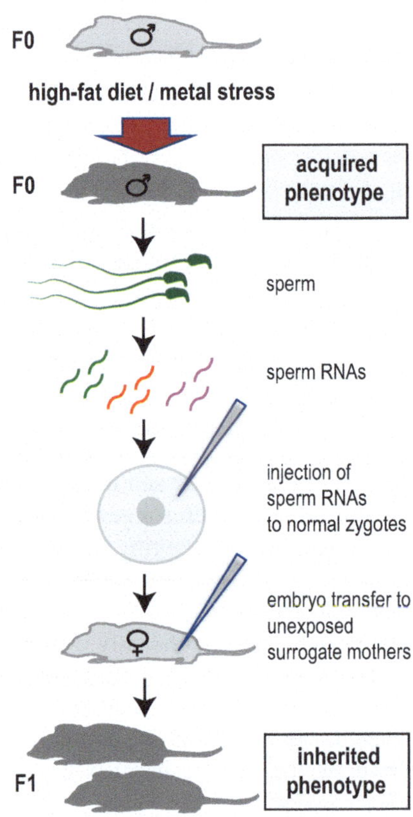

Fig. 11.2 Experimental set-up for the studies of paternal RNA-mediated epigenetic inheritance in mice. Phenotypic changes are induced in male mice, and subsequently, spermatozoa are collected and sperm RNAs are extracted. To study the direct role of sperm RNAs in the transmission of acquired condition to offspring, sperm RNAs from exposed males and control sperm RNAs from unexposed males are injected to normal zygotes that originate from unexposed parents. Early embryos are then implanted in unexposed surrogate mothers, and the next generation is analyzed for the inheritance of acquired phenotypes

epigenetic inheritance was further highlighed by a study on the tRNA methyltransferase DNMT2. The deletion of *Dnmt2* gene in mice altered the sperm RNA profile and prevented HFD-induced elevation of sperm tRF modifications [114]. Importantly, DNMT2 activity was also shown to be required for the small RNA-mediated transmission of HFD-induced metabolic disorders to offspring [114].

In addition to tRFs, sperm miRNAs have also been shown to be sensitive to changes in paternal diet. For example, several members of the let-7 miRNA family known to control lipid and glucose metabolism were downregulated in spermatozoa after exposing mice to low-protein diet before conception [95]. The let-7 miRNAs were also found to be differentially expressed in spermatozoa of HFD-fed rats as well as in the spermatozoa of their offspring [22]. The same study identified let-7c as a potential transgenerational carrier of a HFD-induced metabolic condition. It was shown that the offspring of obese sires developed disturbances in their glucose and lipid metabolism accompanied by an altered expression of let-7c in liver, white adipose tissue, and muscle tissue of adult offspring [22]. miRNAs were also implicated in RNA-mediated epigenetic inheritance in a study of Western-like, high fat, high sugar diet-induced metabolic disorder [43]. In this study, zygotic injection of not only sperm RNAs but also of testicular RNAs from Western-like diet-fed males was able to transmit the metabolic phenotype to offspring [43]. This indicates that testicular germ cells may already contain epigenetic information about the environment and condition of the individual before their release into the epididymis. Western-like diet induced the expression of several miRNAs, and the injection of miR-19b to one-cell embryos was able to induce metabolic alterations in offspring that were similar to the diet-induced alterations [43].

11.5.2 Traumatic Stress and Psychiatric Disorders

Parental exposures to environmental challenges are associated with the increased risk of neuropsychiatric diseases and stress dysregulation in offspring [27, 69, 73]. One well documented example concerns the holocaust survivors; although the children of holocaust survivors did not experience the Holocaust directly, a significant percentage of them have been shown to suffer from psychiatric disorders associated with trauma, such as phobias or depression [99]. Particularly in human, a part of the transmission of trauma obviously occurs through parental behavior. However, the effects of trauma may also be inherited through induced alterations in gene expression, and the current evidence shows that stress and anxiety can affect the epigenome of gametes and be transmitted to offspring [91, 109].

Again, animal models have enabled the dissection of the mechanisms of epigenetic inheritance, and just like in case of metabolic stress, sperm sncRNA are a likely candidate in the transmission of paternal traumatic experiences to offspring. One pioneering study in the field exposed male mice to a modified protocol combining maternal separation with unpredictable maternal stress (MSUS), which yielded male offspring manifesting reduced anxiety levels but greater depressive behavior [34].

The early stress induced altered sperm miRNA profile in exposed compared to non-exposed males. The importance of sncRNAs in the transmission of stress response was established once a population of small RNAs isolated from MSUS-exposed sperm were microinjected into unexposed zygotes and a similar spectrum of anxiety and depression related behaviors in offspring were observed [34].

Neuropsychiatric diseases frequently present with the dysregulation of the hypothalamic–pituitary–adrenal (HPA) stress axis, suggesting a remarkable vulnerability of this system to external perturbations. Six weeks of exposure of male mice to chronic stress before breeding was shown to induce HPA dysregulation detectable by significantly reduced serum corticosterone levels following restraint stress in both female and male offspring [89]. However, the HPA dysregulation was not accompanied by altered offspring behavior. This study identified nine miRNAs that were consistently up-regulated in sperm of mice exposed to chronic variable stress either during adolescence or adulthood. A direct connection between miRNAs and a physiological stress was also demonstrated; zygotic microinjection of these nine miRNAs produced offspring with similar HPA-axis dysregulation and altered hypothalamic transcriptome demonstrating that an overexpression of these miRNAs can in fact impact the physiological response to stress [90].

In order to dissociate the behavioral and physiological aspects of paternal stress, the dysregulation of HPA axis activity was induced by supplementing male mice with a low dose of corticosterone over the course of one spermatogenic cycle. Interestingly, though the dose of the corticosterone and the chosen exposure period were insufficient to alter the anxiety and depression-related measures of the exposed mice, male offspring experienced increased anxiety. The exposed fathers had significant changes in miRNA levels in their spermatozoa [80, 97], demostrating that sustained elevation of glucocorticoids can induce alterations in sperm sncRNA profile and is involved in the transmission of paternal stress-induced traits across generations.

11.6 How Inherited RNAs Can Affect the Development and Health of Offspring?

When the oocyte and sperm merge during fertilization, in addition to providing their genomic content, they both contribute small RNA molecules to the zygote. Mature oocytes and zygotes have very similar miRNA profiles, which indicates that the majority of zygotic miRNAs are maternally inherited [104]. However, another more recent study was able to identify 14 miRNAs that were not found in oocytes, but were present in both wild-type sperm and 2-pronuclei embryos [111], which indicates a paternal contribution to the miRNA content of the embryo. The miRNAs delivered by sperm were also shown to be important for the developmental potential of the embryo; intracytoplasmic sperm injection (ICSI) using miRNA-depleted sperm from germ cell-specific *Dicer1* or *Drosha* knockout mice resulted in embryos with reduced developmental capacity. Importantly, the phenotype was rescued upon injection of small RNAs from wild-type sperm [111]. Injecting miRNA-depleted sperm to miRNA-depleted oocytes also resulted in reduced developmental ability but this phenotype could not be recovered by injecting wild-type sperm small RNAs, indicating the importance of maternal miRNAs for embryonic development [111]. Only few studies exist to date to address the impact of individual paternal miRNAs. The role of spermatozoal miR-34c has been studied with inconclusive results. While it was shown that a zygotic injection of a miR-34c inhibitor disrupted the first embryonic cleavage division [68], another study showed that miR-34c knockout male mice were fertile, and intracytoplasmic injection of their sperm resulted in normal fertilization, normal preimplantation development and normal birth rate [110].

Small RNAs have a well-established role in the control of gene expression [38]. Though the exact mechanisms of transmitting epigenetic data to zygotes remains to be studied further, their well-known functions as posttranscriptional regulators in other cells suggests they may target specific transcripts to regulate gene expression in zygotes. Furthermore, accumulating evidence suggest that they may influence other epigenetic marks in the zygote. Small RNAs are able to, for example, recruit histone modifiers or affect DNA methylation levels at specific loci by interacting with DNA methyltransferases [48, 67]. This active crosstalk between different epigenetic marks implies that the transmission of small RNA-mediated epigenetic signals from the gametes to zygotes may include a complex network of different epigenetic mechanisms.

The studies on the sperm RNA-mediated epigenetic inheritance of diet-induced metabolic disorders have proven the capability of sperm RNAs to modify gene expression in the zygote. Zygotic injection of a pool of sperm tRFs collected from HFD-fed males induced different effects on early embryonic gene expression compared to tRFs from males on normal diet [16]. It was shown that those sperm tRFs that were differentially expressed between normal diet vs. HFD-fed males preferentially matched to gene promoter regions, which could allow direct targeting of promoters. Interestingly, those genes that were downregulated in the zygote after HFD-tRF injections were enriched in the pathways related to metabolic regulation, and it was suggested that these changes in embryonic gene expression might induce phenotypic alterations in adult offspring. Gene expression changes were observed in the offspring islet cells, which could explain the observed inherited metabolic disorder [16].

In another recent study, Sharma et al. demonstrated that the experimental blocking of a specific tRF, tRF-Gly-GCC, that was earlier shown to be altered as a response to low-protein diet, resulted in upregulation of a specific set of genes in cultured embryonic stem cells and zygotes [95]. These upregulated genes are known to be highly expressed in preimplantation embryos and are regulated by the endogenous retroelement MERVL. Zygotic injection of synthetic tRF-Gly-GCC oligo resulted in MERVL target gene repression in two-cell embryos, supporting the hypothesis that a specific tRF in sperm is capable of repressing the MERVL-regulated genes after fertilization. While previous work using the same mouse model has demonstrated heritable effects of paternal low-protein diet on liver transcriptome in offspring [12], it has not been directly shown whether the reported tRF-induced effects in zygotic gene expression have phenotypic consequences in offspring.

Following fertilization, the zygote gains the ability to differentiate into any cell type of the body. The developing embryo begins to produce unique transcripts after embryonic genome activation and relinquishes itself from parental gamete transcripts. This transition includes major gene expression changes that may determine the later development and health of an individual and is a likely target for the small RNA action [102]. Altogether, it appears that small RNAs have all functional requirements and capacity to bring about gene expression changes in the developing embryo and therefore, potentially modify the phenotype of offspring.

11.7 Future Perspectives

During recent years, immense progress has been made in revealing the phenomenon and mechanisms of paternal epigenetic inheritance. However, despite the advances in the field, several burning questions remain to be answered before we can fully understand the scope and the impact of the process in development and health. These include thorough mechanistic studies to elucidate the details of how sperm RNA profile is generated during testicular gametogenesis and posttesticular maturation, and how the environmentally-induced epigenetic changes are transmitted from somatic tissues to the sperm RNA profile. Furthermore, it is critical to understand at which point of the gamete development the environmentally-induced signal can be programmed to the germ cell's epigenome and how permanent the changes in sperm RNAs are, i.e. can they be reversed by specific interventions. The possibility to reverse adverse changes in sperm RNA profile would potentially prevent the transmission of unfavorable acquired condition to offspring. It will also be highly important to fully characterize the mechanisms by which sperm RNAs transfer the epigenetic information about environmentally-induced conditions to developing embryo and to understand the crosstalk between different epigenetic mechanisms that bring about changes in embryonic gene expression and determine the offspring phenotype.

All functional studies on epigenetic inheritance have been conducted using animal models. The phenomenon of epigenetic inheritance is well-recognized in humans as well, and a convincing bunch of epidemiologic studies has proven the association between the parent's acquired preconception conditions and the offspring phenotype. However, transgenerational mechanistic information is still lacking because of the challenge in setting up multigenerational studies in humans. While we still have to wait for mechanistic evidence, it is very likely that the epigenetic inheritance of acquired disorders plays a significant role in the etiology of human complex diseases. Therefore, it will be important to invest a lot of efforts in elucidating the nature and mechanisms of epigenetic inheritance to be able to understand, and one day to prevent the transmission of unfavorable acquired conditions to offspring.

Acknowledgements We thank all Kotaja lab members for stimulating discussions. The research in the lab is supported by the Academy of Finland and Sigrid Jusélius Foundation. T.L. is funded by the Turku Doctoral Program in Molecular Medicine.

References

1. Adhikari D, Liu K. Molecular mechanisms underlying the activation of mammalian primordial follicles. Endocr Rev. 2009;30:438–64.
2. Anderson P, Ivanov P. tRNA fragments in human health and disease. FEBS Lett. 2014;588:4297–304.
3. Anway MD, Cupp AS, Uzumcu M, Skinner MK. Epigenetic transgenerational actions of endocrine disruptors and male fertility. Science. 2005;308:1466–9.
4. Aravin AAA, Sachidanandam R, Bourc'his D, Schaefer C, Pezic D, Toth KFKF, Bestor T, Hannon GJGJ. A piRNA pathway primed by individual transposons is linked to de novo DNA methylation in mice. Mol Cell. 2008;31:785–99.
5. Bao J, Wu J, Schuster AS, Hennig GW, Yan W. Expression profiling reveals developmentally regulated lncRNA repertoire in the mouse male germline. Biol Reprod. 2013;89:107.
6. Belleannée C, Calvo E, Thimon V, Cyr DG, Légaré C, Garneau L, Sullivan R. Role of microRNAs in controlling gene expression in different segments of the human epididymis. PLoS One. 2012;7:e34996.
7. Ben Maamar M, Sadler-Riggleman I, Beck D, McBirney M, Nilsson E, Klukovich R, Xie Y, Tang C, Yan W, Skinner MK. Alterations in sperm DNA methylation, non-coding RNA expression, and histone retention mediate vinclozolin-induced epigenetic transgenerational inheritance of disease. Environ. Epigenetics. 2018;4:dvy010.
8. Björkgren I, Saastamoinen L, Krutskikh A, Huhtaniemi I, Poutanen M, Sipilä P. Dicer1 ablation in the mouse epididymis causes dedifferentiation of the epithelium and imbalance in sex steroid signaling. PLoS One. 2012;7:e38457.
9. Björkgren I, Gylling H, Turunen H, Huhtaniemi I, Strauss L, Poutanen M, Sipilä P. Imbalanced lipid homeostasis in the conditional Dicer1 knockout mouse epididymis causes instability of the sperm membrane. FASEB J. 2015;29:433–42.
10. Brieño-Enríquez MA, García-López J, Cárdenas DB, Guibert S, Cleroux E, Děd L, de Hourcade JD, Pěknicová J, Weber M, del Mazo J. Exposure to endocrine disruptor induces transgenerational epigenetic deregulation of MicroRNAs in primordial germ cells. PLoS One. 2015;10:e0124296.
11. Carmell MA, Girard A, van de Kant HJG, Bourc'his D, Bestor TH, de Rooij DG, Hannon GJ. MIWI2 is essential for spermatogenesis and repression of transposons in the mouse male germline. Dev Cell. 2007;12:503–14.
12. Carone BR, Fauquier L, Habib N, Shea JM, Hart CE, Li R, Bock C, Li C, Gu H, Zamore PD, et al. Paternally induced transgenerational environmental reprogramming of metabolic gene expression in mammals. Cell. 2010;143:1084–96.
13. Chalmel F, Rolland AD. Linking transcriptomics and proteomics in spermatogenesis. Reproduction. 2015;150:R149–57.
14. Chalmel F, Rolland AD, Niederhauser-Wiederkehr C, Chung SS, Demougin P, Gattiker A, Moore J, Patard JJ, Wolgemuth DJ, Jegou B, et al. The conserved transcriptome in human and rodent male gametogenesis. Proc Natl Acad Sci U S A. 2007;104:8346–51.
15. Champroux A, Cocquet J, Henry-Berger J, Drevet JR, Kocer A. A decade of exploring the mammalian sperm epigenome: paternal epigenetic and transgenerational inheritance. Front Cell Dev Biol. 2018;6:50.
16. Chen Q, Yan M, Cao Z, Li X, Zhang Y, Shi J, Feng G, Peng H, Zhang X, Zhang Y, et al. Sperm tsRNAs contribute to intergenerational inheritance of an acquired metabolic disorder. Science. 2016a;351:397–400.

17. Chen Q, Yan W, Duan E. Epigenetic inheritance of acquired traits through sperm RNAs and sperm RNA modifications. Nat Rev Genet. 2016b;17:733–43.
18. Conine CC, Sun F, Song L, Rivera-Pérez JA, Rando OJ. Small RNAs gained during Epididymal transit of sperm are essential for embryonic development in mice. Dev. Cell. 2018;46:470–80.. e3
19. Cornwall GA. New insights into epididymal biology and function. Hum Reprod Update. 2009;15:213–27.
20. Cortés-López M, Miura P. Emerging functions of circular RNAs. Yale J Biol Med. 2016;89:527–37.
21. Czech B, Hannon GJ. One loop to rule them all: the ping-pong cycle and piRNA-guided silencing. Trends Biochem Sci. 2016;41:324–37.
22. de Castro Barbosa T, Ingerslev LR, Alm PS, Versteyhe S, Massart J, Rasmussen M, Donkin I, Sjögren R, Mudry JM, Vetterli L, et al. High-fat diet reprograms the epigenome of rat spermatozoa and transgenerationally affects metabolism of the offspring. Mol Metab. 2016;5:184–97.
23. De Fazio S, Bartonicek N, Di Giacomo M, Abreu-Goodger C, Sankar A, Funaya C, Antony C, Moreira PN, Enright AJ, O'Carroll D. The endonuclease activity of Mili fuels piRNA amplification that silences LINE1 elements. Nature. 2011;480:259–63.
24. de Rooij DG. The nature and dynamics of spermatogonial stem cells. Development. 2017;144:3022–30.
25. Deng W, Lin H. Miwi, a murine homolog of piwi, encodes a cytoplasmic protein essential for spermatogenesis. Dev. Cell. 2002;2:819–30.
26. Denham J, O'Brien BJ, Harvey JT, Charchar FJ. Genome-wide sperm DNA methylation changes after 3 months of exercise training in humans. Epigenomics. 2015;7:717–31.
27. Dietz DM, LaPlant Q, Watts EL, Hodes GE, Russo SJ, Feng J, Oosting RS, Vialou V, Nestler EJ. Paternal transmission of stress-induced pathologies. Biol Psychiatry. 2011;70:408–14.
28. Donkin I, Barrès R. Sperm epigenetics and influence of environmental factors. Mol. Metab. 2018;14:1–11.
29. Donkin I, Versteyhe S, Ingerslev LR, Qian K, Mechta M, Nordkap L, Mortensen B, Appel EVR, Jørgensen N, Kristiansen VB, et al. Obesity and bariatric surgery drive epigenetic variation of spermatozoa in humans. Cell Metab. 2016;23:369–78.
30. Ferguson-Smith AC. Genomic imprinting: the emergence of an epigenetic paradigm. Nat Rev Genet. 2011;12:565–75.
31. Firlit CF, Davis JR. Morphogenesis of the residual body of the mouse testis. J Cell Sci. 1965:s3–106.
32. Friedman RC, Farh KK, Burge CB, Bartel DP. Most mammalian mRNAs are conserved targets of microRNAs. Genome Res. 2009;19:92–105.
33. Gapp K, Bohacek J. Epigenetic germline inheritance in mammals: looking to the past to understand the future. Genes, Brain Behav. 2018;17:e12407.
34. Gapp K, Jawaid A, Sarkies P, Bohacek J, Pelczar P, Prados J, Farinelli L, Miska E, Mansuy IM. Implication of sperm RNAs in transgenerational inheritance of the effects of early trauma in mice. Nat Neurosci. 2014;17:667–9.
35. Gaucher J, Reynoird N, Montellier E, Boussouar F, Rousseaux S, Khochbin S. From meiosis to postmeiotic events: the secrets of histone disappearance. FEBS J. 2010;277:599–604.
36. Ge Z-J, Liang X-W, Guo L, Liang Q-X, Luo S-M, Wang Y-P, Wei Y-C, Han Z-M, Schatten H, Sun Q-Y. Maternal diabetes causes alterations of DNA methylation statuses of some imprinted genes in murine Oocytes1. Biol Reprod. 2013a;88:117.
37. Ge Z-J, Luo S-M, Lin F, Liang Q-X, Huang L, Wei Y-C, Hou Y, Han Z-M, Schatten H, Sun Q-Y. DNA methylation in oocytes and liver of female mice and their offspring: effects of high-fat-diet–induced obesity. Environ Health Perspect. 2013b;122:159–64.
38. Ghildiyal M, Zamore PD. Small silencing RNAs: an expanding universe. Nat Rev Genet. 2009;10:94–108.

39. Girard A, Sachidanandam R, Hannon GJ, Carmell MA. A germline-specific class of small RNAs binds mammalian Piwi proteins. Nature. 2006;442:199–202.
40. Gòdia M, Swanson G, Krawetz SA. A history of why fathers' RNA matters†. Biol Reprod. 2018;99:147–59.
41. Goh WSSWSS, Falciatori I, Tam OHOH, Burgess R, Meikar O, Kotaja N, Hammell M, Hannon GJGJ. piRNA-directed cleavage of meiotic transcripts regulates spermatogenesis. Genes Dev. 2015;29:1032–44.
42. Gou L-T, Dai P, Yang J-H, Xue Y, Hu Y-P, Zhou Y, Kang J-Y, Wang X, Li H, Hua M-M, et al. Pachytene piRNAs instruct massive mRNA elimination during late spermiogenesis. Cell Res. 2015;25:266.
43. Grandjean V, Fourré S, De Abreu DAF, Derieppe M-A, Remy J-J, Rassoulzadegan M. RNA-mediated paternal heredity of diet-induced obesity and metabolic disorders. Sci Rep. 2015;5:18193.
44. Hackett JA, Zylicz JJ, Surani MA. Parallel mechanisms of epigenetic reprogramming in the germline. Trends Genet. 2012;28:164–74.
45. Hess RA, Renato de Franca L. Spermatogenesis and cycle of the seminiferous epithelium. Adv Exp Med Biol. 2008;636:1–15.
46. Hilscher B, Hilscher W, Bülthoff-Ohnolz B, Krämer U, Birke A, Pelzer H, Gauss G. Kinetics of gametogenesis. I Comparative histological and autoradiographic studies of oocytes and transitional prospermatogonia during oogenesis and prespermatogenesis. Cell Tissue Res. 1974;154:443–70.
47. Hirakata S, Siomi MC. piRNA biogenesis in the germline: from transcription of piRNA genomic sources to piRNA maturation. Biochim Biophys Acta - Gene Regul. Mech. 2016;1859:82–92.
48. Holoch D, Moazed D. RNA-mediated epigenetic regulation of gene expression. Nat Rev Genet. 2015;16:71–84.
49. Hou Y-J, Zhu C-C, Duan X, Liu H-L, Wang Q, Sun S-C. Both diet and gene mutation induced obesity affect oocyte quality in mice. Sci Rep. 2016;6:18858.
50. Huypens P, Sass S, Wu M, Dyckhoff D, Tschöp M, Theis F, Marschall S, de Angelis MH, Beckers J. Epigenetic germline inheritance of diet-induced obesity and insulin resistance. Nat Genet. 2016;48:497–9.
51. Idler RK, Yan W. Control of messenger RNA fate by RNA-binding proteins: an emphasis on mammalian spermatogenesis. J Androl. 2012;33:309–37.
52. Ingerslev LR, Donkin I, Fabre O, Versteyhe S, Mechta M, Pattamaprapanont P, Mortensen B, Krarup NT, Barrès R. Endurance training remodels sperm-borne small RNA expression and methylation at neurological gene hotspots. Clin Epigenetics. 2018;10:12.
53. Jodar M, Selvaraju S, Sendler E, Diamond MP, Krawetz SA. The presence, role and clinical use of spermatozoal RNAs. Hum. Reprod. Update. 2013a;19:604–24.
54. Jodar M, Selvaraju S, Sendler E, Diamond MP, Krawetz SA, Reproductive Medicine Network. The presence, role and clinical use of spermatozoal RNAs. Hum Reprod Update. 2013b;19:604–24.
55. Johnson GD, Mackie P, Jodar M, Moskovtsev S, Krawetz SA. Chromatin and extracellular vesicle associated sperm RNAs. Nucleic Acids Res. 2015;43:6847–59.
56. Jones RC. To store or mature spermatozoa? The primary role of the epididymis. Int J Androl. 1999;22:57–67.
57. Kierszenbaum AL, Tres LL. Structural and transcriptional features of the mouse spermatid genome. J Cell Biol. 1975;65:258–70.
58. Kleene KC, Kleene KC. Patterns, mechanisms, and functions of translation regulation in mammalian spermatogenic cells. Cytogenet Genome Res. 2003;103:217–24.
59. Korhonen HM, Meikar O, Yadav RP, Papaioannou MD, Romero Y, Da Ros M, Herrera PL, Toppari J, Nef S, Kotaja N. Dicer is required for haploid male germ cell differentiation in mice. PLoS One. 2011;6:e24821.
60. Kotaja N. MicroRNAs and spermatogenesis. Fertil Steril. 2014;101:1552–62.

61. Krawetz SA, Kruger A, Lalancette C, Tagett R, Anton E, Draghici S, Diamond MP. A survey of small RNAs in human sperm. Hum Reprod. 2011;26:3401–12.
62. Kuramochi-Miyagawa S, Kimura T, Ijiri TW, Isobe T, Asada N, Fujita Y, Ikawa M, Iwai N, Okabe M, Deng W, et al. Mili, a mammalian member of piwi family gene, is essential for spermatogenesis. Development. 2004;131:839–49.
63. Kuramochi-Miyagawa S, Watanabe T, Gotoh K, Totoki Y, Toyoda A, Ikawa M, Asada N, Kojima K, Yamaguchi Y, Ijiri TW, et al. DNA methylation of retrotransposon genes is regulated by Piwi family members MILI and MIWI2 in murine fetal testes. Genes Dev. 2008;22:908–17.
64. Laiho A, Kotaja N, Gyenesei A, Sironen A. Transcriptome profiling of the murine testis during the first wave of spermatogenesis. PLoS One. 2013;8:e61558.
65. Lane M, Zander-Fox DL, Robker RL, McPherson NO. Peri-conception parental obesity, reproductive health, and transgenerational impacts. Trends Endocrinol Metab. 2015;26: 84–90.
66. Lehtiniemi T, Kotaja N. Germ granule-mediated RNA regulation in male germ cells. Reproduction. 2018;155:R77–91.
67. Li L-C. Chromatin remodeling by the small RNA machinery in mammalian cells. Epigenetics. 2014;9:45–52.
68. Liu W-M, Pang RTK, Chiu PCN, Wong BPC, Lao K, Lee K-F, Yeung WSB. Sperm-borne microRNA-34c is required for the first cleavage division in mouse. Proc Natl Acad Sci U S A. 2012;109:490–4.
69. Matthews SG. Early programming of the hypothalamo-pituitary-adrenal axis. Trends Endocrinol Metab. 2002;13:373–80.
70. Messerschmidt DM, Knowles BB, Solter D. DNA methylation dynamics during epigenetic reprogramming in the germline and preimplantation embryos. Genes Dev. 2014;28:812–28.
71. Miller D, Briggs D, Snowden H, Hamlington J, Rollinson S, Lilford R, Krawetz SA. A complex population of RNAs exists in human ejaculate spermatozoa: implications for understanding molecular aspects of spermiogenesis. Gene. 1999;237:385–92.
72. Moazed D. Small RNAs in transcriptional gene silencing and genome defence. Nature. 2009;457:413–20.
73. Nestler EJ, Barrot M, DiLeone RJ, Eisch AJ, Gold SJ, Monteggia LM. Neurobiology of depression. Neuron. 2002;34:13–25.
74. Nilsson EE, Sadler-Riggleman I, Skinner MK. Environmentally induced epigenetic transgenerational inheritance of disease. Environ. epigenetics. 2018;4:dvy016.
75. Ornellas F, Carapeto PV, Mandarim-de-Lacerda CA, Aguila MB. Obese fathers lead to an altered metabolism and obesity in their children in adulthood: review of experimental and human studies. J Pediatr. 2017;93:551–9.
76. Öst A, Lempradl A, Casas E, Weigert M, Tiko T, Deniz M, Pantano L, Boenisch U, Itskov PM, Stoeckius M, et al. Paternal diet defines offspring chromatin state and intergenerational obesity. Cell. 2014;159:1352–64.
77. Ostermeier GC, Dix DJ, Miller D, Khatri P, Krawetz SA. Spermatozoal RNA profiles of normal fertile men. Lancet. 2002;360:772–7.
78. Ostermeier GC, Miller D, Huntriss JD, Diamond MP, Krawetz SA. Reproductive biology: delivering spermatozoan RNA to the oocyte. Nature. 2004;429:154.
79. Ostermeier GC, Goodrich RJ, Moldenhauer JS, Diamond MP, Krawetz SA. A suite of novel human spermatozoal RNAs. J. Androl. 2005;26:70–4.
80. Pang TY, Short AK, Bredy TW, Hannan AJ. Transgenerational paternal transmission of acquired traits: stress-induced modification of the sperm regulatory transcriptome and offspring phenotypes. Curr Opin Behav Sci. 2017;14:140–7.
81. Panzeri I, Pospisilik JA. Epigenetic control of variation and stochasticity in metabolic disease. Mol. Metab. 2018;14:26–38.
82. Peng H, Shi J, Zhang Y, Zhang H, Liao S, Li W, Lei L, Han C, Ning L, Cao Y, et al. A novel class of tRNA-derived small RNAs extremely enriched in mature mouse sperm. Cell Res. 2012;22:1609–12.

83. Pigeyre M, Yazdi FT, Kaur Y, Meyre D. Recent progress in genetics, epigenetics and metagenomics unveils the pathophysiology of human obesity. Clin Sci (Lond). 2016;130:943–86.
84. Radford EJ. Exploring the extent and scope of epigenetic inheritance. Nat Rev Endocrinol. 2018;14:345–55.
85. Rassoulzadegan M, Cuzin F. Nutrition meets heredity: a case of RNA-mediated transmission of acquired characters. Environ. Epigenetics. 2018;4:dvy006.
86. Rassoulzadegan M, Grandjean V, Gounon P, Vincent S, Gillot I, Cuzin F. RNA-mediated non-mendelian inheritance of an epigenetic change in the mouse. Nature. 2006;441:469–74.
87. Reilly JN, McLaughlin EA, Stanger SJ, Anderson AL, Hutcheon K, Church K, Mihalas BP, Tyagi S, Holt JE, Eamens AL, et al. Characterisation of mouse epididymosomes reveals a complex profile of microRNAs and a potential mechanism for modification of the sperm epigenome. Sci Rep. 2016;6:31794.
88. Reuter M, Berninger P, Chuma S, Shah H, Hosokawa M, Funaya C, Antony C, Sachidanandam R, Pillai RS. Miwi catalysis is required for piRNA amplification-independent LINE1 transposon silencing. Nature. 2011;480:264–7.
89. Rodgers AB, Morgan CP, Bronson SL, Revello S, Bale TL. Paternal stress exposure alters sperm MicroRNA content and reprograms offspring HPA stress Axis regulation. J Neurosci. 2013;33:9003–12.
90. Rodgers AB, Morgan CP, Leu NA, Bale TL. Transgenerational epigenetic programming via sperm microRNA recapitulates effects of paternal stress. Proc Natl Acad Sci. 2015;112:13699–704.
91. Rowold d'H E, Schulze L, Van der Auwera S, Grabe HJ. Paternal transmission of early life traumatization through epigenetics: do fathers play a role? Med. Hypotheses. 2017;109:59–64.
92. Saitou M, Barton SC, Surani MA. A molecular programme for the specification of germ cell fate in mice. Nature. 2002;418:293–300.
93. Schorn AJ, Gutbrod MJ, LeBlanc C, Martienssen R. LTR-retrotransposon control by tRNA-derived small RNAs. Cell. 2017;170:61–71.. e11
94. Sendler E, Johnson GD, Mao S, Goodrich RJ, Diamond MP, Hauser R, Krawetz SA. Stability, delivery and functions of human sperm RNAs at fertilization. Nucleic Acids Res. 2013;41:4104–17.
95. Sharma U, Conine CC, Shea JM, Boskovic A, Derr AG, Bing XY, Belleannee C, Kucukural A, Serra RW, Sun F, et al. Biogenesis and function of tRNA fragments during sperm maturation and fertilization in mammals. Science. 2016;351:391–6.
96. Sharma U, Sun F, Conine CC, Reichholf B, Kukreja S, Herzog VA, Ameres SL, Rando OJ. Small RNAs are trafficked from the epididymis to developing mammalian sperm. Dev Cell. 2018;46:481–494.e6.
97. Short AK, Fennell KA, Perreau VM, Fox A, O'Bryan MK, Kim JH, Bredy TW, Pang TY, Hannan AJ. Elevated paternal glucocorticoid exposure alters the small noncoding RNA profile in sperm and modifies anxiety and depressive phenotypes in the offspring. Transl Psychiatry. 2016;6:e837.
98. Sipilä P, Björkgren I. Segment-specific regulation of epididymal gene expression. Reproduction. 2016;152:R91–9.
99. Sorscher N, Cohen LJ. Trauma in children of holocaust survivors: transgenerational effects. Am J Orthopsychiatry. 1997;67:493–500.
100. Soumillon M, Necsulea A, Weier M, Brawand D, Zhang X, Gu H, Barthes P, Kokkinaki M, Nef S, Gnirke A, et al. Cellular source and mechanisms of high transcriptome complexity in the mammalian testis. Cell Rep. 2013;3:2179–90.
101. Spadafora C. The "evolutionary field" hypothesis. Non-Mendelian transgenerational inheritance mediates diversification and evolution. Prog Biophys Mol Biol. 2018;134:27–37.
102. Suh N, Blelloch R. Small RNAs in early mammalian development: from gametes to gastrulation. Development. 2011;138:1653–61.
103. Sun J, Lin Y, Wu J. Long non-coding RNA expression profiling of mouse testis during postnatal development. PLoS One. 2013;8:e75750.

104. Tang F, Kaneda M, O'Carroll D, Hajkova P, Barton SC, Sun YA, Lee C, Tarakhovsky A, Lao K, Surani MA. Maternal microRNAs are essential for mouse zygotic development. Genes Dev. 2007;21:644–8.
105. Watanabe T, Cheng E, Zhong M, Lin H. Retrotransposons and pseudogenes regulate mRNAs and lncRNAs via the piRNA pathway in the germline. Genome Res. 2015;25:368–80.
106. Weick E-M, Miska EA. piRNAs: from biogenesis to function. Development. 2014;141: 3458–71.
107. Western P. Foetal germ cells: striking the balance between pluripotency and differentiation. Int J Dev Biol. 2009;53:393–409.
108. Wong N, Wang X. miRDB: an online resource for microRNA target prediction and functional annotations. Nucleic Acids Res. 2015;43:D146–52.
109. Yeshurun S, Hannan AJ. Transgenerational epigenetic influences of paternal environmental exposures on brain function and predisposition to psychiatric disorders. Mol Psychiatry. 2018;
110. Yuan S, Tang C, Zhang Y, Wu J, Bao J, Zheng H, Xu C, Yan W. Mir-34b/c and mir-449a/b/c are required for spermatogenesis, but not for the first cleavage division in mice. Biol. Open. 2015;4:212–23.
111. Yuan S, Schuster A, Tang C, Yu T, Ortogero N, Bao J, Zheng H, Yan W. Sperm-borne miRNAs and endo-siRNAs are important for fertilization and preimplantation embryonic development. Development. 2016;143:635–47.
112. Zhang P, Kang J-Y, Gou L-T, Wang J, Xue Y, Skogerboe G, Dai P, Huang D-W, Chen R, Fu X-D, et al. MIWI and piRNA-mediated cleavage of messenger RNAs in mouse testes. Cell Res. 2015;25:193–207.
113. Zhang X, Gao F, Fu J, Zhang P, Wang Y, Zeng X. Systematic identification and characterization of long non-coding RNAs in mouse mature sperm. PLoS One. 2017;12:e0173402.
114. Zhang Y, Zhang X, Shi J, Tuorto F, Li X, Liu Y, Liebers R, Zhang L, Qu Y, Qian J, et al. Dnmt2 mediates intergenerational transmission of paternally acquired metabolic disorders through sperm small non-coding RNAs. Nat Cell Biol. 2018;20:535–40.
115. Zimmermann C, Stévant I, Borel C, Conne B, Pitetti J-L, Calvel P, Kaessmann H, Jégou B, Chalmel F, Nef S. Research resource: the dynamic transcriptional profile of sertoli cells during the progression of spermatogenesis. Mol Endocrinol. 2015;29:627–42.

Chapter 12
Future Perspectives in Epigenetic Inheritance

Jonatan Darr

Keywords Epigenetic inheritance · Microbiome · Epitranscriptomics · Chromatin · Circulating RNAs · Prions · Health

12.1 Introduction

The following chapter contains a short summary on known mechanisms of epigenetic inheritance, elaborated at length throughout the book. Following this, novel and speculative mechanisms of epigenetic inheritance are presented. These novel mechanisms of epigenetic regulation remain to be explored but can potentially effect aspects of human health including proper development, disease susceptibility and fertility. Figure 12.1 illustrates in general terms the mechanisms presented within this chapter. The chapter ends with a discussion on the possible implications of recent findings and developments in the field of epigenetic inheritance on human health and wellbeing and a short discussion on open questions in the field that remain to be resolved.

12.2 Current Mechanistic Insights on Epigenetic Inheritance

12.2.1 Paramutation and Epialleles

Our current understanding of the mechanistic underpinnings of '**epigenetic inheritance**' arose from early studies that have been conducted primarily in plants, with observations of epigenetic inheritance and epialleles going back to the beginning of

J. Darr (✉)
Helmholtz Zentum München, German Research Center for Environmental Health, Institute of Experimental Genetics, Neuherberg, Germany

German Center for Diabetes Research (DZD), Neuherberg, Germany
e-mail: jonatan.darr@mail.huji.ac.il

R. Teperino (ed.), *Beyond Our Genes*,
https://doi.org/10.1007/978-3-030-35213-4_12

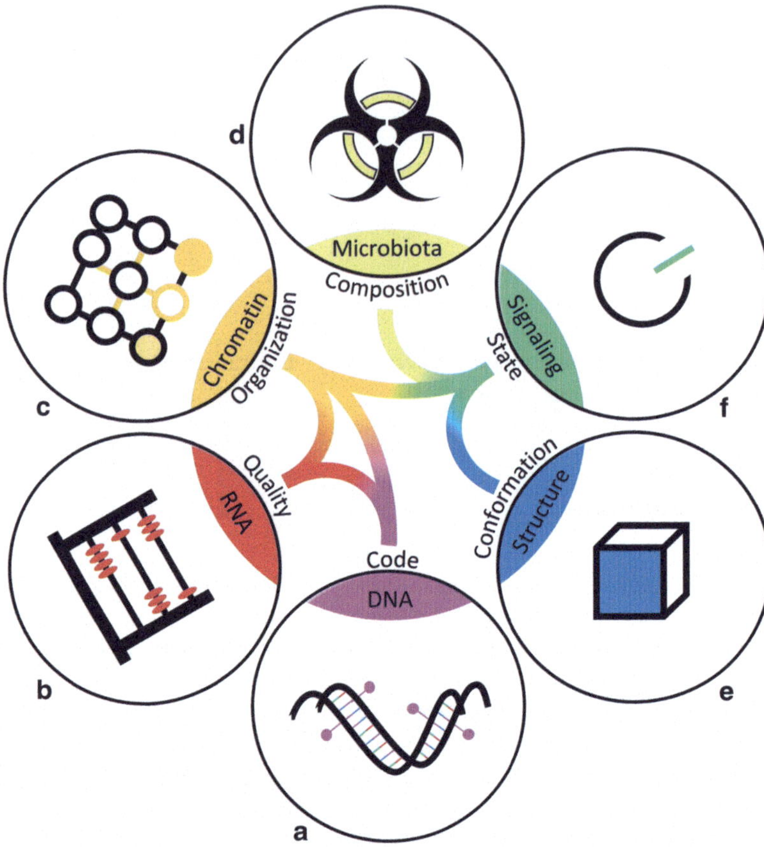

Fig. 12.1 Epigenetic inheritance mechanisms and their possible crosstalk. (**a**) 'DNA code' – apart from the DNA base sequence that is the basis of genetic inheritance, various DNA base modifications are speculated to be involved in epigenetic inheritance. The most characterized modification 5mdC is a well-known epigenetic regulator. (**b**) **'RNA quality'** – describes the RNA transcript sequence, type, abundance and modification within germ cells that can mediate epigenetic inheritance. (**c**) **'Chromatin organization'** – including histone variants, histone tail modifications and nucleosome positioning. (**d**) **'Microbiota composition'** - parental transmission of microbiota to progeny can affect developmental programs and future disease risks. (**e**) **'Structure conformation'** – structural inheritance as with prion particles. Structural inheritance may also include membrane and organelle structural transmission. (**f**) **'Signaling state'** – activation of signaling pathways during early embryonic development or germ cell differentiation can lead to persistent heritable traits. Activators can be parental factors such as growth factors, hormones and nutrients, or environmental factors such as pollutants. Persistent signaling can act as a feedback loop to reinforce transmission of additional molecular mechanisms such as chromatin state or protein conformation

the twentieth century. These early studies focused on highly variable and easily observable genetic traits that in many cases dictated coloring of plant tissues and / or shape of leaves and petals. The mechanistic nature behind these early observations remained enigmatic for the greater part of the twentieth century and has only begun to be elucidated following the development of novel molecular techniques that could address them.

One of the first cases of epigenetic inheritance to be described in the literature is that of '**paramutation**'. The first described instance of paramutation and of a heritable '**epiallele**' was reported in 1915 in the garden pea [1]. A 'rogue' phenotype with narrow leaves and petals was described within crops of wild-type plants. Hybrids gave rise to an intermittent phenotype that progressively became more 'rogue' like and only transmitted the rogue phenotype to progeny. The most studied case of paramutation is of the r1 (red1) [2] locus in the maize with additional loci such as the b1 (booster1) [3], pl1 (purple plant1) [4] and more [5, 6] later described in great detail in the same model plant.

The r1/b1 loci encode for basic helix-loop-helix (b-HLH) transcription factors that together with the pl1 transcription factor activate genes of the anthocyanin biosynthetic pathway to produce red/purple pigments. Even slight changes in the RNA levels of these transcription factors confer prominent changes in the pigmentation of the maize kernels, giving rise to distinct visible phenotypes. A cross between an R-r haplotype, with dark purple seeds, and an R-stippled haplotype, with purple stippled seeds, yielded F1 seeds that upon subsequent crosses with an r1 null haplotype gave rise to an F2 progeny with unaltered R-stippled kernels and an altered phenotype of the R-r haplotype termed R-r'. This haplotype exhibited a reduced pigmentation of the kernels compared to the R-r haplotype which resulted from a lower expression of the transcription factor.

Paramutations seemed to violate Mendel's law of segregation since alleles segregated in the heterozygote influence each other. Additionally, given the high rate of allelic conversion (100%) in F2 and rare instances of reversion in subsequent generations, genetic mutations could not serve as a plausible molecular mechanism. These and additional observed contradictions to commonly held genetic theories of inheritance forced a rethink of what may constitute heritable information.

Detailed genetic and epigenetic studies have found that '**paramutable**' R-r haplotypes in maize share several common features [7, 8]. Genetically, these haplotypes contain a genetic element called S-subcomplex. This element includes a truncated inactive R gene and an inverted repeat comprised from two functional R genes positioned head-to-head. Epigenetically, paramutable and '**neutral**' alleles demonstrate marked hypometylation of Cystein residues compared to '**paramutagenic**' alleles and an increased DNAse sensitivity. In light of these findings it became apparent that the change of R-r allele to R-r′ allele represents an epigenetic change, and the two alleles are in fact epialleles with differential DNA methylation patterns.

It is now clear that paramutation involves trans-communication between homologous sequences. This communication establishes heritable changes in chromatin structure that often correlate with alterations in DNA methylation. Paramutations have even been established synthetically in mouse strains demonstrating a requirement for repeating elements and implicating DNA methylation in the process yet again [9].

12.2.2 DNA Methylation

Epialleles and paramutations have since been observed in multiple species including fungi, plants and animals and in many cases involve differential DNA methylation at CpG sites. Surprisingly, only recently has differential DNA methylation been described also in the rough epiallele [10], solving a century old observation that remained long unaddressed. In mammals, the best-studied epiallele is the agouti variable yellow (A^{vy}) locus [11, 12]. In genetically identical A^{vy} mice, higher expression of the agouti protein leads to pleotropic phenotypes including yellow coat color, obesity, diabetes and increased susceptibility to tumors. These phenotypes are passed to progeny only from the maternal side with a high degree of correlation present between the phenotypic presentations in the progeny to the maternal phenotype. Genetic studies demonstrated that the A^{vy} locus contains an insertion of a IAP retrotransposon upstream to the Agouti gene, where it functions as a cryptic promoter driving Agouti expression. CpG methylation levels in the A^{vy} IAP retrotransposon LTR region are directly correlated between females and their progeny and inversely correlated with Agouti expression.

Another less known example of a metastable mammalian epiallele is the Axin-Fused ($Axin^{Fu}$) allele. Axin is a protein phosphatase that inhibits Wnt signaling [13]. The $Axin^{Fu}$ allele is a dominant gain-of-function with an insertion into intron 6 of a IAP retrotransposon in an antisense orientation. Expression of the allele leads to a characteristic kinked tail presentation whilst silencing leads to a normal tail presentation. Though the phenotype was described already in 1937 [14] the mechanistic nature of the mutation was solved only much later [15]. Similarly to the A^{vy} allele, the LTR region of the IAP retrotransposon insertion demonstrates differential methylation levels between individuals with variable phenotypic presentation, with hypomethylation resulting in a kinked tail presentation. Hypomethylation is associated with the transcription and expression of truncated forms of the Axin protein containing axon 7–10. These truncated forms have been associated with defective axis formation in additional model organisms [13]. In contrast to A^{vy}, $Axin^{Fu}$ is inherited transgenerationally both from the maternal and the paternal side, with methylation pattern of the IAP retrotransposon in sperm reflecting the methylation patter in somatic tissues of the offspring.

Despite these and a few additional examples there is still an ongoing debate as to the extent of methylation based epigenetic inheritance, this in part due to the profound epigenetic reprograming that germ cells undergo [16, 17]. Nonetheless, whilst the bulk of the genome becomes demethylated in primordial germ cells, IAP retrotransposons escape global demethylation [18]. How this class of retrotransposons escapes reprogramming is still unknown, yet IAPs constitute prime candidates in mediating and driving trans-generational inheritance in mammals. If any additional genomic regions are able to escape global reprograming in a context specific manner and what mechanisms may enable this remains to be elucidated.

12.2.3 Small RNA

The first indication that RNA may mediate epigenetic inheritance in mammals came from work on a phenotypic trait arising from a targeted knockout of the Kit gene in mice [19]. The KIT tyrosine kinase receptor has a critical role in several developmental processes including germ cell differentiation, haematopoiesis and melanogenesis. Homozygote knockouts die shortly after birth whilst heterozygotes have a distinctive white tail tip and white feet. In crosses between heterozygotes, wild type progeny exhibited weaker but similar color patterns to heterozygotes. Wild type progeny from crosses between a heterozygote and a wild type also persistently exhibited this color pattern regardless of the parental sex of the heterozygote. Overall, this Kit' paramutation was most prominent in crosses between heterozygotes, but progressively disappeared in following generations.

Unlike most paramutations, no significant changes to either cytosine or histone methylation were observed between wild type, heterozygous and paramutated kit' animals. Surprisingly, levels of polyadenylated Kit mRNA were reduced in Kit' animals, whilst variable forms of Kit mRNA stemming from abnormal arrest and/or initiation of the primary transcription, or abnormal post-transcriptional processing and splicing, accumulated in multiple tissues. In contrast to somatic cells, Kit transcription in sperm was up regulated in heterozygous animals with transcriptional deregulation of both alleles most evident in the late spermatogenic stages. Unexpectedly, substantial amounts of Kit RNA were detected even in mature epididymal sperm of heterozygotes and paramutated mice.

Microinjection of sperm RNA (and surprisingly brain RNA) collected from heterozygotes into one-cell embryos was enough to establish the kit' paramutation in 50% of the litter. Microinjections of Kit targeting microRNAs mir-221 or mir-222 was also sufficient to induce the Kit' paramutation at high frequencies, suggesting that exposure to microRNAs in early embryonic development can induce permanent and heritable epigenetic signatures that effect gene expression in the adult.

Later experiments were able to demonstrated that microinjection of various miRNAs to early embryos can induce developmental phenotypes and pathologies with possible links to various human conditions. One such model induced heritable cardiac hypertrophy following microinjection of mir-1 into one cell embryos [20]. Mice born after mir-1 microinjection had an increased heart and multiple abnormalities including mitochondrial disorganization, myofibrillar disorganization, sarcomere shrinkage and more, with abnormalities arising already at embryonic day 18.5. Cdk9, a putative target of mir-1, was demonstrated to accumulate both in RNA and protein levels in embryonic hearts around E18.5. Indeed, microinjection of

Cdk9 fragments to one cell embryos induced the same observed phenotypes as with mir-1 microinjection. The cardiac abnormalities observed following mir-1 microinjection were heritable, suggesting a similar paramutation phenomena as with the Kit locus.

In a similar set of experiments microinjection of mir-124 led to a heritable 30% increase in body weight of mice at birth and adulthood when compare to control animals [21]. This phenotype was associated with a heritable epigenetic modification of a putative regulatory region found 3 kilobases upstream to the Sox9 locus. This regulatory region was found to be enriched in H3K9me2/me3 methylation that demonstrated a high degree of correlation between founder mice to progeny with increased body weight.

In addition to miRNAs, nuclear RNA interference was also demonstrated to induce a heritable epigenetic trait. In C.elegance nuclear RNA interference is used as an antiviral defense mechanism to silence replication of RNA viruses. A seminal work demonstrated how induction of an antiviral response in C.elegance can induce transgenerational inheritance of antiviral resistance in unexposed progeny [22]. This acquired immune resistance is transmitted through small RNAs called viRNA and requires the RNA-dependent RNA polymerase RRF-1 for maintenance of long term silencing. This work demonstrated for the first time intergenerational inheritance of an acquired adaptive immune response that is beneficial and advantageous for long term survival of the specious. To date, no RNA dependent RNA polymerases have been found in mammals [23], suggesting that the described mode of germ line RNA amplification is limited to specific species.

In addition to micro-RNAs and small interfering RNAs Piwi-interacting RNAs have also been implicated in the establishment of transgenerational epigenetic inheritance [24]. These piRNA regulate transcriptional gene silencing of retrotransposons in germ cells through both epigenetic and post-transcriptional mechanisms including DNA methylation [25, 26], RNAi slicing [27] and transcript deadenylation [28]. Interestingly, in c.elegance PIWI-RNAs can induce a positive feedback loop that promotes their sustained production and transgenerational inheritance through chromatin based processes that require the RNAi machinery for its maintenance [29].

Overall, small RNAs are to date the most recognized molecular mechanisms that mediate epigenetic inheritance. Their ability to direct chromatin based processes such as DNA methylation and histone modification and post translational processes such as transcript splicing or degradation in a targeted sequence dependent manner, together with the ability in specific cases to amplify and perpetuate their own production makes them truly unique.

12.2.4 Chromatin Structure

In addition to DNA methylation, nucleosome occupancy, histone variants and post-transcriptional modifications are central in epigenetic regulation of gene expression and cell identity. Differentiation and maturation of germ cells entails drastic remodeling of the chromatin landscape, giving rise to cells with unique chromatin landscapes that differ substantially from somatic cells. Reprogramming and "re-setting" of epigenetic markers is critical for proper fertilization and formation of a totipotent zygote. This resetting includes extensive demethylation of the DNA [30] and drastic changes in nucleosome occupancy, composition and post-transcriptional modifications of histones [31].

Sperm cells undergo extensive chromatin remodeling so as to allow the compaction of DNA into the head of mature spermatozoa. Histones are replaced during spermatogenesis by protamines that are small arginine rich proteins allowing for the tight compaction of the DNA within sperm heads [32]. This protein exchange leaves only a small fraction of nucleosomes on DNA, roughly 1% in mice and 10% in humans, with retention of nucleosomes mostly in loci critical for development [33]. Retained nucleosomes are composed from unique histone variants that are expressed only during spermatogenesis such as testes-specific histone H2A and H2B (TH2A / TH2B) [34]. Oocytes demonstrate a global reduction in histone acetylation levels which facilitates recruitment of the ATRX chromatin remodeling factor required for chromosome alignment and meiotic spindle organization [35]. In addition, the somatic histone H1 is replaced with the H1Foo (H1 histone family, oocyte-specific) variant during oogenesis [36].

Given the early timing of primordial germ cell differentiation in animals and fundamental chromatin remodeling events during maturation of germ cells, the scientific community rejected for a long time the possibility that epigenetic inheritance, and specifically in animals, is possible. Yet despite these preconceptions H3K4 and H3K9 methyltransferases, as well as the H3K4 and H3K9 lysine-specific demethylase LSD1, have been implicated in transgenerational inheritance in multiple model organisms including S. pombe, C. elegans and mice. Indeed, transcriptional gene silencing mediated by RNAi mechanisms is achieved in part through heterochromatic modifications of target loci [37–41]. These modifications include H3K9m2 and DNA methylation and can lead to a heritable phenotype [42–46].

Deregulation of histone modifications in germ cells can lead to transgenerational defects. In a mutated C. elegance line for the LSD1 homologue spr-5, a steady decline in brood size across 28 successive generations and an increase in the frequency of sterile animals in each brood was noted. This deteriorating phenotype was accompanied by the stable accumulation in PGCs of H3K4me2 at spermatogenesis-expressed genes, leading to defects in oogenesis and spermatogenesis [47].

Environmental stress was also found to induce transgenerational inheritance by modulating levels of H3K9me3 [48]. In C.elegance expression from a daf-21 (Hsp90) promoter::fluorescent protein multi-copy array was induced for 7 generations following growth of a single generation at a higher temperature of 25 °C. This transgenerational inheritance persisted through both maternal and paternal inheritance, with elevated expression evident from the onset of zygotic transcription. Exposure to 25 °C during germline development resulted in depletion of H3K9me3 from the array, with reduced levels of the H3K9me3 repressive mark evident also in the F2 progeny. No changes were observed in levels of the repressive mark

H3K27me3, or of the active markers H3K36me3 and H3K4me2. The histone methyltransferase SET-25 was found to be required for the repression of the array, with its activity reduced under higher temperature. Indeed, it was found that reduced SET-25 activity under higher temperatures results in the derepression of multiple classes of repetitive elements and pseudogenes.

Another study in C. elegance demonstrate the requirement for the JMJD-2 and JMJD-3/UTX-1 demethylase activity in the environmentally induced transgenerational transmission of a fertility defect [49]. A single dose of Bisphenol A (BPA) in F0 led to the expression of a fluorescent reporter in the germline as well as to reproductive dysfunctions that lasted for 5 generations. This single exposure led to a reduction in H3K9me3 and H3K27me3 levels in germline nuclei. Finally, RNAi knockdown of either the JMJD-2 or JMJD-3/UTX-1 demethylases restored H3K9me3 and H3K27me3 levels and alleviates the transgenerational effects, implicating both enzymes in mediating the observed environmentally induced epigenetic inheritances.

The involvement of histone modifications and of the LSD1 demethylase in the establishment and transmission of epigenetic inheritance was also noted in mice [50]. Using transgenic mice with a transient over-expression of LSD1 during spermatogenesis, the authors generated mice with sperm containing reduced H3K4me2 levels. The progeny of these male mice demonstrated severely impaired development and reduced survival rates which persisted for two generations following the transient overexpression of LSD1. No changes in sperm DNA methylation were evident but rather a reduction in H3K4me2 levels and altered RNA content in sperm.

12.3 Prospective Mechanisms

12.3.1 Microbiome

The human '**microbiota**' is a rich community of bacteria, fungi and viruses that live on and in humans as commensal, symbiotic or pathogenic organisms. They are found in multiple organs including skin, mammary glands, oral cavity, uterus, seminal fluids and gastrointestinal tract [51, 52] . The microbiota is highly variable between different organs of a single person and between different people [52, 53], though between family members the microbiota tends to be more similar. Adult and infant microbial communities are quite different with microbial colonization starting in the womb, continuing through delivery, breastfeeding and environmental exposures until it stabilizes and forms the complexity observed in adults at around 3 years of age [54]. Parental transmission of the microbiota influences the developing microbiota of infants and is essential for proper physiological development. Experiments in '**germ free**' (GF) animal models have demonstrated that proper anatomical development of intestinal epithelium and maturation of gut-associated lymphoid tissue are dependent upon proper microbiome colonization and development [55]. Moreover colonization by gut microbiota has been demonstrated to impact mammalian brain development and the '**hypothalamic–pituitary–adrenal**' (HPA) axis, leading to altered adult behavior and stress response [56, 57].

One of the first works to directly demonstrate the involvement of the gut microbiota in stress response compared GF mice to '**Specific Pathogen Free**' (SPF) and '**gnotobiotic**' mice. Restraint stress lead to higher levels of plasma adrenocorticotropic hormone (ACTH) and corticosterone in GF mice compared to SPF mice. GF mice also demonstrated lower expression of neurotrophic factor in the cortex and the hippocampus. Inoculation of GF mice with various bacteria modulated ACTH and corticosterone levels, a finding that directly demonstrated the impact of various gut bacteria on stress response. Another work demonstrated how treatment of SPF mice with antimicrobial factors changed the gut microbiota and led to an increased exploratory behavior and hippocampal expression of the brain-derived neurotrophic factor (BDNF), which are associated with reduced stress behavior [58].

In flies, administration of G418 antibiotics led to several transgenerationaly heritable phenotypes including a delay in larval development [59]. This delay was mediated by a depletion of a commensal Acetobacter bacteria found in the gut [60]. Reintroduction of the bacteria strain to G418 treated adults prior to mating or to developing larva abrogated the delayed larval development, demonstrating that the bacteria are necessary for the establishment of the inheritance of this trait. Use of antibiotics such as Puromycin did not induce a heritable developmental delay and failed to kill gut Acetobacter bacteria, whilst other antibiotics like Ampicillin and Ciprofloxacin established a heritable developmental delay and killed the gut Acetobacter bacteria, demonstrating again that inheritance of the delayed larval development is mediated specifically by the gut Acetobacter bacteria population. Supplementing Riboflavin (B12) to progeny of treated flies could rescue the delayed development, whilst inhibition of Riboflavin synthesis in bacteria with Roseoflavin could induce the phenotype in flies not exposed to antibiotics. These results demonstrated that Riboflavin availability can modulate fly larval development and that commensal gut bacteria supplement Riboflavin to developing larva. Antibiotic interference with the gut microbiome in one generation was recapitulated in its progeny leading to reduced Riboflavin availability and delayed larval development.

Overall**,** an accumulating body of work demonstrates how changes in the relative abundances of gut microbial taxa or **'Dysbiosis',** can be linked to multiple pathologies including inflammatory bowel disease [61], cardiovascular disease, metabolic syndromes [62, 63], obesity [64], depression and more. Dysbiosis following antimicrobial treatment in infants can also lead to the development of multiple pathologies later on in adult life [65, 66]. Critically, in several studies the changes observed in the gut microbiome have a causal relation to the observed pathology [64]. Given these findings it is not surprising that parental transfer of the microbiome to offspring can lead to variable inherited phenotypes that are considered as epigenetic inheritance.

More surprising are works that demonstrate unexpected links between commensal microbiome to germ-cells. One well described example is the relationship between the Wolabachia intracellular bacteria to their arthropod hosts including worms and flies (reviewed extensively in [67]). These bacteria can be considered as reproductive parasites that induce several different phenotypes in the host including; feminization of genetic males, parthenogenesis, selective male killing and cytoplasmic incompatibility of germ cells that prevents infected males from mating with females infected with a different Wolabachia type. The phenotype is dependent upon the Wolabachia strain and the host species and can effectively lead to speciation of infected hosts. Wolabachia can be transferred vertically through the female germ-line as well as horizontally in the population. This transmission can be considered as a form of epigenetic inheritance that leads to phenotypic variability in genetically identical individuals. To date, no such intercellular bacteria is known to exist in vertebrates.

Another seminal work in Drosophila demonstrated how the gut microbiome can directly affect the germ line to allow greater phenotypic variability in the progeny [68]. Elimination of the gut bacteria led to a reduction in the number of oocytes in $F_{0.}$ this reduction could be rescued by specific reintroduction of Acetobacter to GF flies. Embryos of GF flies demonstrated faster developmental progression compared to controls stemming from an expedited '**maternal to zygotic transition**' (MZT) and an increased rate of pre- and post-cellularization development. Larva of GF flies also demonstrated greater phenotypic variability when mutant backgrounds were used. The crosstalk uncovered in this work between the gut microbiome and the germ-line demonstrates the potential for transgenerational influences on developmental stability and disease susceptibility.

12.3.2 RNA Mediated Soma to Germ-Line Information Transfer

Towards the end of the nineteenth century August Weismann, a prominent evolutionary biologist of the time, published his hypotheses on the continuity of the germ-plasm and what later became known as the '**Weismann barrier**'. In simple terms, the first hypothesis proposed that somatic cells cannot give rise to germ cells whilst the second hypothesis stated that information cannot flow from somatic cells to germ cells. Together, these ideas excluded the possibility of inheritance of acquired somatic traits as suggested by Lamarck, and were in contradiction to Darwin's own late idea of '**pangenesis'** in which he proposed that the soma secretes small particles he called '**gemmules**' that migrate to the gametes were they contribute to heritable information. Interestingly though, Weismann did not per-se exclude the possibility of a heritable transfer of acquired changes of the germ-plasm.

Weismann's hypotheses dominated the scientific discourse on hereditary in animals, as it was discovered that in plants germ-cells arise from the vegetative meristem and somatic cells. His ideas progressively led to a gene-centric view of hereditary that was supported by the '**central dogma**' which stated that information flows exclusively from DNA to RNA to proteins. Together, the possibility that a soma to germ-line information transfer can transpire and result in heritable changes to the progeny was strongly rejected.

One of the first works to challenge the Weizmann barriers demonstrated soma-to-germline information transfer in C. elegans [69]. The authors used a transgenic C. elegans line that expressed cytosolic GFP in all somatic tissues and GFP–dsRNA in all neurons. GFP expression was found to be reduced in most somatic tissues and the gametes. The silencing in soma and gametes was found to be dependent upon expression of the SID-1 RNA importer and the RDE-1 argonaute protein but independent of the RNA-dependent RNA polymerase RRF-1. The authors demonstrated how neuronal derived mobile RNAs can induce a persistent transgenerational silencing of GFP expression in subsequent generations. The progeny of worms with a gamete specific GFP expression and neuronal specific GFP-dsRNA expression, continued to demonstrated complete suppression of GFP in the gametes for more than 25 generations, even following loss of the neuronal specific GFP-dsRNA expression construct.

Screens for pathways that are required for the establishment of transgenerational repression identified in addition to SID-1 and RDE-1, the nuclear Argonaute HRD-1 and MUT-7 an RNase D homologue. Taken together, these results suggested that germline silencing due to neuronal dsRNAs relies on the import of dsRNAs through the SID-1 RNA transporter, and subsequent processing within the germline by the primary Argonaute RDE-1 (and additional proteins) to secondary single-stranded small RNAs. Maintenance of transgenerational silencing was found to require HRDE-1 and MUT-7 and independent of SID-1 and RDE-1. HRDE-1 itself was demonstrated to use small RNAs to guide histone 3 lysine 9 trimethylation (H3K9me3), which suggests that initiation and maintenance of transgenerational silencing is a H3K9me3 dependent process.

A seminal works implicated sperm tRNA fragments in transmission of environmentally induced epigenetic inheritance in mice and raised the possibility that these tRNA fragments are transported to sperm from somatic cells [70]. In this work IVF generated progeny of male mice subject to a low-protein diet, demonstrated elevated levels of the cholesterol biosynthesis enzyme Sqle in the liver at 3 weeks of age when compared to the IVF generated progeny of control male mice. Low protein diet effected levels of multiple small RNAs in the sperm of challenged fathers including tRNA fragments that were found to be highly abundant and among the most differentially regulated RNA species in the sperm. Intriguingly, low levels of tRNA fragments were detected in sperm isolated directly from the testes, but the relative fraction and abundance of these tRNA fragments increased along the epididymal track where sperm are stored after maturation.

Since maturation of sperm was known to depend upon fusion with extracellular vesicles called epididymosomes that deliver various proteins essential for sperm maturation, the authors tested the hypothesis that these epididymosomes carry tRNA fragments into maturing sperm and indeed found a high correlation between the small RNA content within epididymosomes and the sperm small RNA content. This finding suggested that tRNAs may be loaded into maturing sperm along the epididymal track. The authors continued and demonstrated that tRNA fragments may exert their regulatory effect through genes that are regulated by the long terminal repeats of the endogenous MERVL retrotransposon.

These works may be the first in a line of works demonstrating RNA mediated soma-to-germline information transfer. The presence of stable cell free circulating RNAs in bio-fluids of multicellular organisms including mice and humans [71, 72] suggests that RNA mediated cell to cell communication and possibly soma-to-germline information transfer may be more common than previously thought and poses a tantalizing future research field.

12.3.3 Non-canonical Base Modifications

Apart from the canonical DNA methylation of deoxycytidine (5mdC) in CpG sites which is observed in multiple species and that is already associated with epigenetic inheritance, DNA carries multiple non-canonical DNA base modifications that potentially serve as epigenetic regulators. Active DNA demethylation is achieved in part with the aid of dedicated TET enzymes, which mediate the sequential oxidation of 5mdC to 5-carboxy-deoxycytidine (5cadC) through the intermittent oxidation states 5-hydroxymethyl-deoxycytidine (5hmdC) and 5-formyl-deoxycytidine (5fdC) [73, 74]. 5hmdC, 5fdC and 5cadC have been recognized as epigenetic markers by their own right with dedicated binding proteins and possible regulatory functions. 5fC and 5caC have been demonstrated to inhibit transcriptional elongation by RNA polymerase II [75]. 5fC may also affect the structure of the DNA double helix [76]. In addition to deoxycytidine methylation, N6-methyl-deoxyadenosine (m6dA) has been recently recognized in vertebrates including mice and humans [77]. M6dA is a recognized epigenetic marker in bacteria where it regulates multiple pathways including DNA replication [78], repair [79], transcription and transposition [80]. In mice, m6dA was shown to accumulate in the prefrontal cortex (PFC) following chronic stress, where it was preferentially deposited on LINE transposons within intergenic regions. These LINE transposons exhibited transcriptional downregulation as a result of the methylation [81]. Though work is still underway to precisely elucidate the nature and function of various non-canonical DNA base modifications, it is clear that these modifications represent a tantalizing mechanism for establishment and transmission of epigenetic inheritance.

In addition to DNA base modification, RNA bases also display a rich collection of base modification with over 100 base modification known to date [82]. From these non-canonical base modifications cytosine methylation of RNA has already been implicated in epigenetic inheritance. The RNA metyltransferase Dnmt2 was found to be required for the establishment and maintenance of the mouse kit' paramutation [83]. Dnmt2 catalyzes cytosine methylation of tRNA to regulate stability and steady state levels [84]. Using homozygous knockouts for Dnmt2 the authors could demonstrate its requirement for kit' paramutation establishment. Another work demonstrate the requirement of Dnmt2 for the establishment and intergenerational transmission of acquired metabolic disorders induced by a paternal high fat diet (HFD) challenge [85]. Here, RNA derived from sperm of Dnmt2 knockout HFD fed males and wildtype HFD / control diet fed males was used for zygotic injection. Development of metabolic phenotypes transpired only following zygotic injection of sperm RNA derived from the wildtype HFD fed males. Importantly, natural breading still led to a metabolic phenotype in progeny of Dnmt2 knockout HFD males, implicating multiple mechanisms in the establishment of epigenetic inheritance. In wild-type males HFD lead to a significant increase of m5C and m2G in the 30–40 bp RNA fraction that was not evident in Dnmt2 knockout males. Taken together, the authors postulated that RNA hypomethylation may lead to increased tRNA fragmentation and perhaps alter the biology of tRNA derived fragments.

12.3.4 Prions and Structural Inheritance

'**Prions**' are misfolded proteins that have the surprising capacity to self-propagate in an infectious manner. An interaction between a native protein to the misfolded prion protein induces misfolding of the former into a prion particle that can propagate itself further [86]. The discovery of prions is tightly linked to research of infectious and fatal neurodegenerative disease such as '**Creutzfeldt–Jakob**' and '**Kuru**', where prions were discovered to be the underlying molecular mechanism leading to tissue damage and cell death [87].

A long lasting enigma in yeast research related to non-Mendelian segregation was resolved with the realization that prions are the heritable material responsible for the observed phenotypes. Indeed, in yeast several prions can dictated phenotypes and be inherited from mother to daughter cells in what constitutes as a form of epigenetic inheritance. Though the diversity of prion proteins and of prion variants in yeast is surprising, the relevance to epigenetic inheritance in humans is not clear.

One early prion example, described initially in the 70's is the [URE3+] prion that was identified in a mutagenesis screen of S. cerevisiae. Starting from aspartate transcarbamylase mutant lines that require ureidosuccinic acid (USA) supplementation to the growth media, the researchers looked for mutants that were able to take up USA even in the presence of large quantities of glutamic acid, which normally inhibits uptake of USA [88]. Backcrossing of the [URE3+] strain demonstrated dominant non-mendalian segregation with all resulting daughter cells able to utilize USA. Later works demonstrated the [URE3+] phenotype to occur at a low frequency in haploid yeast strains, to be reversible and to be tightly depend on the expression of the Ure2 protein with higher levels of protein expression resulting in higher rates of spontaneous occurrence. Eventually it was recognized that the [URE3+] phenotype arises from a prion form of the Ure2 protein [89]. The Ure2 protein itself acts as an inhibitor of the Gln3 transcription factor under favorable nitrogen conditions (as with presence of glutamic acid in the media). This inhibition leads to repression of nitrogen catabolic processes including downregulation of the Dal5 Allantoic acid transporter. USA can be effectively transported by the Dal5 transporter when it is expressed given some structural resemblance to Allantoic acid. In contrast, the prion form of Ure2 cannot inhibit Gln3, allowing for uptake and usage of USA when glutamic acid is present in the media.

Though generally speaking protein misfolding and prion states lead to loss of function such as with [URE3+] prion form of Ure2 that cannot inhibit Gln3, there is at least one described case of a gain of function. The S. cerevisiae [PIN+] prion facilitates and promotes the formation of new prions de novo including the [URE3+] prion [90]. [PIN+] is a prion form of the Rnq1 protein which acts as a functional amyloid [91]. These are proteins that in their amyloid form (such as with the [PIN+] prion form of Rnq1), play a positive role in modulation of cell physiology.

Prions may confer not only distinct phenotypic state but also phenotypic heterogeneity. Some proteins can exist in several structurally and functionally distinct heritable prion forms without any changes of the primary amino acid sequence. These prion strains or variants can generate distinct phenotypes, as with the mammalian PrP prion strains that cause neurodegenerative diseases which vary in incubation periods, clinical symptoms, and neural degeneration patterns [92]. In S. cerevisiae one such protein is Sup35. The [PSI+] prion variants of the Sup35 protein are the most intensively studied prions whose study revealed many basic properties of prion biology in general. Sup35 links GTP hydrolysis to the release of the nascent polypeptide from the peptidyl-tRNA at the ribosomal P site. A complete loss of Sup35 function in yeast is lethal, but even a significant reduction in protein levels are tolerated well and result in partial read-through of termination codons and a nonsense suppression phenotype. [PSI+] variants can be classified into strong, moderate and weak variants, which define the strength of the nonsense suppression phenotype and its mitotic stability [16].

Several additional prions have been identified in Saccharomyces cerevisiae and other yeast and fungi species. To date mammalian prions are associated only to neurodegenerative disease and no developmental phenotypes have been associated to prions. There still remains however an intriguing possibility that prions could be transgenerationally inherited to dictate variable phenotypes including disease susceptibility.

Prions represent the most studied and best understood example of structural inheritance. This mode of inheritance is the transmission of an epigenetic trait by a self-perpetuating spatial structures. In yeast, non-prion proteins have also been demonstrated to acquire altered conformations that elicit variable phenotypic traits that persist for multiple generations and whose inheritance was non Mendelian [93]. Over-expression of every single ORF of the yeast proteome lead to the identification of fifty proteins that do not contain any known prion domain nor have any N/Q-rich regions. Instead, these proteins have intrinsically disordered regions that are evolutionary conserved. Most of these proteins are transcription factors or RNA binding proteins whose variable folding conferred beneficial adaptive traits. This finding extends the possibilities for protein structural inheritance beyond proteins containing prion domains.

Additional mechanisms that may represent a form of structural inheritance include inheritance of the cortical structure of the surface of the ciliates [94], inheritance of organelle membranes, such as the mitochondria or chloroplasts [95] and transmission of proteins that are essential for their own assembly like Hsp60p [96].

12.3.5 Cellular Signaling and Feedback Loops

Multiple signaling pathways that regulate gene expression through modulation of chromatin structure can potentially lead to environmentally induced epigenetic inheritance. These signaling pathways can act directly on the germ-cells themselves, or effect very early stages of fertilization and embryonic development to induce phenotypic variations later in life. Hormones, growth factors, small molecules and endocrine disruptors have all been implicated in multiple species as environmental inducers of acquired traits that in specific cases have been demonstrated to be epigenetically transmitted to non-exposed progeny.

One interesting example for a hormonal induced epigenetic inheritance was observed in A. thaliana following exposure to herbivory stress [97]. **Jasmonates** are essential **phytohormones** that mediate signaling pathways essential for plant development and survival. Jasmonic acid (JA) signaling (and signaling of relate metabolites) triggers the production of antiherbivore defenses, whilst the volatile metabolite methyl jasmonate (MeJA) allows for rapid propagation of the defense signal to other parts of the stressed plant (and possibly neighboring plants). Challenging of A. thaliana with caterpillar predation or with MeJA were found to upregulated antiherbivore defense mechanisms in progeny of challenged plants, which result in a substantial reduction of the growth rate of predatory caterpillars. This upregulation of antiherbivore defense mechanisms persisted for two generations following the initial challenge. Perception of the JA metabolite JA-Ile was found to be essential for the transmission of the predatory resistance to successive generation, as parental plants knocked out for the COI1 gene, an intracellular receptor of JA-Ile, were unable to induce epigenetic inheritance. COI1 expression in the progeny generation was however not required for inherited resistance. The authors also demonstrate that siRNA signaling is required for the observed heritable effects, as parental lines lacking nuclear RNA polymerases required for the synthesis of small interfering RNAs (nrpd2a nrpd2b double mutants) or defective in siRNA processing (dcl2 dcl3 dcl4 triple mutants) did not induce resistance in successive progeny.

Surprisingly, induction of epigenetic inheritance by ‘hormonal’ signaling is not limited to plants and has been observed in animals. The induction of helmet formation in Daphnia cucullata by predator secreted ‘**kairomones**’, was found to be inherited to unexposed F1 progeny [98]. This helmet structure helps protect the Daphnia from various predators. Daphnia grown in the presence of kairomones gave rise to an F1 progeny that consistently demonstrated longer helmet structures when compared to control progeny, effectively transmitting an acquired adaptive trait that reduces the predation rate of the unexposed progeny.

In mammals, the possible involvement of hormonal signaling in the establishment of transgenerational epigenetic inheritance was made clear through studies conducted on the pathophysiological effects of various polluting compounds that are used in industrial, agricultural or medical contexts. Specifically, studies into the effect of various polluting ‘**endocrine disrupting chemicals**’ (EDCs) on male fertility led to novel models of transgenerational epigenetic inheritance. One such study conducted on rats studied the effect of methoxychlor and vinclozolin, a pesticide and a fungicide with antiandrogenic activities [99]. In-utero or early post-natal exposure of male rats to these compounds influences sexual differentiation, gonad formation, and fertility. Looking at transgenerational effects following a transient exposure in gestating females, the study found that both vinclozolin and methoxychlor induced transgenerational defects in spermatogenesis and reduced sperm viability that were evident as much as four generations following the initial exposure. This transgenerational effect was found to be transmitted through the male germ line and associated with altered heritable DNA methylation. Transgenerational effects of other EDC have been described also in fish [100]. Exposure of medaka (Oryzias latipes) fish embryos to BPA and 17α-ethinyl estradiol (EE2), led to lower fertility rates and higher embryo mortality in F2 and onward, compared to progeny of unexposed fish. The induction of transgenerational effects following exposure to endocrine disruptors can potentially be devastating to wild populations that are routinely exposed to these and other similar chemicals [101].

In light of the findings on the transgenerational effects of synthetic EDCs on fertility and health it is perhaps less surprising that various small molecules, growth factors, interleukins and signaling molecules present at early stages of fertilization and embryonic development impact multiple phenotypic traits in the progeny. In females, nutritional availability has been demonstrated to directly influence DNA methylation levels in developing embryos [102–105]. Nutritional and environmental cues also affect the microenvironment of the early preimplantation embryo to exert phenotypic variability in progeny (see [106, 107] for a comprehensive review). Some of these maternal effects are well documented and known.

Lately, it has been acknowledged that seminal fluids can also impact early embryonic development. In addition to sperm multiple signaling factors are present in seminal fluids [108, 109]. These can potentially interact with and reshape female reproductive track physiology to promote embryonic development [110–112]. One effect these factors seem to facilitate is an immune tolerance within the female reproductive tract to the male gametes and to the embryo.

The effect seminal fluids have on metabolic traits of the progeny have been demonstrated in several works. One work excised the seminal vesicles of male mice and noted an increase in body weight and total fat mass in adult progeny compared to control [113]. Adult male progeny also demonstrated impaired glucose clearance, increased plasma leptin levels and elevated blood pressure. The authors also noted reduced fecundity of excised males as a result of a reduced embryo implantation rate in mated females. In addition, implanted embryos demonstrated placental hypertrophy. Using IVF and embryo transfer experiments the authors demonstrated that the exposure of the female's reproductive track to seminal fluids effect very early embryonic development by inducing Csf2, Lif, and Il6 expression in the oviduct.

Though the precise extent to which seminal fluids are influenced by environmental cues is unclear, there are evidence to suggest this is the case. In humans temporal fluctuations in the levels of various growth factors and cytokines were noted in seminal fluids [114, 115]. In mice, high fat diet (HFD) induced obesity led to higher levels of insulin and leptin and decreased levels of estradiol in the seminal vesicle fluid [116]. The seminal microbiome was also demonstrated to change in response to a HFD [117].

Taken together, cell signaling may induce positive feedback loops to sustain signaling even in the absence of the initial signaling trigger. When signaling leads to chromatin based changes in gene expression these may persist through mitosis to affect phenotype of future progeny. Signaling may be triggered by various environmental factors such as with EDCs, or endocrine signaling factors present in the microenvironment of the developing germ cells or of the early embryo.

12.4 Implications on Health and Well-being

12.4.1 Environmental Health and Protection

Overall, the literature regarding epigenetic inheritance seems to suggest that humanity may be facing a greater '**environmental health**' concern than is currently perceived [118]. Drastic changes in life style and dietary habits during the last century combined with a continued widespread use of antibiotics and environmental pollution, may affect health of future generations in unforeseen ways through epigenetic mechanisms (Fig. 12.2).

The majority of studies into the impact of various environmental factors on human health focus on the effects observed and measured in the exposed individuals, in effect disregarding any inter- or transgenerational effects. This methodological choice probably stems from the lack of any viable options to study multi-generational impacts in human populations other than with association studies (for example [119–121]). Regardless the implications are clear; First, the scientific community may be blind to developing health risks and trends. One current example is obesity and diabetes which have grown to epidemic proportions over the last decades [122, 123]. Though the rise in obesity rates in the adult population can be partially attributed to environmental and life style changes, the alarming rates of childhood obesity and the emergence of obesity and metabolic disorders earlier on in life cannot be attributed to genetic and environmental factors alone. It is now evident that epigenetic inheritance mechanisms are at least partially driving this epidemiological spread. Indeed, one model suggests that exposure to environmental '**obesogens**' can predispose individuals and their progeny to metabolic syndromes and obesity [124, 125], with multiple animal experiments supporting this epidemiological model (see [125] for an extensive review). Similarly, future health trends in coming generations may be dictated by the exposure of present day individuals to current environmental factors. A second implication stemming from the limitations of current research into epigenetic inheritance in humans is that there may be a systemic and widespread underestimation of the health risks imposed by various factors, as these risks may reveal themselves progressively in future generations.

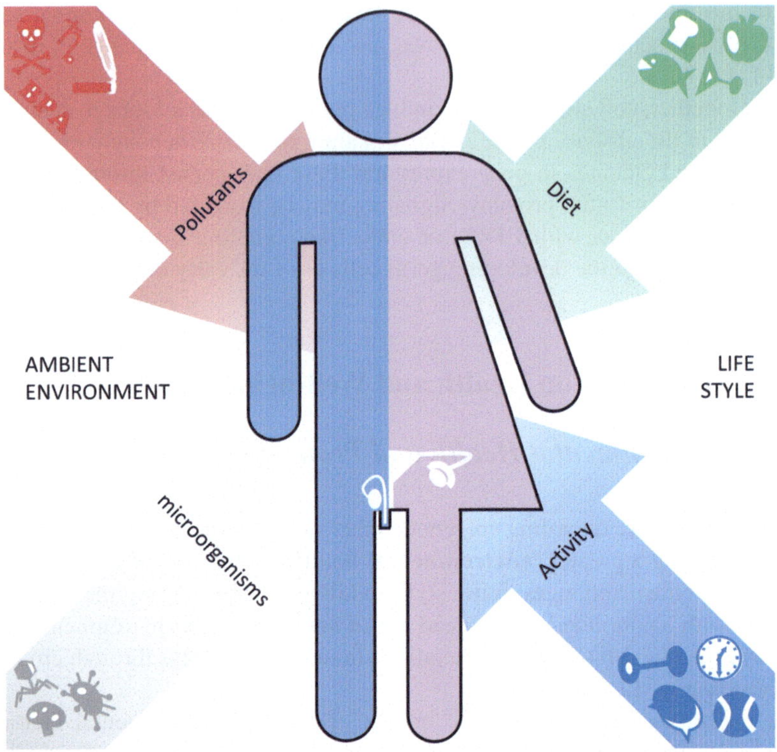

Fig. 12.2 Changes in ambient environment and life style can lead to altered disease susceptibility of future generations**.** Multiple environmental and life style factors can induce maternal and paternal inter/transgenerational epigenetic inheritance through epigenetic modulation of the germ cells. Additional changes in seminal fluid composition and / or changes in the oviduct and uterine microenvironment can effect early embryonic development to induce inter/ transgenerational inheritance. Factors that have been demonstrate to induce inter/transgenerational epigenetic inheritance in various species include pollutants, microorganisms, nutritional availability and multiple activities including; physical activity, circadian activity, stress levels, and social interactions. In humans these can be divided to life style factors and ambient environment factors

In addition to environmental health, works on the transgenerational effects of various environmental pollutants including EDC suggest that multi-generational effects observed in the lab may have devastating consequences on the ecosystem and by extension on human population. Specifically, observations on transgenerational effects on fertility, fecundity and progeny survival rates may necessitates new estimations for sustainable resource management and development of additional strategies for biodiversity conservation.

Collectively, a greater focus within the field of environmental health and protection should be placed on impacts and repercussions that extend beyond a single generation with a focus on the effects of various environmental factors on an offspring's disease risk and susceptibility. Novel insights into generational implications of environmental factors can help predict future disease trends and risks and to better assess the health and environmental risks imposed by various pollutants. Together with a better understanding of the underlying molecular mechanisms of transgenerational inheritance, novel intervention strategies can be devised in environmental health and environmental protection to protect both humans and the ecosystem.

12.4.2 Pharmacovigilance

As with environmental health, a rethink is needed concerning the methodologies of pharmacological safety studies and more broadly within the field of **pharmacovigilance**. In relation to reproduction safety, pre-clinical and clinical studies monitor adverse drug effects on the fertility of exposed individuals, both male and female. Pre-clinical animal studies (and those clinical studies that are relevant to pregnant women) also monitor the health of progeny from exposed pregnant females, whilst pregnancy exposure registries track the health of new born babies of women that use pharmacological drugs during pregnancy. These studies focus on the **teratogenic** potential of pharmaceutical drugs during in-utero exposure and on adverse effects through breast feeding exposure, and normally monitor the offspring health for a relative short period of time. One conclusion drawn from the current drug safety guidelines is that these seem to ignore and neglect any inter/transgenerational and/or paternal inheritance that may adversely affect human health following drug use. The new findings put forth throughout the book concerning epigenetic inheritance may necessitate a revisal of the guidelines for drug safety studies.

As for phase IV clinical studies and routine pharmacovigilance measures, current technological and legislative limitations hinder the assessment of inter- and transgenerational effects of pharmacological compounds in human populations. These limitations may be better addressed in the future with the tools and databases that are being developed through the **digital health** revolution. Indeed, proper acquisition, management and storage of health records together with increasing use of digital health monitoring can aid future pharmacovigilance studies and in particular in matters related to epigenetic inheritance. These same tools would be indispensable for basic research into the scope and mechanisms of epigenetic inheritance in human populations.

12.4.3 ART and Preconception Care

Studies in humans have already demonstrated a higher rate of children born with Beckwith-Wiedemann syndrome following '**in-vitro fertilization**' (IVF) treatments, with associated hypomethylation of the differentially methylated region of the imprinted genes LIT1 and H19 [126]. Additional studies revealed associations between various assisted reproduction technologies (ARTs) to **imprinting disorders** including Angelman syndrome, Prader-Willi syndrome and Silver-Russell syndrome [127]. These correlations suggest ARTs can effect epigenetic imprinting and lead to developmental defects.

In mice, IVF was demonstrated to impair health [128] and metabolism [129] and lead to higher fasting glucose levels, impaired glucose tolerance and shortened life span compared to natural conception mice. In addition, liver, adipose tissue, muscle and pancreatic islets demonstrate altered gene expression [130]. These studies suggest that ARTs can negatively affect human health in a manner extending beyond developmental defects. In light of these finding and our growing knowledge on the effects of the early preimplantation embryonic microenvironment on embryonic development and future disease risk, new IVF protocols need to be established to assure minimizing disease risk.

Another interesting concept that stems from the findings described throughout the book concerns preconception care during family planning [131], with better practices and measures that might be taken to assure the future health of the population and to minimize disease risk. Whilst today preconception care is focused on women with the aim of modifying biochemical, behavioral and social risks to maximize health of new born children, it is becoming clear that men have a greater contribution than previously thought of to their child's health. As sperm and seminal fluids carry multiple epigenetic modifications and signaling factors that can potentially shape and effect early embryonic development and future disease risks, men should also be advised whilst planning a family on proper measures to reduce disease risk of prospective children.

12.5 Outstanding Questions

How are epigenetic marks, specifically histone modifications and variants, maintained through mitotic and meiotic cell divisions?

What are the molecular mechanisms that mediate conversion of environmental signals to epigenetically encoded information in the gametes?

How relevant is transgenerational inheritance to humans?

What are the disease risks that can be environmentally modulated in a transgenerationally?

What is the function of non-canonical DNA and RNA bases and how common are they?

Glossary

The **Central dogma** was proposed by Francis Crick in the 50's and stated that information generally flows from DNA to proteins via intermittent RNA molecules but cannot flow back from proteins to DNA.

Cretzfeldt-Jakob disease (CJD) is a fatal brain disease caused by a prion particle of the PrP protein encouded by the PRNP gene. Can appear as familial, sporadic or acquired disease.

Digital health aims at improving healthcare and allow for a more personalized care through the use of information and communication technologies.

Dysbiosis is a shift in the relative abundances of different microbial taxa comprising the microbiota of a sick organisms compared to taxa observed in healthy organisms.

Endocrine disrupting chemicals are synthetic chemicals that can interfere with the endocrine signaling system. Can cause infertility and lead to the development of cancer, metabolic disorders, birth defects and developmental disorders.

Environmental health is defined as a branch of public health that addresses the direct pathological impacts of chemicals, radiation and some biological agents, and the effects on health and wellbeing of the broad physical, psychological, social and aesthetic environment, which includes housing, urban development, land use and transport.

Environmental protection aims to conserve natural resources, biodiversity and the ecosystem through the implementation of various policies and procedures that protect the environment for the benefit of humans and the ecosystem.

Epialleles are genetically identical but variably expressed alleles. This variable expression is due to epigenetic modifications that are established during early development.

Epigenetic inheritance is defined as mitotically and/or meiotically heritable changes in gene expression that do not result from changes in the DNA sequence. Several different phenomena observed across different kingdoms can be described as epigenetic inheritance.

Gemmules are hypothetical particles suggested by Charles Darwin as part of his Pangenesis theory. These particles are emitted from every tissue and migrate to germ cells where they mediate inheritance. These particles were suggested to contribute to the development of the embryo and for phenotypic traits of progeny. In effect Darwin suggested that acquired somatic traits can be carried on to the offspring via the gemmules.

Germ-free mice are mice grown in sterile conditions which allow to maintain the mice void of any microbiota.

Gnotobiotic mice are mice with a defined set of bacteria and microorganisms.

The **Hypothalamic-pituitary-adrenal axis** is a major neuroendocrine system that controls stress reactions and regulates multiple processes including digestion, energy storage and expenditure, immunity, mood, sexuality and more. Comprised from the intricate interactions between the hypothalamus, the pituitary gland, the adrenal glands and their respective endocrine signaling factors.

In-vitro fertilization is the process of oocyte fertilization with sperm outside the body and in cell culture.

Imprinting disorders are a collection of congenital disorders that stem from miss-regulation of imprinted loci, which are normally expressed from only one parental allele (maternal or paternal). This miss-regulation can stem from deletions, sense mutations and hyper/hypo-methylation.

Jasmonates are lipid based plant hormones that regulate a wide range of processes including growth, photosynthesis and reproductive development.

Kairomones are signaling molecules that mediate interspecies interactions in a way that benefits the recipient of the signal rather than the secreting party.

Kuru is another fatal neurodegenerative disorder that was common among the Fore people of New- Guinea. As with CJD, a prion strain of the PrP protein caused the disease which was transmitted within the tribe through ritualistic cannibalism.

The **Maternal to zygotic transition** is an early stage of embryonic development during which the zygotic genome because transcriptionally active and maternaly deposited transcripts are degraded.

The **Microbiota** is the collection of bacteria, fungi and viruses that live as commensal, symbiotic or pathogenic microorganisms of multicellular organisms.

Obesogens are chemicals that disrupt normal development and homeostatic controls of adipogenesis, lipid metabolism and energy balance in such a way that can induce obesity.

Pangenesis is Charles Darwin's theory of inheritance published in his book 'The Variation of Animals and Plants under Domestication' published at 1868. Pangenesis complemented Darwin's theory of natural selection in that in proposed a mechanisms for inheritance and development namely Gemmules.

Paramutation involves trans-communication between homologous sequences. This communication establishes heritable changes in chromatin structure that often correlate with alterations in DNA methylation. Loci implicated in paramutation can have three types of allele: alleles that do not participate in paramutation are termed neutral or non-paramutagenic; sensitive alleles are termed paramutable; and alleles that induce the change are paramutagenic.

Pharmacovigilance is the study field related to detection, assessment and prevention of adverse effects caused about by pharmaceutical products with the overall aim of minimizing these risks.

Phytohormones are plant hormones that can control multiple aspects of development starting with reproduction and embryogenesis, through pathogen defense and stress tolerance and more. Phytohormones can be secreted from all cells of the plant and act locally or distally.

Prions are misfolded proteins that have the surprising capacity to self-propagate in an infectious manner. An interaction between a native protein with the misfolded prion protein induces misfolding of the former into a prion particle that can propagate itself further.

Specific pathogen free mice are mice grown free of specific pathogenic infections and certified to be kept under these conditions.

Teratogenic compounds are compounds that cause developmental malformations, a well-known example is thalidomide.

Weismann barrier is a central concept proposed by August Weismann which maintains that germ cell are strictly separated from somatic cells and can only arise from an immortal germ cell lineage. This concept requires genetic information to flow in a unidirectional manner from germ cells to soma, basically excluding any type of soma to germ line information transfer.

References

1. Bateson W, Pellew C. On the genetics of "rogues" among culinary peas (Pisum sativum). J Genet. 1915;5(1):13–36.
2. Brink RA. A genetic change associated with the R locus in maize which is directed and potentially reversible. Genetics. 1956;41(6):872–89.
3. Coe EH. A regular and continuing conversion-type phenomenon at the B locus in maize. Proc Natl Acad Sci U S A. 1959;45(6):828–32.
4. Hollick JB, Patterson GI, Coe EH Jr, Cone KC, Chandler VL. Allelic interactions heritably alter the activity of a metastable maize pl allele. Genetics. 1995;141(2):709–19.
5. Pilu R, Panzeri D, Cassani E, Cerino Badone F, Landoni M, Nielsen E. A paramutation phenomenon is involved in the genetics of maize low phytic acid1-241 (lpa1-241) trait. Heredity (Edinb). 2009;102(3):236–45.
6. Sidorenko LV, Peterson T. Transgene-induced silencing identifies sequences involved in the establishment of paramutation of the maize p1 gene. Plant Cell. 2001;13(2):319–35.
7. Walker EL, Robbins TP, Bureau TE, Kermicle J, Dellaporta SL. Transposon-mediated chromosomal rearrangements and gene duplications in the formation of the maize R-r complex. EMBO J. 1995;14(10):2350–63.
8. Panavas T, Weir J, Walker EL. The structure and paramutagenicity of the R-marbled haplotype of Zea mays. Genetics. 1999;153(2):979–91.
9. Herman H, Lu M, Anggraini M, Sikora A, Chang Y, Yoon BJ, et al. Trans allele methylation and paramutation-like effects in mice. Nat Genet. 2003;34(2):199–202.
10. The Pea (*Pisum sativum* L.). Rogue paramutation is accompanied by alterations in the methylation pattern of specific genomic sequences. Epigenomes. 2017;1(1):6.
11. Duhl DM, Vrieling H, Miller KA, Wolff GL, Barsh GS. Neomorphic agouti mutations in obese yellow mice. Nat Genet. 1994;8(1):59–65.
12. Morgan HD, Sutherland HG, Martin DI, Whitelaw E. Epigenetic inheritance at the agouti locus in the mouse. Nat Genet. 1999;23(3):314–8.
13. Zeng L, Fagotto F, Zhang T, Hsu W, Vasicek TJ, Perry WL 3rd, et al. The mouse fused locus encodes Axin, an inhibitor of the Wnt signaling pathway that regulates embryonic axis formation. Cell. 1997;90(1):181–92.
14. Reed SC. The inheritance and expression of fused, a new mutation in the house mouse. Genetics. 1937;22(1):1–13.

15. Rakyan VK, Chong S, Champ ME, Cuthbert PC, Morgan HD, Luu KV, et al. Transgenerational inheritance of epigenetic states at the murine Axin(Fu) allele occurs after maternal and paternal transmission. Proc Natl Acad Sci U S A. 2003;100(5):2538–43.
16. Derkatch IL, Chernoff YO, Kushnirov VV, Inge-Vechtomov SG, Liebman SW. Genesis and variability of [PSI] prion factors in Saccharomyces cerevisiae. Genetics. 1996;144(4):1375–86.
17. Hackett JA, Zylicz JJ, Surani MA. Parallel mechanisms of epigenetic reprogramming in the germline. Trends Genet. 2012;28(4):164–74.
18. Seisenberger S, Andrews S, Krueger F, Arand J, Walter J, Santos F, et al. The dynamics of genome-wide DNA methylation reprogramming in mouse primordial germ cells. Mol Cell. 2012;48(6):849–62.
19. Rassoulzadegan M, Grandjean V, Gounon P, Vincent S, Gillot I, Cuzin F. RNA-mediated non-mendelian inheritance of an epigenetic change in the mouse. Nature. 2006;441(7092):469–74.
20. Wagner KD, Wagner N, Ghanbarian H, Grandjean V, Gounon P, Cuzin F, et al. RNA induction and inheritance of epigenetic cardiac hypertrophy in the mouse. Dev Cell. 2008;14(6):962–9.
21. Grandjean V, Gounon P, Wagner N, Martin L, Wagner KD, Bernex F, et al. The miR-124-Sox9 paramutation: RNA-mediated epigenetic control of embryonic and adult growth. Development. 2009;136(21):3647–55.
22. Rechavi O, Minevich G, Hobert O. Transgenerational inheritance of an acquired small RNA-based antiviral response in C. elegans. Cell. 2011;147(6):1248–56.
23. Stein P, Svoboda P, Anger M, Schultz RM. RNAi: mammalian oocytes do it without RNA-dependent RNA polymerase. RNA. 2003;9(2):187–92.
24. Malone CD, Hannon GJ. Small RNAs as guardians of the genome. Cell. 2009;136(4):656–68.
25. Aravin AA, Sachidanandam R, Bourc'his D, Schaefer C, Pezic D, Toth KF, et al. A piRNA pathway primed by individual transposons is linked to de novo DNA methylation in mice. Mol Cell. 2008;31(6):785–99.
26. Kuramochi-Miyagawa S, Watanabe T, Gotoh K, Totoki Y, Toyoda A, Ikawa M, et al. DNA methylation of retrotransposon genes is regulated by Piwi family members MILI and MIWI2 in murine fetal testes. Genes Dev. 2008;22(7):908–17.
27. Reuter M, Berninger P, Chuma S, Shah H, Hosokawa M, Funaya C, et al. Miwi catalysis is required for piRNA amplification-independent LINE1 transposon silencing. Nature. 2011;480(7376):264–7.
28. Rouget C, Papin C, Boureux A, Meunier AC, Franco B, Robine N, et al. Maternal mRNA deadenylation and decay by the piRNA pathway in the early Drosophila embryo. Nature. 2010;467(7319):1128–32.
29. Luteijn MJ, Ketting RF. PIWI-interacting RNAs: from generation to transgenerational epigenetics. Nat Rev Genet. 2013;14(8):523–34.
30. Popp C, Dean W, Feng S, Cokus SJ, Andrews S, Pellegrini M, et al. Genome-wide erasure of DNA methylation in mouse primordial germ cells is affected by AID deficiency. Nature. 2010;463(7284):1101–5.
31. Sasaki H, Matsui Y. Epigenetic events in mammalian germ-cell development: reprogramming and beyond. Nat Rev Genet. 2008;9(2):129–40.
32. Balhorn R. The protamine family of sperm nuclear proteins. Genome Biol. 2007;8(9):227.
33. Hammoud SS, Nix DA, Zhang H, Purwar J, Carrell DT, Cairns BR. Distinctive chromatin in human sperm packages genes for embryo development. Nature. 2009;460(7254):473–8.
34. Kimmins S, Sassone-Corsi P. Chromatin remodelling and epigenetic features of germ cells. Nature. 2005;434(7033):583–9.
35. De La Fuente R, Viveiros MM, Wigglesworth K, Eppig JJ. ATRX, a member of the SNF2 family of helicase/ATPases, is required for chromosome alignment and meiotic spindle organization in metaphase II stage mouse oocytes. Dev Biol. 2004;272(1):1–14.
36. Teranishi T, Tanaka M, Kimoto S, Ono Y, Miyakoshi K, Kono T, et al. Rapid replacement of somatic linker histones with the oocyte-specific linker histone H1foo in nuclear transfer. Dev Biol. 2004;266(1):76–86.
37. Irvine DV, Zaratiegui M, Tolia NH, Goto DB, Chitwood DH, Vaughn MW, et al. Argonaute slicing is required for heterochromatic silencing and spreading. Science. 2006;313(5790):1134–7.

38. Volpe TA, Kidner C, Hall IM, Teng G, Grewal SI, Martienssen RA. Regulation of heterochromatic silencing and histone H3 lysine-9 methylation by RNAi. Science. 2002;297(5588):1833–7.
39. Buker SM, Iida T, Buhler M, Villen J, Gygi SP, Nakayama J, et al. Two different Argonaute complexes are required for siRNA generation and heterochromatin assembly in fission yeast. Nat Struct Mol Biol. 2007;14(3):200–7.
40. Wassenegger M, Heimes S, Riedel L, Sanger HL. RNA-directed de novo methylation of genomic sequences in plants. Cell. 1994;76(3):567–76.
41. Mette MF, Aufsatz W, van der Winden J, Matzke MA, Matzke AJ. Transcriptional silencing and promoter methylation triggered by double-stranded RNA. EMBO J. 2000;19(19):5194–201.
42. Gu SG, Pak J, Guang S, Maniar JM, Kennedy S, Fire A. Amplification of siRNA in Caenorhabditis elegans generates a transgenerational sequence-targeted histone H3 lysine 9 methylation footprint. Nat Genet. 2012;44(2):157–64.
43. Burton NO, Burkhart KB, Kennedy S. Nuclear RNAi maintains heritable gene silencing in Caenorhabditis elegans. Proc Natl Acad Sci U S A. 2011;108(49):19683–8.
44. Watanabe T, Tomizawa S, Mitsuya K, Totoki Y, Yamamoto Y, Kuramochi-Miyagawa S, et al. Role for piRNAs and noncoding RNA in de novo DNA methylation of the imprinted mouse Rasgrf1 locus. Science. 2011;332(6031):848–52.
45. Ashe A, Sapetschnig A, Weick EM, Mitchell J, Bagijn MP, Cording AC, et al. piRNAs can trigger a multigenerational epigenetic memory in the germline of C. elegans. Cell. 2012;150(1):88–99.
46. Zhai J, Liu J, Liu B, Li P, Meyers BC, Chen X, et al. Small RNA-directed epigenetic natural variation in Arabidopsis thaliana. PLoS Genet. 2008;4(4):e1000056.
47. Katz DJ, Edwards TM, Reinke V, Kelly WG. A C. elegans LSD1 demethylase contributes to germline immortality by reprogramming epigenetic memory. Cell. 2009;137(2):308–20.
48. Klosin A, Casas E, Hidalgo-Carcedo C, Vavouri T, Lehner B. Transgenerational transmission of environmental information in C. elegans. Science. 2017;356(6335):320–3.
49. Camacho J, Truong L, Kurt Z, Chen YW, Morselli M, Gutierrez G, et al. The memory of environmental chemical exposure in C. elegans is dependent on the Jumonji demethylases jmjd-2 and jmjd-3/utx-1. Cell Rep. 2018;23(8):2392–404.
50. Siklenka K, Erkek S, Godmann M, Lambrot R, McGraw S, Lafleur C, et al. Disruption of histone methylation in developing sperm impairs offspring health transgenerationally. Science. 2015;350(6261):aab2006.
51. Javurek AB, Spollen WG, Ali AM, Johnson SA, Lubahn DB, Bivens NJ, et al. Discovery of a novel seminal fluid microbiome and influence of estrogen receptor alpha genetic status. Sci Rep. 2016;6:23027.
52. Human Microbiome Project C. Structure, function and diversity of the healthy human microbiome. Nature. 2012;486(7402):207–14.
53. Qin J, Li R, Raes J, Arumugam M, Burgdorf KS, Manichanh C, et al. A human gut microbial gene catalogue established by metagenomic sequencing. Nature. 2010;464(7285):59–65.
54. Koenig JE, Spor A, Scalfone N, Fricker AD, Stombaugh J, Knight R, et al. Succession of microbial consortia in the developing infant gut microbiome. Proc Natl Acad Sci U S A. 2011;108(Suppl 1):4578–85.
55. Chung H, Pamp SJ, Hill JA, Surana NK, Edelman SM, Troy EB, et al. Gut immune maturation depends on colonization with a host-specific microbiota. Cell. 2012;149(7):1578–93.
56. Diaz Heijtz R, Wang S, Anuar F, Qian Y, Bjorkholm B, Samuelsson A, et al. Normal gut microbiota modulates brain development and behavior. Proc Natl Acad Sci U S A. 2011;108(7):3047–52.
57. Sudo N, Chida Y, Aiba Y, Sonoda J, Oyama N, Yu XN, et al. Postnatal microbial colonization programs the hypothalamic-pituitary-adrenal system for stress response in mice. J Physiol. 2004;558(Pt 1):263–75.
58. Bercik P, Denou E, Collins J, Jackson W, Lu J, Jury J, et al. The intestinal microbiota affect central levels of brain-derived neurotropic factor and behavior in mice. Gastroenterology. 2011;141(2):599–609.. e1-3

59. Stern S, Fridmann-Sirkis Y, Braun E, Soen Y. Epigenetically heritable alteration of fly development in response to toxic challenge. Cell Rep. 2012;1(5):528–42.
60. Fridmann-Sirkis Y, Stern S, Elgart M, Galili M, Zeisel A, Shental N, et al. Delayed development induced by toxicity to the host can be inherited by a bacterial-dependent, transgenerational effect. Front Genet. 2014;5:27.
61. Sartor RB. Microbial influences in inflammatory bowel diseases. Gastroenterology. 2008;134(2):577–94.
62. Vijay-Kumar M, Aitken JD, Carvalho FA, Cullender TC, Mwangi S, Srinivasan S, et al. Metabolic syndrome and altered gut microbiota in mice lacking toll-like receptor 5. Science. 2010;328(5975):228–31.
63. Larsen N, Vogensen FK, van den Berg FW, Nielsen DS, Andreasen AS, Pedersen BK, et al. Gut microbiota in human adults with type 2 diabetes differs from non-diabetic adults. PLoS One. 2010;5(2):e9085.
64. Ridaura VK, Faith JJ, Rey FE, Cheng J, Duncan AE, Kau AL, et al. Gut microbiota from twins discordant for obesity modulate metabolism in mice. Science. 2013;341(6150):1241214.
65. Leclercq S, Mian FM, Stanisz AM, Bindels LB, Cambier E, Ben-Amram H, et al. Low-dose penicillin in early life induces long-term changes in murine gut microbiota, brain cytokines and behavior. Nat Commun. 2017;8:15062.
66. Neuman H, Forsythe P, Uzan A, Avni O, Koren O. Antibiotics in early life: dysbiosis and the damage done. FEMS Microbiol Rev. 2018;42(4):489–99.
67. Werren JH, Baldo L, Clark ME. Wolbachia: master manipulators of invertebrate biology. Nat Rev Microbiol. 2008;6(10):741–51.
68. Elgart M, Stern S, Salton O, Gnainsky Y, Heifetz Y, Soen Y. Impact of gut microbiota on the fly's germ line. Nat Commun. 2016;7:11280.
69. Devanapally S, Ravikumar S, Jose AM. Double-stranded RNA Made in C. elegans neurons can enter the germline and cause transgenerational gene silencing. Proc Natl Acad Sci U S A. 2015;112(7):2133–8.
70. Sharma U, Conine CC, Shea JM, Boskovic A, Derr AG, Bing XY, et al. Biogenesis and function of tRNA fragments during sperm maturation and fertilization in mammals. Science. 2016;351(6271):391–6.
71. El-Mogy M, Lam B, Haj-Ahmad TA, McGowan S, Yu D, Nosal L, et al. Diversity and signature of small RNA in different bodily fluids using next generation sequencing. BMC Genomics. 2018;19(1):408.
72. Weber JA, Baxter DH, Zhang S, Huang DY, Huang KH, Lee MJ, et al. The microRNA spectrum in 12 body fluids. Clin Chem. 2010;56(11):1733–41.
73. Carell T, Kurz MQ, Muller M, Rossa M, Spada F. Non-canonical bases in the genome: the regulatory information layer in DNA. Angew Chem Int Ed Engl. 2018;57(16):4296–312.
74. Wu X, Zhang Y. TET-mediated active DNA demethylation: mechanism, function and beyond. Nat Rev Genet. 2017;18(9):517–34.
75. Wang L, Zhou Y, Xu L, Xiao R, Lu X, Chen L, et al. Molecular basis for 5-carboxycytosine recognition by RNA polymerase II elongation complex. Nature. 2015;523(7562):621–5.
76. Raiber EA, Murat P, Chirgadze DY, Beraldi D, Luisi BF, Balasubramanian S. 5-Formylcytosine alters the structure of the DNA double helix. Nat Struct Mol Biol. 2015;22(1):44–9.
77. Koziol MJ, Bradshaw CR, Allen GE, Costa ASH, Frezza C, Gurdon JB. Identification of methylated deoxyadenosines in vertebrates reveals diversity in DNA modifications. Nat Struct Mol Biol. 2016;23(1):24–30.
78. Lu M, Campbell JL, Boye E, Kleckner N. SeqA: a negative modulator of replication initiation in E. coli. Cell. 1994;77(3):413–26.
79. Laengle-Rouault F, Maenhaut-Michel G, Radman M. GATC sequence and mismatch repair in Escherichia coli. EMBO J. 1986;5(8):2009–13.
80. Roberts D, Hoopes BC, McClure WR, Kleckner N. IS10 transposition IS regulated by DNA adenine methylation. Cell. 1985;43(1):117–30.
81. Yao B, Cheng Y, Wang Z, Li Y, Chen L, Huang L, et al. DNA N6-methyladenine is dynamically regulated in the mouse brain following environmental stress. Nat Commun. 2017;8(1):1122.

82. Boccaletto P, Machnicka MA, Purta E, Piatkowski P, Baginski B, Wirecki TK, et al. MODOMICS: a database of RNA modification pathways. 2017 update. Nucleic Acids Res. 2018;46(D1):D303–D7.
83. Kiani J, Grandjean V, Liebers R, Tuorto F, Ghanbarian H, Lyko F, et al. RNA-mediated epigenetic heredity requires the cytosine methyltransferase Dnmt2. PLoS Genet. 2013;9(5):e1003498.
84. Tuorto F, Liebers R, Musch T, Schaefer M, Hofmann S, Kellner S, et al. RNA cytosine methylation by Dnmt2 and NSun2 promotes tRNA stability and protein synthesis. Nat Struct Mol Biol. 2012;19(9):900–5.
85. Zhang Y, Zhang X, Shi J, Tuorto F, Li X, Liu Y, et al. Dnmt2 mediates intergenerational transmission of paternally acquired metabolic disorders through sperm small non-coding RNAs. Nat Cell Biol. 2018;20(5):535–40.
86. Collinge J. Mammalian prions and their wider relevance in neurodegenerative diseases. Nature. 2016;539(7628):217–26.
87. Prusiner SB, Scott MR, DeArmond SJ, Cohen FE. Prion protein biology. Cell. 1998;93(3):337–48.
88. Lacroute F. Non-Mendelian mutation allowing ureidosuccinic acid uptake in yeast. J Bacteriol. 1971;106(2):519–22.
89. Wickner RB. [URE3] as an altered URE2 protein: evidence for a prion analog in Saccharomyces cerevisiae. Science. 1994;264(5158):566–9.
90. Derkatch IL, Bradley ME, Hong JY, Liebman SW. Prions affect the appearance of other prions: the story of [PIN(+)]. Cell. 2001;106(2):171–82.
91. Sondheimer N, Lindquist S. Rnq1: an epigenetic modifier of protein function in yeast. Mol Cell. 2000;5(1):163–72.
92. Ghaemmaghami S. Biology and Genetics of PrP Prion Strains. Cold Spring Harb Perspect Med. 2017;7(8).
93. Chakrabortee S, Byers JS, Jones S, Garcia DM, Bhullar B, Chang A, et al. Intrinsically disordered proteins drive emergence and inheritance of biological traits. Cell. 2016;167(2):369–81.. e12
94. Beisson J, Sonneborn TM. Cytoplasmic inheritance of the Organization of the Cell Cortex in Paramecium Aurelia. Proc Natl Acad Sci U S A. 1965;53:275–82.
95. Hirt RP, Horner DS. Organelles, genomes, and eukaryote phylogeny: an evolutionary synthesis in the age of genomics. Boca Raton: CRC Press; 2004. 388 p. p.
96. Cheng MY, Hartl FU, Horwich AL. The mitochondrial chaperonin hsp60 is required for its own assembly. Nature. 1990;348(6300):455–8.
97. Rasmann S, De Vos M, Casteel CL, Tian D, Halitschke R, Sun JY, et al. Herbivory in the previous generation primes plants for enhanced insect resistance. Plant Physiol. 2012;158(2):854–63.
98. Agrawal AA, Laforsch C, Tollrian R. Transgenerational induction of defences in animals and plants. Nature. 1999;401:60–3.
99. Anway MD, Cupp AS, Uzumcu M, Skinner MK. Epigenetic transgenerational actions of endocrine disruptors and male fertility. Science. 2005;308(5727):1466–9.
100. Bhandari RK, vom Saal FS, Tillitt DE. Transgenerational effects from early developmental exposures to bisphenol A or 17alpha-ethinylestradiol in medaka, Oryzias latipes. Sci Rep. 2015;5:9303.
101. Manikkam M, Guerrero-Bosagna C, Tracey R, Haque MM, Skinner MK. Transgenerational actions of environmental compounds on reproductive disease and identification of epigenetic biomarkers of ancestral exposures. PLoS One. 2012;7(2):e31901.
102. Dolinoy DC, Huang D, Jirtle RL. Maternal nutrient supplementation counteracts bisphenol A-induced DNA hypomethylation in early development. Proc Natl Acad Sci U S A. 2007;104(32):13056–61.
103. Waterland RA, Dolinoy DC, Lin JR, Smith CA, Shi X, Tahiliani KG. Maternal methyl supplements increase offspring DNA methylation at Axin fused. Genesis. 2006;44(9):401–6.

104. Waterland RA, Jirtle RL. Transposable elements: targets for early nutritional effects on epigenetic gene regulation. Mol Cell Biol. 2003;23(15):5293–300.
105. Dominguez-Salas P, Moore SE, Baker MS, Bergen AW, Cox SE, Dyer RA, et al. Maternal nutrition at conception modulates DNA methylation of human metastable epialleles. Nat Commun. 2014;5:3746.
106. Leese HJ, Hugentobler SA, Gray SM, Morris DG, Sturmey RG, Whitear SL, et al. Female reproductive tract fluids: composition, mechanism of formation and potential role in the developmental origins of health and disease. Reprod Fertil Dev. 2008;20(1):1–8.
107. Hochberg Z, Feil R, Constancia M, Fraga M, Junien C, Carel JC, et al. Child health, developmental plasticity, and epigenetic programming. Endocr Rev. 2011;32(2):159–224.
108. Jodar M, Sendler E, Krawetz SA. The protein and transcript profiles of human semen. Cell Tissue Res. 2016;363(1):85–96.
109. Poiani A. Complexity of seminal fluid: a review. Behav Ecol Sociobiol. 2006;60(3):289–310.
110. Sharkey DJ, Tremellen KP, Jasper MJ, Gemzell-Danielsson K, Robertson SA. Seminal fluid induces leukocyte recruitment and cytokine and chemokine mRNA expression in the human cervix after coitus. J Immunol. 2012;188(5):2445–54.
111. Tremellen KP, Seamark RF, Robertson SA. Seminal transforming growth factor beta1 stimulates granulocyte-macrophage colony-stimulating factor production and inflammatory cell recruitment in the murine uterus. Biol Reprod. 1998;58(5):1217–25.
112. Robertson SA. Seminal plasma and male factor signalling in the female reproductive tract. Cell Tissue Res. 2005;322(1):43–52.
113. Bromfield JJ, Schjenken JE, Chin PY, Care AS, Jasper MJ, Robertson SA. Maternal tract factors contribute to paternal seminal fluid impact on metabolic phenotype in offspring. Proc Natl Acad Sci U S A. 2014;111(6):2200–5.
114. Sharkey DJ, Tremellen KP, Briggs NE, Dekker GA, Robertson SA. Seminal plasma transforming growth factor-beta, activin a and follistatin fluctuate within men over time. Hum Reprod. 2016;31(10):2183–91.
115. Sharkey DJ, Tremellen KP, Briggs NE, Dekker GA, Robertson SA. Seminal plasma proinflammatory cytokines interferon-gamma (IFNG) and C-X-C motif chemokine ligand 8 (CXCL8) fluctuate over time within men. Hum Reprod. 2017;32(7):1373–81.
116. Binder NK, Sheedy JR, Hannan NJ, Gardner DK. Male obesity is associated with changed spermatozoa Cox4i1 mRNA level and altered seminal vesicle fluid composition in a mouse model. Mol Hum Reprod. 2015;21(5):424–34.
117. Javurek AB, Spollen WG, Johnson SA, Bivens NJ, Bromert KH, Givan SA, et al. Consumption of a high-fat diet alters the seminal fluid and gut microbiomes in male mice. Reprod Fertil Dev. 2017;29(8):1602–12.
118. Prüss-Üstün A, Wolf J, Corvalán C, Bos R, Neira M, World Health Organization, et al. Preventing disease through healthy environments: a global assessment of the burden of disease from environmental risks. 2nd ed. Geneva, Switzerland: World Health Organization; 2016. xxiii. p. 147.
119. Bygren LO, Kaati G, Edvinsson S. Longevity determined by paternal ancestors' nutrition during their slow growth period. Acta Biotheor. 2001;49(1):53–9.
120. Kaati G, Bygren LO, Edvinsson S. Cardiovascular and diabetes mortality determined by nutrition during parents' and grandparents' slow growth period. Eur J Hum Genet. 2002;10(11):682–8.
121. Pembrey ME, Bygren LO, Kaati G, Edvinsson S, Northstone K, Sjostrom M, et al. Sex-specific, male-line transgenerational responses in humans. Eur J Hum Genet. 2006;14(2):159–66.
122. Wang Y, Beydoun MA. The obesity epidemic in the United States--gender, age, socioeconomic, racial/ethnic, and geographic characteristics: a systematic review and meta-regression analysis. Epidemiol Rev. 2007;29:6–28.
123. Wang Y, Lobstein T. Worldwide trends in childhood overweight and obesity. Int J Pediatr Obes. 2006;1(1):11–25.
124. Grun F, Blumberg B. Environmental obesogens: organotins and endocrine disruption via nuclear receptor signaling. Endocrinology. 2006;147(6 Suppl):S50–5.

125. Janesick AS, Shioda T, Blumberg B. Transgenerational inheritance of prenatal obesogen exposure. Mol Cell Endocrinol. 2014;398(1–2):31–5.
126. DeBaun MR, Niemitz EL, Feinberg AP. Association of in vitro fertilization with Beckwith-Wiedemann syndrome and epigenetic alterations of LIT1 and H19. Am J Hum Genet. 2003;72(1):156–60.
127. Cortessis VK, Azadian M, Buxbaum J, Sanogo F, Song AY, Sriprasert I, et al. Comprehensive meta-analysis reveals association between multiple imprinting disorders and conception by assisted reproductive technology. J Assist Reprod Genet. 2018;35(6):943–52.
128. Rexhaj E, Paoloni-Giacobino A, Rimoldi SF, Fuster DG, Anderegg M, Somm E, et al. Mice generated by in vitro fertilization exhibit vascular dysfunction and shortened life span. J Clin Invest. 2013;123(12):5052–60.
129. Chen M, Wu L, Zhao J, Wu F, Davies MJ, Wittert GA, et al. Altered glucose metabolism in mouse and humans conceived by IVF. Diabetes. 2014;63(10):3189–98.
130. Feuer SK, Liu X, Donjacour A, Lin W, Simbulan RK, Giritharan G, et al. Use of a mouse in vitro fertilization model to understand the developmental origins of health and disease hypothesis. Endocrinology. 2014;155(5):1956–69.
131. Lane M, Robker RL, Robertson SA. Parenting from before conception. Science. 2014;345(6198):756–60.

Index

R. Teperino (ed.), *Beyond Our Genes*,
https://doi.org/10.1007/978-3-030-35213-4

The manufacturer's authorised representative in the EU is Springer Nature Customer Service Centre GmbH, Europaplatz 3, 69115 Heidelberg, Germany. If you have any concerns regarding our products, please contact ProductSafety@springernature.com

Printed and bound by CPI Group (UK) Ltd, Croydon, CR0 4YY
23/09/2026
02221975-0002